Charles Bagge Plowright

# A Monograph of the British Uredineae and Ustilagineae

Charles Bagge Plowright

**A Monograph of the British Uredineae and Ustilagineae**

ISBN/EAN: 9783337121631

Printed in Europe, USA, Canada, Australia, Japan

Cover: Foto ©ninafisch / pixelio.de

More available books at **www.hansebooks.com**

# A MONOGRAPH

OF THE

# BRITISH UREDINEÆ AND USTILAGINEÆ

*WITH AN ACCOUNT OF THEIR BIOLOGY INCLUDING THE METHODS OF OBSERVING THE GERMINATION OF THEIR SPORES AND OF THEIR EXPERIMENTAL CULTURE*

BY

CHARLES B. PLOWRIGHT, F.L.S., M.R.C.S.

SENIOR SURGEON TO THE WEST NORFOLK AND LYNN HOSPITAL
HONORARY MEMBER OF THE NORFOLK AND NORWICH NATURALISTS' SOCIETY
AND OF THE WOOLHOPE NATURALISTS' FIELD CLUB
CORRESPONDING MEMBER OF THE SOCIETÀ CRITTOGAMOLOGICA ITALIANA
SOCIÉTÉ MYCOLOGIQUE DE FRANCE, OF THE SCOTTISH CRYPTOGAMIC SOCIETY
OF THE BIRMINGHAM NATURAL HISTORY AND MICROSCOPICAL SOCIETY
AND MEMBER OF THE SCIENTIFIC COMMITTEE OF THE ROYAL HORTICULTURAL SOCIETY

ILLUSTRATED WITH WOODCUTS AND EIGHT PLATES

LONDON
KEGAN PAUL, TRENCH & CO., 1, PATERNOSTER SQUARE
1889

# PREFACE.

Of the many interesting problems involved in the elucidation of the life-history of the various members of the fungus kingdom, there are none more important to the agriculturist than those connected with the development of the Uredineæ and Ustilagineæ. Not alone to the cultivator of plants are these phenomena of interest, but also to the student of botany and microscopy.

The work of Tulasne in France, and of De Bary in Germany, during the past thirty years, laid the foundation of our knowledge of the biology of these fungi. More recently the continued researches of De Bary, of Schröter, the late Dr. Winter, Wolff, Kühn, Magnus, Woronin, and Hartig in Germany, of Oersted, Rostrup, and Nielsen in Denmark, of Fischer Von Waldheim in Russia, and of Farlow and Thaxter in America, have so increased our information as to render necessary a re-classification of our British species.

The object of the present work is to supply the British student, not only with improved descriptions of these parasites, but also with an account of their life-history as far as this is at present known.

With this object, the materials from which the following pages are compiled have been collected from the various handbooks, transactions, pamphlets, and periodicals in which the several authors have published their work. This has been a task of some little difficulty, and has taken several years to accomplish. It has been my endeavour in all cases to acknowledge the individual work of each author, but has not always been possible for me to do so.

The descriptions of the species are in the main those of the late Dr. Winter in the new edition of Rabenhorst's "Kryptogamen-Flora," although I have not scrupled to amend them in various ways, partly from my own observations, and partly from the writings of others. The arrangement of the species is that first proposed by Dr. J. Schröter and employed by Winter in the above-named work, to which I must refer the student for the descriptions of those European species which are not known to be British, as well as for the full details of the synonymy of those described in the present work. As the present generation of British students owe their knowledge almost exclusively to the writings of Dr. M. C. Cooke, the synonymy of his two works, "The Handbook of British Fungi," and "Microscopic Fungi," is given in full. The synonymy of the older British botanists has, however, not been omitted.

The biology of the Uredineæ is, in the main, the work of the late Professor De Bary, and of the Ustilagineæ that of Von Waldheim, Brefeld, and Schröter. During the past seven years I have devoted much time to these two groups of fungi, and have made between nine hundred and a thousand experimental cultures, with the object of working out the life-history of those species of which it was unknown, as well as of testing the accuracy of the statements of other botanists. The results of my cultures have been from time to time published in various botanical periodicals of this and other countries, and in the Transactions of the Linnean and Royal Societies.

It was at one time hoped, when the æcidiospores and uredospores were shown not to be distinct species, that continued biological investigation would materially lessen the number of species of the Uredineæ. I am convinced, however, that modern botanists have erred in grouping together forms on account of the similarity of the teleutospores; for instance, the Pucciniæ on the Compositæ, Labiatæ, Umbelliferæ, etc. The proper limitation of species of these parasites cannot be effected on purely morphological grounds; it can only be accomplished when the morphological characters are supplemented by a knowledge of the life-history of each individual species.

I am greatly indebted to my friend, Professor J. W. H. Trail, of Aberdeen, for his kind assistance in the revision of the proof-sheets.

I have also to thank numerous friends and correspondents, both in this country and abroad, for information and material for experiment, amongst whom I must mention Professor Farlow, of Harvard University; Mr. W. B. Grove; Rev. Dr. Keith; Dr. P. Magnus, of Berlin; Mr. W. Phillips, F.L.S.; Mr. H. T. Soppitt; Professor P. Sorauer, of Proskau; Rev. Dr. Stevenson; Rev. J. E. Vize; the late Dr. Winter, and many others.

# CONTENTS.

# A MONOGRAPH

OF THE

# BRITISH

# UREDINEÆ AND USTILAGINEÆ.

## CHAPTER I.

### BIOLOGY OF THE UREDINEÆ—INTRODUCTORY REMARKS.

THE Uredineæ constitute a large and important group of parasitic fungi, affecting flowering plants and ferns, representatives of which are familiar to every student of nature, practical agriculturist, and working microscopist. In whatever part of the world phanerogams appear, these parasitic cryptogams accompany them.

Our present object is to obtain an insight into the life-history and the structure of those species which are known to occur in Great Britain. Much has been elucidated concerning these points during the past twenty years, but still much remains to be discovered.

They all grow as parasites upon some living plant, independently of which they cannot exist, and consist of two essential elements—spores and mycelium.

The spores every one sees; the mycelium, on account of its inconspicuous nature, and because it cannot be seen without some little effort, no one takes the trouble to look for. The vegetative mycelium of most of the Uredineæ is very similar, although the spores are very diverse. The mycelium, being present in the appropriate host-plant, may give origin to several different kinds of spores, according to the nature of the Uredine under examination, each kind of which was regarded by the older botanists as being a distinct genus (*Æcidium, Uredo*). While it is true that the relationship between the various spore-forms was suspected as early as the beginning of the present century, yet its actual demonstration has been accomplished only within the last thirty years. From the investigations of Tulasne and De Bary, we now know that each spore-form has a life-history of its own, that they are often products of the same mycelium grown among different environments, and that they all arise from some antecedent spore-form. The ultimate condition in which all the Uredineæ are encountered is the teleutospore, which, after a longer or shorter period of quiescence, manifests its vitality by germinating and producing a body—the promycelial spore. The promycelial spore may then be regarded as the beginning of the series of spore-forms of which the teleutospore is the end.

The actual number of spore-forms intervening between the promycelial spore and the teleutospore is subject to considerable variation in different species.

1. The promycelial spore, after emitting a germ-tube, which enters the tissues of its appropriate host-plant, may give rise to a mycelium, which produces teleutospores exactly similar to the teleutospore from which it originated (Leptopuccinia, Micropuccinia).

2. The promycelial spore in a second group of the Uredineæ, in like manner, gives rise to a mycelium which produces uredospores, and subsequently teleutospores (Hemipuccinia).

3. The promycelial spore may give rise to a mycelium which produces, first, spermogonia, then æcidiospores, then uredospores, and finally teleutospores (Auteupuccinia).

Certain minor variations in the life-cycle occur; for example, the uredospores may be suppressed, so that we have only æcidiospores and teleutospores (Pucciniopsis), or the spermogonia and æcidiospores may occur on one plant, and the uredospores and teleutospores upon another of a totally distinct kind (Heteropuccinia).

Each of the above spore-forms has its own peculiarity of structure, as well as origin, so that it will be better, for clearness, to describe them separately.

The spore relations may be thus tabulated :—

PROMYCELIAL SPORE.

| | | | |
|---|---|---|---|
| Teleutospore | Uredospore | Spermogonium | Spermogonium |
| | Teleutospore | Uredospore | Æcidiospore |
| | | Teleutospore | Uredospore |
| | | | Teleutospore |

## CHAPTER II.

### MYCELIUM OF THE UREDINEÆ.

THIS is, of course, common to all the different spore-forms, inasmuch as it is that part of the fungus which develops them. It consists of a number of hyaline tubes that extend themselves principally between the cells of the host-plant. In some instances these tubes send short branches into the cells (haustoria), but the haustoria are not so common nor so well developed as in some other parasitic fungi (Peronosporeæ, Ustilagineæ). In the tropical genus Hemileia * the haustoria are unbranched, thin-stemmed vesicles, very like those of Cystopus. Bagnis† has figured the haustoria of *Puccinia malvacearum*, but his figure is of a doubtful character; and Barclay‡ has depicted an arborescent haustorium on the mycelium of *Æcidium urticæ*, var. *himalayense*.

The mycelial tubes (Plate I. Figs. 1 and 2) consist of hyaline, membranous walls, containing usually a colourless watery fluid. They are rendered more distinct by the

* Marshall Ward, "On Hemileia vastatrix," *Linn. Soc. Jour. Bot.*, vol. xix., and *Quart. Jour. Micr. Science*, new series, vol. xxi.

† Bagnis, "Obs. Vita et Morphol. Funghi Uredinei," t. i. fig. 11.

‡ Barclay, "Scientific Memoirs by Medical Officers of the Indian Army" (1887), t. iv. fig. 4.

action of caustic potash (KHO). The tubes themselves are rather irregular in their outline, and branch at frequent intervals. These branches unite with other mycelial tubes, so as to form an anastomosing irregular network, which pervades more or less widely the tissues of the affected plant. At rare intervals transverse septa are seen. The function of this mycelial network is to utilize the elaborated material which the host has prepared for its own use, and to turn it into a suitable pabulum for the sustenance of the fungus.

The extent to which the mycelium permeates the tissues of the host-plant varies; in the majority of cases it is localized and confined to a limited area. The germ-tube from a single spore having once entered the tissues of a leaf, the tendency of the mycelium thereby produced is to spread equally in all directions in a centrifugal manner. Many causes, however, come into operation which tend to prevent the ultimate spore products being equidistant from the centre. This may be due, in part, to the mycelial hyphæ growing more luxuriantly in one direction than in another; but to a great extent it arises from a want of uniformity in the tissues of the host-plant. Still, however, we are often able to observe that the fructification of those Uredines which have a limited mycelial growth, is arranged either in a circle or in more or less circular manner (*Puccinia lychnidearum*, *Uromyces scillarum*, *Cæoma orchidis*, *Æcidium zonale*). In those cases in which the mycelium is developed in a leaf with strongly marked venation, this tends to exert a directive influence upon its extension; for example, the primary uredospores of *Triphragmium ulmariæ* and *Uromyces alchemillæ*. The same directive influence of the tissues of the host-plant is seen in the linear arrangement of the sori of *Puccinia magnusiana*, *graminis*, *rubigo-vera*, etc., especially when they occur upon the stems or cauline sheaths of their graminaceous host-plants.

The presence of the mycelial hyphæ generally acts as a local stimulant to the tissues in which they are present, as is evinced by the increased thickness which is often associated with a concave arching or vaulting of the affected places (*Æcidium grossulariæ, berberidis, Ræstelia cancellata*). When the stems are affected, this is usually shown very markedly by the development of swellings and distortions; as a rule these are more or less fusiform, and often induce considerable bending of the attacked stem (*P. difformis*, Kze.; *Ur. trifolii*). Even upon these cauline tumefactions may often be traced the concentric arrangement of the sori. The stimulation of the affected part may be carried to such an extent as to kill the invaded tissues; thus we often find the older leaves of *Malva sylvestris* and *Althæa rosea* with numerous circular holes, punched as it were out of them. Each of these has been the seat of the very localized mycelium of *Puccinia malvacearum*, which has by its presence killed a circumscribed portion of the entire thickness of the leaf tissue, so that, as the leaf itself expands, the dead area above described becomes separated round its circumference and falls out, leaving a circular hole. The above-described perforated foliage is most observable after a period of drought; in rainy weather the reproduction of the parasite is so rapid that the entire leaf tissue is quickly invaded by the fungus and totally destroyed. The same dropping out of mycelial areas occurs upon the stem; here, however, the hyphæ only penetrate the external parts, so that when the affected spots drop out an elongated or fusiform wound is left, at the bottom of which the central woody part of the stem is exposed. Schröter * has pointed out that a somewhat

* Schröter, "Cohn's Beitrage," vol. ii. p. 88.

similar circumscribed dead area surrounds those sori of *P. annularis* which are formed late in the year, and that it is by means of teleutospores contained in these sori that the fungus is reproduced in the following spring.

The localized mycelium of the æcidiospores especially, has a powerful influence upon the chlorophyll, causing it to lose its green colour, and, as De Bary * has pointed out with *R. cancellata*, making it disappear entirely. The affected tissue is found to contain an immense number of minute starch granules. To such an extent does the development of starch take place, that in the Himalayas the natives eat the hypertrophied stems of *Urtica parvifolia*, which are affected with *Æc. urticæ*, on account of the abundant nutritive starchy material they contain. The hypertrophies are eaten just before the æcidia open, and are said to resemble cucumber in flavour.† The affected places generally assume some tint of yellow or reddish yellow, more or less bright. In the Polygoneæ the spots are bright red or purplish (*Æc. rumicis, Uredo bifrons*). In other cases they are yellow, surrounded by a purplish or reddish border (*Æc. zonale, behenis*). In a few instances all colour is more or less discharged, and the spots appear whitish (*Æc. albescens, leucospermum*).

The mycelium of all species is by no means thus localized. With some Uredines, on the contrary, it pervades the whole plant with the exception of its roots—stem, leaves, petioles, peduncles, and the upper part of the root-stock. De Bary ‡ has shown that when the mycelium of a Uredine can be traced into the perennial parts of the host-plant, it is itself perennial; thus with

* De Bary, "Brandpilze," p. 73.

† Surgeon-Major A. Barclay, "On Æcidium Urticæ," "Scientific Memoirs by Medical Officers of the Indian Army" (1887), p. 2.

‡ De Bary, "Neue Untersuchungen über Uredineen" (1865), pp. 20, 21.

*Endophyllum euphorbiæ* it is found more particularly in the pith and in the inner parenchyma of the bark of the affected plants. In plants of *Anemone nemorosa*, it may be found in the leaves, stems, and vascular bundles of the upper part of the rhizome, as is likewise the case with the host-plants bearing *Uredo suaveolens* and *Æc. tragopogonis.*

# CHAPTER III.

## SPERMOGONIA AND THE SO-CALLED SPERMATIA.

THESE bodies are produced by the mycelium, which arises directly from the entrance of the germ-tube of a promycelial spore, or from a perennial mycelium permeating the tissues of the host-plant.

The hyphæ which are destined to form a spermogonium become interwoven into an inextricable network in the subepidermal tissues of the host-plant. They pass for the most part between the cells, and are found more frequently septate where a spore-bed is about to be formed than when they are encountered elsewhere in the tissues of the host-plant. Their contents are watery and transparent. From this tangled mesh, immediately beneath the epidermis, are given off a great number of branches of much smaller diameter, as a rule about 2 or 3$\mu$. The general direction of these finer branches is towards the epidermis. They incline towards a central point, however, and so come together, forming a pyriform or subglobose body, the upper part of which is covered only by the cuticle of the host-plant. This body soon assumes a pyriform or flask-shaped contour (Plate I. Fig. 3). The neck of the flask bursts through the epidermis as a minute conical point, which can be seen to consist of a vast number of straight hyphæ, parallel to one another in the main, but all sloping upwards, converging to the apex of the cone. The outer walls of this body are embedded between the cells of the host-plant, and are likewise composed of similar fine hyphæ, placed side by side. A section of the spermogonium at this time shows its interior to consist of similar converging parallel hyphæ. Very soon those hyphæ which constitute the emerging apex separate from one another, so as to appear as a brush of stiff hair like bodies (paraphyses); in the centre of this brush is a minute canal, which passes downward to the interior of the body of the spermogonium (Plate I. Fig. 4). The apices of those converging hyphæ which occupy the lower part of the spermogonial interior, are now seen to be surmounted by very minute irregularly oval or rounded bodies (Plate I. Fig. 5)—the so-called spermatia. These spermatia vary in size, not only in the different species, but also in the same spermogonium. In those I have examined, they were from 5 to 8$\mu$ long, and 3 or 4 or even 6$\mu$ wide.

Tulasne * gives the measurement of the spermatia of *Triphragmium ulmariæ* and *Puccinia fusca* as from 5 to 6$\mu$ long, while those of most the Æcidia and of *Cæoma pingue* and *C. ribesii* are rarely more than 4$\mu$ long. The spermatia are produced in linear series from the apices of the hyphæ (sterigmata), which fill the interior of the spermogonium. These are held together by a viscid, gelatinous substance, which at first fills the bottom of the canal; but as more spermatia are produced, gradually the whole canal becomes full, and eventually the mass oozes out at its upper end in the form of a globule. The cause of the expulsion of the spermogonial mass is, as De Bary † has shown, the imbibition of moisture, which causes the investing gelatinous material to swell. As the spermogonium advances in maturity, its flask-like neck opens out, so that, instead

* Tulasne, "2e Mémoire," p. 118.
† De Bary, "Brandpilze," p. 60.

of resembling a flask with a bristly mouth, it comes to be a cup-shaped depression on the surface of the leaf, surrounded by a hedge of stiff bristles (Plate I. Fig. 4). The function of these bristly paraphyses appears to be that of preventing the exuded mass of spermatia and jelly from being bodily washed off the surface of the leaf by rain. This is the more necessary, because it is during wet weather, as we have seen, that the spermatia, from the imbibition of moisture by their investing jelly, are brought to the surface of the leaf at all. The paraphyses occur with all spermogonia except those of the Phragmidia.

Whatever their functions may be, whether as a spore-form or as spermatia properly so called, it is obvious that, by being held together by a viscid substance, their chance of dissemination by currents of air, etc., is but small. To a certain extent their diffusion over a limited area might take place in very wet weather. Ráthay * has, however, shown that the spermatial mass contains a certain amount of some saccharine material mixed with it—a substance which has the power of reducing Fehling's solution; and further, that, as a matter of fact, insects do visit spermogonia for the sake of this saccharine matter, and are thereby unwittingly the agents for the distribution of the spermatia. Of this latter point, he obtained actual demonstration in the following manner. He had standing upon his window-sill some plants of *Euphorbia amygdaloides*, upon which a large number of spermogonia of *Endophyllum euphorbiæ* were in the act of exuding their contents. One day, as he approached the window in question, he noticed some flies, which were busy upon these leaves, fly away from the leaves and alight upon the window pane. Closer scrutiny showed that the flies had left their wet footmarks upon the

* Ráthay, "Untersuchungen über die Spermogonien der Rostpilze." Wien: 1882.

glass. These footmarks were found, upon microscopic examination, to contain the spermatia of the Endophyllum. Further investigation of the subject led him to the conclusion that this sugary matter in the spermogonial contents acts as a bait to attract insects, and he has observed some 135 species, of which 31 were Coleoptera, 32 Hymenoptera, 64 Diptera, and 8 Hemiptera, thus visiting various spermogonia. The examination of the spermogonia of *Uromyces pisi*, *Puccinia suaveolens*, *fusca*, *tragopogonis*, *pimpinellæ*, *Endophyllum euphorbiæ*, *Gymnosporangium sabinæ*, and *G. juniperinum*, gave rise to copious deposits of suboxide of copper, when the washings of the affected leaves in distilled water were treated with Fehling's solution; less was obtained from *Uromyces dactylidis*, *Puccinia graminis*, *coronata*, *rubigo-vera*, *sylvatica*, *violæ*, *Gymnosporangium clavariæforme*, and *Æcidium magelhænicum*, and least of all with *Puccinia poarum* and *Æcidium clematidis*. Some of the spermogonial contents actually taste sweet when the tongue is applied to them, as those of *G. sabinæ* and *juniperinum*. The spermogonial contents of *G. sabinæ* were further found to contain dextrose and lævulose, the latter predominating.

Insects are not only attracted to the spermogonia by their saccharine contents, but also by the powerful odours which many of them possess. Persoon * long ago noticed the penetrating odour of the spermogonia of *Puccinia suaveolens*, which precede its uredospores; hence he called the Uredo, *suaveolens*. In a paper upon "Mimicry in Fungi," † I pointed out the probability of this odour being for the purpose of attracting insect visitors, mimicking as it does the perfume of the flowers of *Ænothera biennis*. Sowerby had already observed the fact that flies

* Persoon, "Synopsis Fungorum," p. 221.

† Plowright, "Mimicry in Fungi," "Grevillea," vol. x. pp. 1–14.

visited this plant. Tulasne* remarked the odour of the spermogonia of *Æcidium pini* and of *Uredo serratulæ*, which he compared to that of the pollen of willows. Léveillé † noticed the odour of *Æcidium tragopogonis*, which he said resembled the perfume of some flower. Ráthay indicates that the possession of the pleasant odours is a special characteristic of those Uredines which have a perennial mycelium. The plants thus invaded are further rendered more conspicuous to insects by the manner in which the affected leaves or shoots grow. As he says, these have a strong negative geotropism. Not only are they more erect, standing above the other leaves or shoots, but they are further characterized by an alteration in their foliage; while they are, as a rule, not only paler in colour, but also modified in form, being either more attenuated or the reverse. The Uredines with a short-lived mycelium, on the contrary, have their spermogonia produced upon the upper surface of the leaves on brilliantly coloured spots, which contrast more or less strikingly with the green colour of the healthy foliage. These spots are generally bright yellow or orange, often with a tinge of red. Sometimes they are white (*Æc. fabæ*) or purple-red (*Æc. rumicis*). The negative geotropism is not, however, confined to the species with the perennial mycelium; I have noticed it very markedly with some leaves of *Senecio jacobæa*, upon which I had produced the æcidium of *Puccinia schæleriana*; as soon as the æcidial spots became sufficiently developed to produce spermogonia, the leaves which bore them became almost erect, while the unaffected leaves remained horizontal.

In either case the darker paraphyses surmounting each conceptacle finally attract the insect to the sugary globule.

* Tulasne, "2e Mémoire," p. 118.

† Léveillé, in D'Orbigny's "Dict. Univ. d'Hist. Nat.," tome xii. p. 175, sub. Uredines.

Although the paraphyses themselves, when seen singly as transparent objects, appear hyaline or yellowish, yet *en masse* they are almost black in many species, and in all distinctly darker than the surrounding tissues. Their function, however, is not confined to the attraction of the insect visitors; they also (as has been already stated) maintain the spermatial globule in its place upon the leaf, and prevent it being bodily washed away. With regard to the presence of some saccharine matter in the spermogonial masses, I may add that I have found, in confirmation of Ráthay's statements, that the spermogonial contents of *Puccinia obscura* both reduced Fehling's solution and also gave the reaction with the indigo-carmine test.

The so-called spermatia were long regarded as not possessing the faculty of germination. Cornu,* however, found that when they were placed in water, in which a little white cane sugar had been dissolved, and exposed to the free action of the air, they were capable of germination. During the past five years I have repeated M. Cornu's experiments, at first employing white sugar as the nutrient material. Under these conditions the individual spermatia at first assume a more regular outline than one commonly observes them to have as they are naturally exuded from the spermogonial receptacles (Plate I. Figs. 6, 7). They at one extremity give out a minute prolongation. This prolongation does not develop into a germ-tube, but gradually increases in size, until it acquires the size and form of its parent spermatium; at the end of twenty-four hours many of these twin bodies, joined end to end, will be observed in the culture (Plate I. Figs. 8, 9). Resembling Saccharomyces spores as these bodies do, I at first thought some stray yeast-spores had accidentally gained admission

* Cornu, '*Bulletin de la Société Bot. de France* (1876), tome xxiii. pp. 120, 121. Compt. rendus, January 21, 1875.

into my culture; in fact, De Vauréal * has suggested that yeast-spores are simply the spermatia of Uredines. By frequent repetition the result was always the same—the so-called spermatia budded in the same manner as Saccharomyces spores. Mr. A. Lister has observed the same budding with the spermatia of *Æcidium albescens*, of which process he sent me an excellent figure made by him in 1873 (Plate I. Fig. 10). In 1883, I made a number of cultures in which honey instead of sugar was employed; with this substance the germination is much more active and prolific, the budding spores often remaining in yeast colonies, attached by their ends in chains of half a dozen or more. I have germinated the spermatia of the æcidiospores on *Bellis perennis*, *Ranunculus bulbosus*, *R. ficaria*, *Anemone coronaria*, *Lapsana communis*, and some others. The older germinating spores present two nuclei of noticeable size. Not only are the spores budded in linear series, but also from these chains young spores are developed laterally, from the point of union where two older ones meet (Plate I. Figs. 12–16). This budding goes on continuously, but not with the same rapidity as with Saccharomyces, nor, as far as my observations go, to the complete exhaustion of the saccharine material. Probably these changes in the fungus are accompanied by the production of alcohol, but I was unable to detect it in any of my cultures. When the Uredine yeast-spore falls to the bottom of the fluid, it is excluded from the air and does not further change; whereas if a small number of Saccharomyces spores be placed in a saccharine fluid, they rapidly multiply themselves until all the sugar disappears, while the Uredine spermatial cultures, at the end of several days, retain their sweet taste. Attempts to produce alcohol from brewers'

* De Vauréal, "Schutzenberger on Fermentation" (1876), p. 61. International Scientific Series, vol. xx.

wort in a Pasteur's flask, with Uredine yeast-spores, were with me unsuccessful. The spermatia do not germinate in pure water, but only in the presence of sugar. There is, however, often enough saccharine matter in the investing jelly of the exuded globule for a few of them to bud, as Mr. Lister's figure shows, and as I have several times observed.

Rebentisch * noticed the small black points which the spermogonia present in the spots upon pear leaves, on which *Ræstelia cancellata* are produced. Unger † figures these structures, and describes them as a distinct species under the name *Æcidiolum exanthematicum;* he regarded them as a peculiar exanthem of the affected plant. Meyen ‡ thought they played the part of the male element in the reproduction of the Uredineæ, with which they were associated. Tulasne § considered they were spore-forms developed from the same mycelium as their accompanying spore-forms, but, like De Bary, was unable to observe their germination.

The spermatia of the Uredineæ occur with all the spore-forms. They almost invariably precede and accompany the æcidiospores, being produced from the same mycelium, but generally occupying the upper surface of the leaf, while the æcidial cups occur on the lower. The first formed spermogonia are produced in the centre of a spot; the next more externally, and so on centrifugally. In many cases, however, they occur on the same surface and between the æcidia. When the æcidiospores are cauline, the accompanying spermogonia are often

* Rebentisch, "Prodrom. floræ Neomarch."

† Unger, "Die Exantheme der Pflanzen," t. iii. figs. 18, 19. Wien: 1836.

‡ Meyen, "Pflanzenpathologie" (1841), pp. 143–147.

§ Tulasne, Comptes rendus, tome xxxii. p. 472; and tome xxxvi. p. 1093.

arranged circumferentially. Morphologically, they present a great similarity to the spermogonia of the lichen-fungi, especially to those of Collema.

Since our knowledge of the life-history of the lichen-fungi has been increased, from the facts added to it by Schwendener, De Bary, Stahl, and others, many botanists have come to look upon the Uredine spermatia as fertilizing bodies, and to consider the Æcidiomycetes as being nearly allied to the Ascomycetes. There are, however, certain facts which cannot be overlooked. In the first place, the faculty which the Uredine spermatia have of multiplying themselves by budding in saccharine solutions in exactly the same manner as the spores of Saccharomyces, and also as Brefeld has shown the conidia of the Ustilagineæ do in his *nährlösung*, points rather to their being conidia than spermatia.

Then, again, it is not asserted that all spore-forms of the Uredineæ are sexual—this is claimed as probable for the æcidiospores alone ; true it is that the æcidia are almost always accompanied by spermogonia, but this is not invariably the case. De Bary * cultivated a single plant of Sempervivum affected with *Endophyllum sempervivi* which bore no spermogonia, but the æcidial cups were perfectly developed, and their spores germinated in the normal manner. In the autumn of 1883, I transplanted into my garden a wild plant of *Tragopogon pratensis* affected with the Æcidium. During the spring and summer of 1884, this plant continued to produce a succession of æcidia, the spores from which were used for infecting seedlings of Tragopogon. This they successfully did, causing the development of the teleutospores, with their scanty accompaniment of uredospores. I was never able to find any spermogonia upon this Tragopogon,

* De Bary, "Morphol. und Physiol.," 1st edit. p. 169.

although I repeatedly examined it for them. These two cases, the Sempervivum and the Tragopogon, in both of which perfectly normal, and in the latter instance functionally active, æcidiospores were produced without the presence of spermogonia, show, at any rate, that each individual æcidial cup was not the result of a spermatial fecundation. It may have been that in both instances the plant bore spermogonia before they came under observation, and that they both bore within them an already fecundated mycelium; but this is only a supposition, and, even if it be admitted as possible, yet it does away entirely with the analogy of the sexually produced spore-beds of the lichen-fungi and Polystigma, in which each spore-bed is the result of a separate spermatial impregnation of the trichogyne. Nor is the case of the Uredineæ with short-lived mycelia more tenable. If with them each æcidial cup is a sexual product, then it cannot arise in the first instance without a spermatial fecundation. But Schröter* has shown that when he produced the æcidium of *P. porri* (Sow.) by infecting the *Allium schœnoprasum* with the teleutospores, the resulting æcidia were always unaccompanied by spermogonia, as was also the case when they occurred on onion (*A. cepa*); and, further, he found while the vernal specimens of the æcidiospores of *Uromyces ervi* on *Ervum hirsutum* and of *Puccinia galii* on *G. aparine* were accompanied by spermogonia, yet those produced later in the year were invariably without this accompaniment. If the spermatia were necessary to the development of the æcidia in spring, how could they be dispensed with by the same formation in autumn? The fact, however, stands that spermogonia almost always accompany æcidiospores; but not only is this the case with the true Æcidia, but also with the analogous Cæomata—analogous in as

* Schröter, "Cohn's Beitrage," vol. ii. p. 83.

far that they originate from promycelial infection of their respective host-plants, and that their spores are produced in linear basipetal series. In fact, the Cæomata only differ from the Æcidia in the want of a peridium. Yet we do not find spermogonia exclusively associated with these two spore-forms; on the contrary, they occur not very rarely with the uredospores—a spore-form which does not present any morphologically biological parallel with the two above named, inasmuch as its spores are produced singly on separate and distinct spore-forming hyphæ; nor has any sexuality ever yet been claimed for uredospores. As instances in which spermogonia accompany uredospores may be quoted *Triphragmium ulmariæ* and the Brachy-pucciniæ, *Puccina hieracii* (Schum), *P. suaveolens* (Pers.), *P. bullata* (Pers.), *P. oreoselini* (Strauss), and *Uromyces terebinthi* (D.C.). Nor is this all, for spermogonia also occur with the teleutospores, as in *P. silphii* (Schw.),* *P. falcariæ* (Pers.), *P. liliacearum* (Duby), and *P. fusca* (Relh.).† It has been suggested that the association of the spermogonia with the teleutospores in the last-named species might be only an exceptional case in which the corresponding female spore occurred upon some other plant—in other words, that it is an heterœcious species; but while admitting the full weight of this argument in the abstract, yet of the Pucciniæ in question, if *P. fusca* be a Pucciniopsis, and have for its æcidiospores *Æcidium leucospermum*, an æcidium so markedly accompanied by spermogonia that their presence was recognized by the older botanists, we must consider it to be an exceptionally well-provided species.

The escaped spermatia are very commonly found scattered amongst the æcidiospores, and often adhering to their outsides. Mr. W. G. Smith ‡ has represented this condition,

* Trelase, "Parasitic Fungi of Wisconsin." 1884.

† Tulasne, "2ᵉ Mémoire," p. 116, note.

‡ W. G. Smith, "Diseases of Field and Garden Crops," p. 166, fig. 86.

and suggested that this may in itself be the fecundative act, but the same objection obtains to this latter suggestion as to the former.

As far as I can judge, the balance of evidence is against the supposition that the spermatia are sexual organs. The other supposition, that they are conidia, is more plausible; but one would have thought, had such been the case, it would have been capable of demonstration. I have attempted various experimental cultures with these bodies, but uniformly without result. These consisted of the application of the spermogonia in active germination in honey and water—

1. To the foliage of the plant upon which the spermogonia occur. Thus the spermatia of *Æcidium ranunculi repentis* were applied to the healthy foliage of *Ranunculus repens* (Exp. 123);* of *Ræstelia cornuta* to *Sorbus aucuparia* (Exp. 124); of *Æcidium bellidis* to *Bellis perennis* (Exp. 244, 247).

2. To the corresponding host-plant bearing the uredospores and teleutospores. The spermatia of *Æcidium berberidis* were applied to wheat (Exp. 392).

3. Remembering the fact that the spermatia are carried by insects, and that they germinate so freely in honey, they were applied to the stigmata of certain flowers. I was further induced to try these experiments from the frequency with which the fruit of so many plants in this country are attacked by the æcidiospores of some of the Uredineæ; for instance, the Mahonia berries, the barberry fruit, the gooseberry, and the fruit of the hawthorn. The spermatia of *Æc. bellidis* were applied to the stigmata of *Bellis perennis* (Exp. 267), and those of *Æc. ficariæ* to *Ranunculus ficaria* (Exp. 274), but no result was obtained; the infected plants produced in due course perfectly normal ripe seeds.

* The numbers of these experiments refer to my private note-book of experimental cultures.

# CHAPTER IV.

## ÆCIDIOSPORES.

THE mycelium originated in the tissues of the host-plant by the entrance of the germ-tube from a promycelial spore, after it has produced a certain number of spermogonia, proceeds in the course of its development to give rise to the æcidiospores. This spore-form is of considerable interest, not only on account of the manner in which it arises, but also by reason of its attractive appearance in the mature state. The first appearance of the æcidiospores consists in the formation of the receptacle (pseudoperidium) in which they are contained. The mycelium destined to give rise to an æcidium becomes more frequently branched and interwoven. At certain points, not immediately beneath the epidermis, but pretty deeply placed in the parenchyma of the plant, the mycelial hyphæ compact themselves into spherical bodies, which at first are not larger than the parenchymatous cells of the host-plant. The bodies (the primordia of De Bary) gradually increase in size by the addition of fresh hyphæ from the mycelium. It can soon be observed that the interior and bulk of these bodies consist of a cellular structure, and that they bear the greatest possible resemblance to perithecia (Plate II. Fig. 1). They are globose, but rather flattened upon their upper surface—that is, the surface nearest to the epidermis.

A section through one of them shows it to consist of an external envelope of cells, which encloses the body on all sides, above, below, and laterally. The base of the body is a circular disc (hymenium), which bears upon its upper surface a number of closely packed, erect hyphæ (basidia), each of which supports a linear series of spores. The hymenium increases in width as the æcidium develops, but whether from new basidia arising between those at first formed, or circumferentially, is not known. Surrounding the circumference of the hymenium is a circle of sterile cells (the peridial cells; Plate II. Fig. 2). Spore-formation in the æcidium proceeds from above downwards, in the following manner. The cells destined to become spores are enclosed in a hyaline tube (the mother-cell); they are at first colourless, but soon within each, from above downwards, a number of granules appear, which, becoming invested with a delicate membrane, rapidly augment in size, from an increase of the enclosed protoplasm, until they touch the walls of the mother-cell. The granular contents are by this time orange (*Æcidium*), or brown (*Ræstelia*), or remain white in a few species, such as *Æc. rumicis, vincæ*, etc.*

The growth of the cell continues until it is indistinguishable from the mother-cell. By the mutual pressure of the neighbouring cells the young spores become polygonal. As this process begins above and continues downwards, or, as it is now termed, in a basipetal manner, it follows that the ripe spores are uppermost. As soon as they have attained their full maturity, they separate from one another and are blown away. In some species (*Ræstelia*, æcidio-

* Mr. G. Massee considers the whole Æcidium to be a sexual product resulting from the conjugation of two dilated mycelial hyphæ in the tissues of the host-plant (*Annals of Botany*, June, 1888, vol. ii. pp. 47–51, plate iv. figs. 1–7).

spores of *Chrysomyxa*) the alternate cells remain sterile, so that instead of a long continuous series of spores, we encounter a series of perfect spores alternating, with variously shaped abortive cells, which may be shrivelled into fibres as in *Ræstelia* (Plate II. Fig. 9), or variously flattened as in *Chrysomyxa* (Plate II. Fig. 8). The further development of the spores consists in the cuticularization of their exterior. How this cuticle arises is not clear, whether from a metamorphosis of the mother-cell, or from a separate development. The cuticle is in almost all species variously roughened by a number of minute points, spines, or very minute warty protuberances. By treatment with caustic potash these prominences disappear. According to De Bary's * most recent views, he would appear to maintain that all æcidiospores are developed with alternate abortive cells.

The pseudoperidium of the æcidia is formed in the same manner as the spore-series, only the cells are sterile and empty, instead of being filled with coloured protoplasm (endochrome). They are, however, developed in linear series, the larger and more mature being uppermost, the younger and smaller below; in fact, we may look upon the peridium as being developed from a circle of basidia surrounding the hymenium. The pseudoperidial cells are held pretty firmly together by an intercellular substance. They are variously altered by mutual pressure, being flattened, oval, oblong, square or rhomboidal, and always larger than the spores. The peridial cell-series are continued upwards until they arch over the spores and meet those of the opposite side, so that they form a complete investment above and around the æcidium. As this body grows it comes nearer the surface, until at last it is covered by the epidermis alone, through which it may be seen

* De Bary, "Vergleichende Morphologie und Biologie der Pilze," p. 76.

shining as a small yellow tubercle. The epidermis becomes ruptured by the pressure of the young growing æcidium from below; and, either before or soon afterwards, the pseudoperidium itself gives way at its summit, exposing the ripe æcidiospores. The ruptured peridium now becomes recurved, and, as seen from above, the æcidium is cup-shaped. The white peridium contrasts with the golden yellow of the spores very strikingly. The peridial cell-series still show their linear origin by separating into teeth, the attachment being stronger from above downwards than laterally. This recurved, toothed margin of the peridium is highly characteristic of the æcidia. The æcidium and its peridium consist at this stage of a few fully developed spores and cells above, and of a vast number of undeveloped spores below. As these latter mature they push the old ripe spores upwards and outwards, and they are carried away by air currents, rain, and by any cause that shakes the host-plant. If the affected plant be grown indoors or continually covered by a bell glass, so as to prevent any injury to the æcidium itself or any agitation of the host-plant, the æcidium elongates itself in a cylindrical manner. This was strikingly shown in a specimen of *Æc. berberidis* which I cultivated and exhibited at the Biological Laboratory of the International Health Exhibition, 1884, in which the peridia developed themselves into long curved cylindrical bodies. The above illustra-

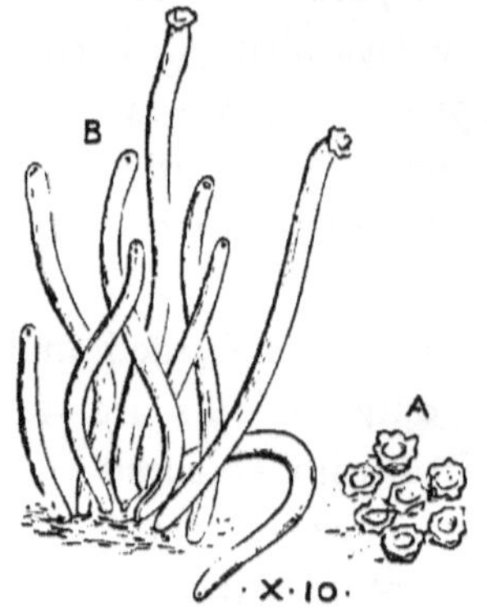


Fig. 1.—*Æcidium berberidis.* A, with normal pseudoperidia; B, with abnormally elongated pseudoperidia. Figured by Mr. Worthington G. Smith from an experimental culture of the author's in the Biological Laboratory of the International Health Exhibition, 1884, for the *Gardener's Chronicle*, September 6, 1884, p. 308.

tion, by Mr. W. G. Smith, was taken from the living plant, and is therefore reliable, although my own impression was that the elongation of the pseudoperidia was not quite so pronounced. In some species the cells of the peridia naturally adhere more firmly, as in *Æc. ornamentale*, *Peridermium*, *Rœstelia*. In *Æcidium pini* the peridium opens at the apex, but the margin shows very little tendency to become recurved. In *R. cancellata* the peridia do not open at the apex. This remains entire for a long time, and surmounts the peridium as a conical cap. The peridial cell-series, however, below this separate from one another laterally, so that the spores escape through a number of longitudinal fissures—a sort of lattice-work.

The membrane of the æcidiospore is not of uniform strength, inasmuch as at certain points, when the spore germinates, there are pushed through it prolongations or outgrowths of the endospore. These potential openings (germ-pores) are exceedingly difficult to observe in the mature spore. When, however, the spore begins to germinate, they become obvious. De Bary * gives the number of germ-pores in *Æc. tragopogonis* as three, and those of *Æc. asperifolii* † as four. From my own observations, I should say the æcidiospores of *Puccinia graminis* and *poarum* have six germ-pores. When these spores begin to germinate they become somewhat quadrangular, with a germ-pore at each angle; but in addition to these a central opening or thin place appears in the centre of that side of the spore which is uppermost, and as, of course, both the upper and under sides of a spore cannot be seen in the same preparation at the same time, it

* De Bary, "Champ. parasit," *Ann. des Sci. Nat.*, 4ᵉ sér. tome xx., Reprint, p. 76, t. ii. fig. 7.

† De Bary, "Neue Untersuchungen über Uredineen," vol. ii. p. 209, Berlin: 1866.

follows that if all the spores in the field present this central germ-pore, then of course there must be a corresponding pore on the opposite side (Plate II. Figs. 4–7 and 10), so that when germination commences, the spores assume a rounded cuboid form, with a germ-pore at each angle. The process of germination continues from one of these germ-pores, rarely from more than one. The outgrowth of the endospore progresses, until it assumes the form of a blunt-pointed, cylindrical tube. Into this tube the protoplasm of the spore passes, and with it the orange granules which give the colour to the spore. The tube grows onwards for a considerable distance—ten or more times the diameter of the spore. Contemporaneously with the growth of the germ-tube, the protoplasmic contents are continuously passed onwards to its extremity; so that the longer the germ-tube grows, the further are the orange granules removed from the spore in which they were originally contained (Plate II. Figs. 6, 7, 10, and 14). The further development of the germ-tube consists in its continued growth; the walls, however, become not only distinctly thinner, but give off various branches, mostly at a right angle to the axis of its growth (Plate II. Fig. 7). While these changes have been taking place, the whole germ-tube has been making a series of convolutions, sometimes from right to left, sometimes from left to right (Plate II. Fig. 6). Often a germ-tube will turn in one direction during the first part of its course, and in the opposite during the latter part. These circumnutatory movements were observed by Tulasne * in *Æc. crassum*, *violæ*, etc. This circumnutation of the germ-tubes I have only observed in the Uredineæ, and it is quite distinct from the torsion one observes in other mycelial hyphæ, when they extend themselves into a dry atmosphere. This has been

* Tulasne, "2e Mémoire," p. 128.

confounded with circumnutation by some authors,* but is simply a hygrometric phenomenon. The distal extremity of the germ-tube of an æcidiospore gains an entrance into the tissues of the host-plant through one of the stomata. This is effected by either the end of the germ-tube or one of its principal branches insinuating itself into the cleft (Plate II. Fig. 10). When an entrance has been achieved, further development takes place in the air-spaces below the stoma, by the branches of the germ-tube growing between the cells of the host-plant; these soon develop into a mycelium, which in due course produces the spore-form proper to the species.

The time which the æcidiospores retain their germinative faculty is stated by De Bary † to be some weeks. From my observations, I should say it is rather a matter of hours. Much depends, however, upon the temperature and the amount of dryness to which the spores are exposed. If they be placed in a very moist atmosphere, they germinate at once; if, on the contrary, in a perfectly dry one, they almost as rapidly die. But if they be kept slightly moist and cool, they will remain uninjured for a much longer time. I have, however, seldom found them germinate after forty-eight hours, and then only a small proportion will do so. It is only the mature spores at the top of the cup, that have already separated themselves from the spore-series, which will germinate. The process begins within a few hours after they have been placed in water.

* Cooke, "Circumnutation in Fungi," *Quekett Journal* (1884), vol. i. 2nd series, p. 309.

† De Bary, "Vergleich.," p. 369.

## CHAPTER V.

### UREDOSPORES.

THE uredospores may be developed from mycelia produced from the entrance of the germ-tube of an æcidiospore, a promycelial spore, or another uredospore. The uredospores are developed singly on the ends of separate mycelial hyphæ (sterigmata, basidia). Their development is as follows:—the mycelial hyphæ in the affected host-plant at certain points become interwoven and intertwined, and the hyphæ which constitute these favoured points are more richly branched and more freely septate than is the case in other parts of the host-plant. In them may be observed a number of orange granules. The hyphæ are so crowded together in the spore-bed, that to a great extent they lose their individual outline. Spore-formation takes place, first towards the centre of the bed, and then extends centrifugally. The mycelial aggregations take place just beneath the cuticle of the plant, to which their upper surfaces are parallel; they are termed spore-beds, or sori (sporenlager, stroma, hymenia, clinodes). Each spore-bed consists of an accumulation of hyphæ, which ramify in all directions; but those hyphæ that are nearest to the cuticle of the host-plant give off a number of branches parallel to each other, at right angles to the spore-bed below, and the host-plant cuticle above. Each of these branches is destined to produce a uredospore. At first they are of uniform diameter, but soon the upper end dilates so as to become club-shaped (Plate III. Figs. 2, 5). Into this dilated extremity the protoplasmic contents of the hyphæ are gradually emptied; and thus by their continued accumulation it becomes almost spherical. This dilated tube-end is now full of granular protoplasm, towards the centre of which appear a few orange granules (in the orange-spored species), while externally it becomes invested with a thin cell-wall of its own (Plate III. Fig. 3). The cell increases in size, its walls in thickness, and its contents become more and more distinctly coloured. The orange granules, which consist of reddish-yellow, oleaginous particles, are at first confined to the centre of the cell, and gradually increase in number from the centre outwards; so that we frequently observe immature spores, which are orange only in their centre, having the centre surrounded by a hyaline zone. The spore has at this stage of its development two coats—the outer, which is extremely thin and is the dilated mycelial hypha; and the inner, the proper wall of the spore. The latter increases somewhat in thickness, although it always constitutes a comparatively thin investment. These two coatings become closely applied to one another. The exospore, which is at first smooth, becomes, in most species at their maturity, variously roughened by the appearance of minute projections from its surface (Plate III. Fig. 4). These may be in the form of short, fine spines, when the spore is said to be finely echinulate, or the prominences may be shorter and less acutely pointed (verrucose). De Bary has shown that the spores of many of the Uredineæ owe their roughness to minute, densely crowded, prismatic, staff-like processes between which similar smaller processes are closely packed. The irregularities of the exospore disappear under the action of caustic potash; and they are much less easily observed when the spores are examined in water than when they are seen dry. This is equally true of the æcidiospores.

In the endospore are two or more openings (germ-pores, oscules) through which the germ-tubes emerge when the spores germinate. They can sometimes be made out pretty clearly in perfectly ripe spores, but not often. There is no difficulty, however, in observing them in those spores in which the process of germination has commenced. In the globose uredospores they are arranged in a circle round the equator; in the oval spores, also midway between the poles. Whether it be correct to regard them as openings is doubtful; they would be more correctly described as thin places, which become holes. The number of germ-pores varies in different species; they are never less than two. Their variation in number and position is but slight: thus De Bary* gives for *U. linearis*, four; *U. fabæ*,† three; *U. phaseoli*,‡ two; and *U. suaveolens*,§ three.

Recently De Bary ‖ has stated that in the uredospores of Puccinia and Uromyces the germ-pores are sharply defined, circular holes in the endospore, closed externally by the exospore; but this hardly accords with my observations. The spore-bed continues to produce uredospores for some considerable time; at length it ceases to do so. If it be examined in this condition it will be found to consist of little else than barren basidia, with here and there one bearing a spore. The uredospores vary in colour; most are some shade of orange, many are brown. De Bary ¶ has pointed out that in those species with brown spores the contents are colourless, *e.g.* in *U. phaseoli*, *rumicis*,

* De Bary, "Brandpilze," p. 33.
† De Bary, "Champ. paras.," p. 74.
‡ Ibid., p. 76.
§ De Bary, "Brandpilze," p. 33.
‖ De Bary, "Vegl.," p. 109.
¶ De Bary, "Brandpilze," p. 31.

*trifolii*, etc., the brown colour being due to the spore-wall. The colouring matter of the uredo and æcidiospores has been spectroscopically investigated by Bachmann.* He examined the æcidiospores of *Gymnosporangium juniperinum* and *Puccinia coronata*, and the uredospores of *Melampsora farinosa*, *Triphragmium ulmariæ*, and *Uromyces alchemillæ*. Combined as it is with some oleaginous material in granules, and enclosed within the cell walls of the spores, it is exceedingly difficult to extract it with ether. But by adopting the saponification process of Kühne † and Hansen ‡ he was able to arrive at the following conclusions. The above-named fungi give very similar spectra, namely two narrow absorption bands, one between *b* and F, the other between F and G, showing that the pigment is the same in all cases; that it is very similar to, if not identical with, the colouring matter of most of the yellow phanerogamous blossoms (the anthoxanthin of Hansen) although combined differently, and allied to the xanthophyll group. It is soluble in alcohol, ether, chloroform, bisulphide of carbon, and benzol, and is coloured green by potassium iodide.

As soon as the uredospore has arrived at its maturity it becomes separated below from the hypha which produced it (basidium, sterigma). This separation takes place by the basidia breaking off either close to the spore (Uredo) or at a short distance below, in which case the upper part of the basidium remains attached to the spore (Trichobasis). This separation is facilitated, if not caused, by the spore being pushed off by the continued formation of other

* E. Bachmann, "Spektroskopische Untersuchungen von Pilzfarbstoffen" (1886), pp. 21-23, figs. 27-31.

† W. Kühne, "Ueber lichtbestandige Farben der Netzhaut" ("Untersuch. d. physiol. Inst. d. Univ. Heidelberg"), bd. i. hft. iv. p. 347.

‡ F. A. Hansen, "Der Chlorophyll Farbstoff" ("Arbeiten des botan Inst. zu Wurzburg"), bd. iii. hft. i. p. 126.

spores in the spore-bed. Whether this be the simple passive process we have hitherto regarded it or not is undetermined. It is quite possible that it may be aided by a projective force, as Brefeld * has shown in the Agaricini. In them the basidia become by degrees fuller of protoplasm, until a certain state in the tension of the elastic walls takes place. When this has attained its full extent, the upper part of the basidium, immediately below the spore, gives way and becomes split off all round, the spore being projected from its sterigma at the point of its attachment. A similar process occurs in Pilobolus, only in this case the rupture is lower down, and not at the exact point of junction of the spore with its basidium.

As soon as each uredospore is ripe it is capable of germinating, and when placed in a sufficiently damp environment it does so in a few hours. This process is accomplished in the same manner as has been already described under the æcidiospores. It consists in the protrusion of a germ-tube through one or more of the germ-pores, which branches, elongates, circumnutates, and receives the protoplasm from the interior of the spore, and passes it onwards to its peripheral extremity (Plate III. Figs. 11, 12, 13). In the same way its extremity, or the extremity of one of its branches, enters into one of the stomata of the host-plant, and in its tissues develops a fresh mycelium (Plate III. Fig. 15).

Under certain circumstances, when the germ-tube cannot enter a stoma, instead of growing in the mode described, it dilates in a bulbous manner † at its extremity into a spherical dilatation, into which the orange granules accumulate; or it may give off one or more lateral out-

* Brefeld, "Schimmelpilze."

† Plowright, "Germination of Uredines;" "Grevillea," vol. ix. pl. 159, figs. 10, 11, 12.

growths, which become globose and filled with endochrome (reserve spores) (Plate III. Fig. 14). These abnormal phenomena show how nearly allied the different spore-forms of the Uredineæ are; and are especially interesting from an evolutionary point of view. They show that although the uredospore has attained a definite mode of germination proper to itself, yet it has not entirely lost its capability of reverting to, and simulating the mode of growth of other spore-forms.*

In certain species (Lecythea) are found bodies known as paraphyses, or cystidea. They are in most cases arranged around the circumference of each spore-bed, arching over it and arising from basidia in the same manner as the spores themselves. The paraphyses contain no coloured endochrome. In shape they may be globose, pyriform, subcylindrical, or capitate (Plate III. Figs. 16, 17). Although they cannot be regarded as undeveloped spores, inasmuch as they are pretty much confined to particular species, and do not occur indiscriminately with all, yet from their mode of origin their affinity is clearly with the pseudoperidial cells of the æcidiospores, and to a certain extent they are protective organs in the same way as the latter. From my observations it appears that the presence of paraphyses with certain species greatly depends upon some special condition of the fungus, as they may be absent, or nearly absent, according to special circumstances. I find them constantly present with the uredospores of *P. perplexans*, when these have arisen not directly, but rather at a considerable distance, from the æcidiospores. On the other hand, when the uredo arises directly from the æcidiospore, they are hardly present at all; this looks very much as if they were an indication of exhaustion of vital energy

* Tulasne figures a similar condition in the germination of the uredospores of Cronartium ("2e Mémoire," tab. xi. figs. 3, 8, 9).

on the part of the fungus, which was combated by protective efforts on the part of the parasite in conserving those spores which it does produce, but when full of vigour and fresh from the æcidiospore it is less careful of its spores. When it begins to feel the effect of exhaustion, and is unable to develop such energetic spores, it takes more care of those which are produced.

The same is true of the quantity of uredospores themselves. When the uredospores are produced directly from the æcidiospores they are much less copious than when they originate from other uredospores; especially is this so when they have arisen from a long series of uredospores. The converse is observable in such species as *Puccinia tragopogonis*, where the teleutospore occurs often on the same leaf as the æcidiospore, and the uredospores are very few indeed. We have striking illustrations of the contrary condition with many heterœcious Uredines. In Australia, where the barberry is not an indigenous plant, and occurs only in gardens and shrubberies, the agriculturists complain not of mildew (*P. graminis*) as destroying their wheat crops, but of rust. Some years ago I received specimens of the affected wheat plants from New South Wales, Queensland, and South Australia, all of which showed a profuse development of uredospores in proportion to the teleutospores, quite out of all parallel to that which obtains in England.* Mr. (C.) J. Arthur informs me that this is equally true of those districts in the United States from which barberries are absent. The same occurs in this country with *P. rubigo-vera*, the uredospores of which are extremely abundant on our wheat crops in spring; but the autumnal Æcidium is a very infrequent fungus, partly because the host-plants are none of them very abundant, and partly because, occurring

* Plowright, "Reproduction of Heterœcious Uredines," *Jour. Linn. Soc. Botany*, vol. xxi. p. 368.

as it does in September and October, the frosts of winter soon destroy their foliage, so that the uredospores have to reproduce themselves throughout the winter and spring months on the wheat plant. Rostrup * has remarked the same fact with regard to *Coleosporium senecionis*, that when it occurs in localities from which fir-trees are absent it consists almost wholly of uredospores. Independently of this, however, the uredospores of some species are much more abundant than of others; in *P. oblongata*, for instance, they occur in great quantities, while in *P. hydrocotyles* the teleutospores are very few in number, and occur in the same spore-beds as the far more numerous uredospores.

Although uredospores have their vitality so easily destroyed by heat and dryness, they can withstand a considerable amount of cold, for freshly developed spore-beds of *P. rubigo-vera* can be found almost at any time, on wheat, during the winter months. Of course, it may be that these have been developed by mycelium produced from a spore-infection effected at an earlier date. I found that the uredospores of this species which had been exposed to several nights of frost, when the thermometer registered 23° F. (— 5° C.), germinated with the greatest freedom. Dietel found the uredospores of *Phragmidium obtusum*, which had been covered by snow from December 18 to January 28, germinated in a day and a half in a warm room.†

* Rostrup, "Heterœciske Uredineen" (1884), p. 6.

† P. Dietel, "Beiträge zur Morphol. und Biol. der Uredineen" (1887), p. 9.

## CHAPTER VI.

### TELEUTOSPORES.

AT some period in the life of all Uredines * a spore is produced from one of the before-mentioned mycelial developments, to which the appellation teleutospore has been given. In the strict sense of the term teleutospore means the last-formed spore—that is to say, it is formed later in point of time than the æcidiospores and uredospores ; but while this is true in the majority of cases, yet there are many species of the Uredineæ in which no other spore-form than the teleutospore occurs. Still, the name has become so familiarized to us by long usage that it is unadvisable to change it. These spores are formed in a similar manner to the uredospores from a stroma (spore-bed, clinode, hymenium) produced by an aggregation and entanglement of mycelial hyphæ placed just beneath the cuticular structures of the host-plant and parallel to them. Perpendicularly from this stroma are given off erect branches, which, becoming dilated at their free ends, are soon in-

* De Bary ("Vergl.," p. 305) points out that *Uredo symphyti* has no other spore-form than the Uredo, and indicates this species as being a degenerate type, which, having lost its other spore-forms, is capable of existing without them. Without questioning the truth of this statement, one cannot fail to remember that until quite recently the other spore-forms of many Uredines were unrecognized, for instance, *Melampsora cerastii ;* so it is quite possible that the teleutospores of *U. symphyti* may exist, although we at present do not know of them.

vested with a thick membrane (Plate III. Figs. 18, 19). Their subsequent development is similar to that which occurs in the uredospores. The investing membrane, however, undergoes considerable thickening, and becomes darker in colour. At the apex of the spore this thickening is most marked, but it does not always follow that here the depth of its colour is most noticeable, although it generally is so. This is, to a certain extent at any rate, owing to the fact that the apex of the spore is perforated by a small tubular canal, which is, however, by no means easy to observe in the perfect spore, but whose presence is obvious enough in those spores which have already germinated, or are in the act of germinating. The interior of this membrane is lined by a very thin endospore. This can be brought into view by the action of undiluted sulphuric ($H_2SO_4$) or nitric acid ($HNO_3$). These reagents exert no influence on the investing membrane, but cause the endospore to shrink away from it. The endospore encloses a finely granulated protoplasmic mass, near the middle of which is usually a vacuole. The teleutospore remains for a longer or shorter period attached to the spore-bed by the lower part of its spore-forming hypha, which contains a hyaline or watery material, and constitutes a stem (pedicel, peduncle) to the teleutospore.

Teleutospores may be simple (Uromyces, Melampsora) or compound (Puccinia, Triphragmium, Chrysomyxa).

The teleutospores of Uromyces are developed in the manner above described. The germinal canal is always at the apex of the spore. The membrane is generally smooth, but in some species, as *Ur. alchemillæ*, it is studded with prominent wart-like tubercles. Similar tubercles are observed on some of the species affecting the Leguminosæ, but instead of remaining as discrete tubercles they become in part or wholly confluent, so as to form elevated ridges. As an accidental variety, one sometimes sees a few bicellular spores in the true Uromyces spore-bed; this has been observed with *Uromyces trifolii* on *Vicia sepium*.

Fig. 2.—Teleutospore of *Uromyces fabæ* germinating, showing the germ-canal through which the promycelium has been protruded, and the development of the promycelial spores from above downwards. (Tulasne.)

The teleutospores of Puccinia are compound. They consist of two distinct spores borne on one pedicel. The dilated upper extremity of the hypha, which is destined to become a Puccinia spore, soon assumes a fusiform shape, and is uniformly filled with granular protoplasm, while the lower part of the hypha contains only a hyaline or watery material (Plate III. Fig. 18). Soon a delicate transverse septum appears near the middle of the swollen part, dividing it transversely into two nearly equal compartments. The spore now becomes invested by a stouter membrane (Fig. 19), which gradually increases in thickness so that the body now can be seen to consist of two distinct accumulations of protoplasm. As the thickening of the outer membrane goes on it becomes darker. This thickening is most marked at the apex of the upper compartment. The granular protoplasm at first usually shows two or three vacuoles (Fig. 20), but these become reduced in number, so that in the mature spore only one is observable (Fig. 21). How this vacuole is formed is not altogether clear, but it probably is produced by the protoplasm of each compartment accumulating more towards the walls of the cell than elsewhere. Each compartment, then, consists of a thick outer membrane, lined by a thin one (the endospore) which encloses the protoplasm and its vacuole.

In the American species *P. amorphæ*, Curtis, the mother-cell remains distinct from the spore-wall of the teleutospore, separated by a gelatinous substance. When the teleutospores of this species are placed in water the mother-cell swells in a very remarkable manner, and can be seen surrounding the teleutospore (Plate IV. Figs. 15, 16, 17). The germinal canal penetrates the membrane only, being closed below by the endospore. The germinal canal of the lower spore is placed laterally immediately below the septum. In most species the exterior of the spore is smooth, but in several it is tuberculate, papillose, or verrucose. In one of the American species, *P. aculeata*, these protuberances are nearly cylindrical and curved. In *P. fusca, pruni*, etc., the tubercles are small, but exceedingly numerous, covering the whole exterior of the spore. In *P. smyrnii* they are few, discrete, and conical. *P. coronata* is characterized by a crown of elongated, variously bent, cylindrical processes, surmounting the upper segment. Here the superior germ canal is not situated at the apex of the spore, but towards one side (Plate IV. Fig. 3), just as it normally occurs in the lower cell of other species. The general outline of the Puccinia spore varies not only in different species, but also in individual spores from the same spore-bed. In *P. pruni* and *fusca* the teleutospores consist of two distinct superimposed globose bodies, flattened at the point of contact; these easily separate from one another by rough manipulation. Not infrequently, only one segment of the spore becomes developed, the other, usually the lower, remaining abortive. In some instances numerous single spores (mesospores) are produced, either mixed with the normal bicellular spores or without them, as in *P. obscura, scirpi, convolvuli, porri*, and *sonchi*. These mesospores are morphologically analogous to the teleutospores of Uromyces. *P. rubigo-vera*, when it occurs on various species of Hordeum, consists almost entirely of mesospores (=*P. anomala*, Rost). An American species produces bicellular spores in the summer (= *P. vexans*, Farl.), and subsequently mesospores only (=*Uromyces brandegei*, Peck). The brown colouration of the membrane of *P. graminis* is sometimes abnormally confined to the pedicel, the spore remaining colourless. This condition I first observed in some specimens sent me by Mr. W. Marshall, from near Ely. I have also seen it in Australian specimens. Upon the other hand, instead of spores being bicellular, they are sometimes variously complex, triseptate, or even quadriseptate; or the upper or lower cell may be longitudinally or obliquely septate, which is the normal condition of the spores of Triphragmium. Greville * figures this condition as *P. variabilis.* This septation is purely accidental, occurring in one spore-bed, but not in any of the others on the same plant. Mr. Vize and Professor Trail have met with this condition in *P. bullata*, and the Rev. Dr. Keith in *P. lychnidearum.* I have seen it in one of the graminaceous species on *Lolium perenne*, and Mr. Grove in *Puccinia betonicæ.*

The teleutospores of Phragmidium and Xenodochus differ from those of Puccinia in being composed of a greater number of elementary spores, which are developed from above downwards (basipetal formation). The dilated upper end of the hypha becomes full of coarsely granular protoplasm; delicate septa appear from above downwards. The protoplasm thus differentiated is soon seen to be surrounded by a cell-wall (endospore), which at this stage of its development is considerably smaller than the compartment in which it is situated; so that the young spores are obviously, in structure, individually distinct from one another. The uppermost spore is usually

* Greville, "Scot. Crypt. Flora," t. 75.

surmounted by a prominent conical point. The membrane soon becomes dark and thickened, and each spore, enveloped by its endospore, pretty well fills each compartment. In the mature state the entire teleutospore is clothed by a colourless transparent cuticle, which is generally tuberculate. That this cuticle is distinct from the thick membrane is shown by warming a spore in caustic potash, when the former disappears, while the thick membrane remains unaffected.

From two to four lateral germ canals or pores occur in the membrane, as Tulasne first showed (Plate IV. Fig. 5). These are arranged equatorially, and appear to be about four in number. Dietel,* however, remarks that in *Ph. obtusum*, Strauss, the germ-pores are placed as in Puccinia, namely, the superior one at the apex of the upper cell, and those of the lower cells laterally immediately below each of the septa, and that there is only one germ-pore in each cell. A similar condition, he states, exists in the Australian species *Ph. barnardi*, Plow. Longitudinal septation very rarely happens.†

In Coleosporium a similar condition exists, but in Triphragmium the divisions are multiple and longitudinal. In Cronartium the teleutospores are arranged in the form of a solid pillar or column, that projects perpendicularly to the spore-bed, and is surrounded at its base by a nest of uredospores. In Chrysomyxa the teleutospores occur in waxy masses, and consist of a series of superimposed spores, each of which has a single germ-pore placed as in the lower cell of Puccinia, not equatorially, but laterally near the upper end. In Gymnosporangium the spores are held together by a gelatinous matrix; they are shaped

* *Dietel*, "Beitrage zur Morphol. und Biolog. der Uredineen," *Botan. Centralblatt.*, bd. xxxii. (1887), tab. ii. figs. 1 and 2, reprint, p. 9.

† Eysenhardt, "Linnæa," vol. iii. 1828.

like Puccinia spores, but have, in the European species, generally two germ-pores in each cell, and these are placed

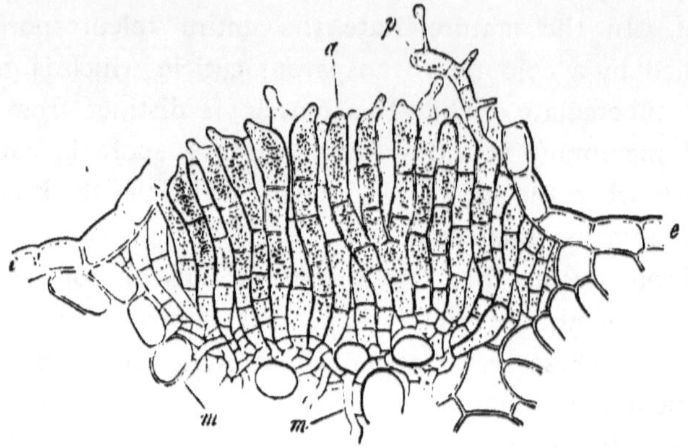


Fig. 3.—*Chrysomyxa rhododendri.* Section of spore-bed showing the compound teleutospores (*a*), one of which is in the act of germination, having emitted a promycelium (*p*) from upper corner of its superior compartment. In most of the other spores germination has commenced. *e*, The epidermal cells of the host-plant (*R. hirsutum*); *m*, mycelial hyphæ. (De Bary.)

at that end of the cell which is nearest the septum (Plate III. Figs. 22, 23; and Plate IV. Figs. 11, 13, and 14).

*Germination.*—The period at which germination takes place in teleutospores is subject to variation with the different species of Uredines. In the greater number of cases this process ensues only after the teleutospore has remained for some considerable time in a state of quiescence or rest. Generally this resting period extends from the summer or autumn of one year until the ensuing spring, corresponding, in fact, with the conditions of the host-plant, which of course, in the case of deciduous plants, is without suitable foliage during the winter months into which the fungus can gain an entrance. On the other hand, there are several species in which teleutospores germinate as soon as they are mature, without undergoing any resting period (Leptopuccinia, Lepturomyces); while there are others which have only a very short resting period (*P. rubigo-*

*vera, poarum*). In either case the germinative process is the same. This consists in the protrusion of a germ-tube from the teleutospore through the germ-pore above mentioned, which tube is really an extension of the endospore through the germ-pore. This tube consists of a hyaline, homogeneous membrane, into which the protoplasm contained within the spore passes. This tube (the promycelium) does not, like the germ-tubes of the uredospores and æcidiospores, elongate indefinitely, but, after attaining a certain length, ceases to grow onwards and terminates in a blind, blunt extremity. Into the promycelial tube the protoplasm passes from the spore, and accumulates towards the peripheral extremity, so that the end nearest the spore becomes empty. The distal end now exhibits one or more septa, which are formed from above downwards. At the same time, from each compartment thus divided off there arises a single, short, pointed branch; these at their points dilate. The dilated end becomes the receptacle for the protoplasm originally contained in each compartment, and rapidly assumes an oval or subreniform outline (the promycelial spore), which in the course of a few hours becomes abstricted and falls off (Plate IV. Figs. 2–10). The promycelial spores, when placed in water or upon any damp surface, forthwith germinate (Plate III. Fig. 24). This process consists in the outgrowth, from some point of their surface, of a short but acutely pointed tube. Placed upon the cuticle of a living leaf of the proper host-plant, this germ-tube turns its point downwards, and, piercing the epidermal cell-walls, enters the tissue of the leaf (Plate IV. Fig. 1). Having thus gained an entry into its host, the protoplasm contained in the promycelial spore passes downwards into the germ-tube, leaving the spore empty. The empty spore-cell falls off, and the minute opening through which the germ-tube has entered the epidermal cell ceases to be visible; this is probably due to the elasticity of the epidermal cell-wall. The germ-tube itself continues its onward growth; soon branching, it insinuates itself between the cells of the host, where it gives rise to a mycelial development, similar to that which arises from the æcidiospore or uredospore germ-tubes.

With the Leptopuccinia and Lepturomyces, the germ-tube of the promycelial spore does not bore its way through the epidermal cells, but enters, as De Bary * first showed with *P. dianthi*, through the stomata, or, as Ráthay † points out with *P. malvacearum*, it enters, like the germ-tube of *Tubercinia trientalis*, between the epidermal cells, and then pierces laterally into the side of the adjacent cell.

If the germination of the promycelial spores takes place in water, and not upon their proper host-plant, the germ-tubes sometimes become swollen at their extremities, so as to form reserve spores in the same manner as has been previously described with the uredospores (Plate III. Fig. 24).

The mycelium produced from a promycelial spore in the tissues of its proper host-plant in due time gives rise to a fresh spore-development. This spore-development, however, varies in different cases.

1. In certain species the mycelium produces spores which are exact counterparts of the original teleutospores (Micropuccinia).

2. It may give rise to teleutospores similar in general appearance as to its progenitors, but endowed with the faculty of immediate germination (Leptopuccinia, etc.).

3. It may give rise to a crop of uredospores, which may

* De Bary, "Champ. paras.," *Ann. Sc. Nat.*, 4ᵉ sér. tome xx. p. 84 (reprint).

† Ráthay, "Ueber d. Eindringen. Puccinia Malvacearum," *Verhandl. d. Kk. Zool. bot. Ges.*, band xxxi. (1881), t. 1.

be typical of the species in question. There are instances, however, in which, instead of being truly typical, these uredospores depart in some degree from those subsequently found in the life-history of the Uredine. For example, in *Triphragmium ulmariæ* the primary uredospores do not differ in their individual form and size from the secondary, but the former occur in large dusty sori principally on the petioles and midribs of the host-plant, while the latter are much smaller and scattered over the under-surface of the leaves. A more common deviation, however, consists in the association of the primary uredospores with spermogonia (Brachypuccinia).

4. In other instances the mycelium produces spermogonia and æcidiospores. The æcidiospores may be upon the same species of host-plant upon which the teleutospores occur (Auteupuccinia), or upon some plant of a totally diverse nature (Heteropuccinia)

The teleutospores of Endophyllum, although they are produced in the same way and closely resemble æcidiospores, yet in germination they behave like the teleutospores of Puccinia (Plate IV. Fig. 7). In Coleosporium (Plate IV. Fig. 9) each cell produces a single promycelial spore, while in Melampsora (Plate IV. Fig. 8) and Chrysomyxa (Plate IV. Fig. 10) three or four are developed from each promycelium, as is also the case with Cronartium.

## CHAPTER VII.

### HETERŒCISM.

*History.*—The fact that a certain number of Uredines possess the faculty of passing a part of their lives upon one plant, and the remainder of it upon another and a totally different one, is so remarkable that until quite recently there were persons who declined point-blank to believe it. It is quite unnecessary now to undertake *seriatim* to answer the theoretical objections which have been raised against the fact that heterœcism does occur, for the simple reason that these objections are only theoretical. The process of simply placing the spores of the fungi in question upon their various host-plants is so easy, that any one wishing to appeal to Nature herself can do so with very little trouble.

The first Uredine in which this peculiarity of development was observed was *Puccinia graminis*, the wheat mildew. The mildew of wheat has, as a blight, probably been known from remote antiquity. The Romans held a festival on April 25—the Robigalia, or Rubigalia—with the object of protecting their fields from mildew. The sacrifices offered on this occasion consisted of the entrails of a dog and a sheep, accompanied with frankincense and wine.* In Wickliffe's Bible, the suggestion is made that

* Smith, "Smaller Dictionary of Greek and Roman Antiquities," 5th edit. (1863), p. 322.

the seven years of famine with which the Egyptians were afflicted were caused by mildew. This, of course, does not prove that the ancient Egyptians had any acquaintance with mildew, but it does show that in early English times the disease was not only known as an affection of cereals, but also that it was regarded as an agency sufficiently powerful to cause famine. In Shakspeare's time, mildew must have existed, for we read of how "The foul fiend Flibbertigibbet mildews the white wheat." * The fungoid nature of the mildew was not known, however, until the latter half of the last century, for Tull,† writing in 1733, attributes it to the attack of small insects, "brought (some think) by the east wind," which feed upon the wheat, leaving their excreta as black spots upon the straw, "as is shown by the microscope"! Felice Fontana,‡ some thirty years later, published an account of the fungus, with figures. Persoon,§ in 1797, gave it the name it still bears (*Puccinia graminis*), and also figured it, as did Sowerby || in 1799, under the name of *Uredo frumenti*.

With regard to the fact that the barberry in some way favours the growth of mildew upon wheat, there is no doubt that it was well known to practical farmers during the eighteenth century, for we find in America, as early as 1760, in the state of Massachusetts, an Act was passed by the legislature, compelling the inhabitants to extirpate all barberry bushes.¶ In New England a similar law existed, which is referred to by Schöpf.** In our own country;

* Shakspeare, "King Lear," act iii. sc. iv.

† Jethro Tull, "Horse Howeing Husbandry," 3rd edit. (1751,) p. 151.

‡ Felice Fontana, "Osservazioni sopra la Ruggine del Grano." Lucca: 1767.

§ Persoon, "Tentamen Dispos. Method. Fungorum" (1797), p. 39, t. iii. fig. 3.

|| Sowerby, "English Fungi," vol. ii. (1799), t. 140.

¶ For "Barberry Law of Massachusetts," see p. 302.

** J. D. Schöpf, "Reise durch die mittleren und südlichen vereinigten Nordamerikanischen Staaten," theil i. p. 56. Erlangen: 1788.

especially in Norfolk, this belief also existed, for we find Marshall,* writing in 1781, says, "It has long been considered as one of the first vulgar errors among husbandmen that the barberry plant has a pernicious quality (or rather a mysterious power) of blighting wheat which grows near it.

"This idea, whether it be erroneous or founded on fact, is nowhere more strongly rooted than among the Norfolk farmers; one of whom mentioning, with a serious countenance, an instance of this malady, I very fashionably laughed at him. He, however, stood firm, and persisted in his being in the right, intimating that, so far from being led from the cause to the effect, he was, in the reverse, led from the effect to the cause; for, observing a stripe of blasted wheat across his close, he traced it back to the hedge, thinking there to have found the enemy; but being disappointed, he crossed the lane into a garden on the opposite side of it, where he found a large barberry bush in the direction in which he had looked for it. The mischief, according to his description, stretched away from this point across the field of wheat, growing broader and fainter (like the tail of a comet) the further it proceeded from its source. The effect was carried to a greater distance than he had ever observed it before, owing, as he believed, to an opening in the orchard behind it to the south-west, forming a gut or channel for the wind.

*    *    *    *    *    *

"Being desirous of ascertaining the fact, be it what it may, I have inquired further among intelligent farmers concerning the subject. They are, to a man, decided in their opinion as to the fact, which appears to have been so long established in the minds of the principal farmers that

* Marshall, "Rural Economy of Norfolk," 2nd edit. vol ii. p. 19. London: 1795.

it is now difficult to ascertain it from observation, barberry plants having (of late years more particularly) been extirpated from the farm hedges with the utmost care and assiduity. One instance, however, of mischief this year I had related to me, and another I was myself eye-witness to. Mr. William Barnard, of Bradfield, says, that this year, seeing a patch of his wheat very much blighted, he looked round for a barberry bush, but seeing none conspicuous in the hedge, which was thick, he with some difficulty got into it, and there found the enemy. He is clearly decided as to the fact. Mr. William Gibbs, of Rowton, telling me that a patch of his wheat was blighted in the same manner, and that he believed it to proceed from some sprigs of barberry which remained in the neighbouring hedge (which a few years ago was weeded from it), I went to inspect the place, and true it is that near it we found three small plants of barberry, one of which was particularly full of berries. The straw of the wheat is black, while the rest of the piece is of a much superior quality.

"These circumstances are undoubtedly strong evidence, but do not by any means amount to proof."

On October 16, 1782, Marshall writes *—

"To endeavour to ascertain the truth of this opinion, I had a small bush of the barberry plant set in February or March last, in the middle of a large piece of wheat.

"I neglected to make any observations upon it until a little before harvest, when a neighbour (Mr. John Baker, of Southrepps) came to tell me of the effect it had produced.

"The wheat was then changing, and the rest of the piece (about twenty acres) had acquired a considerable degree of whiteness (white wheat), while about the barberry bush there appeared a long but somewhat oval-shaped

* Marshall, *loc. cit.*, p. 359.

stripe of a dark livid colour, obvious to a person riding on the road at a considerable distance.

"The part affected resembled the tail of a comet, the bush itself representing the nucleus, on one side of which the sensible effect reached about twelve yards, the tail pointing towards the south-west, so that probably the effect took place during a north-east wind.

"At harvest, the ears near the bush stood erect, handling soft and chaffy; the grains slender, shrivelled, and light. As the distance from the bush increased the effect was less discernible, until it vanished imperceptibly.

"The rest of the piece was a tolerable crop, and the straw clean, except on a part which was lodged, where the straw nearly resembled that round the barberry; but the grain on that part, though lodged, was much heavier than it was on this, where the crop stood erect.

"The grain of the crop, in general, was thin-bodied; nevertheless ten grains, chosen impartially out of the ordinary corn of the piece, took twenty-four of the barberried grains, chosen equally impartially, to balance them."

In 1784, Marshall repeated his experiment at Statfold, in Staffordshire, with the same result. He says *—

"Upon the whole, although I have not from this year's experience been able to form any probable conjecture as to the cause of the injury, it nevertheless serves to fix me still more firmly in my opinion that the barberry is injurious to wheat."

Withering, writing in 1787 of *Berberis vulgaris*, says,† "This shrub should never be permitted to grow in cornlands, for the ears of wheat near it never fill, and its influence in this respect has been known to extend as far as three hundred or four hundred yards across a field."

* Marshall, "Rural Economy of the Midland Counties" (1790), vol. ii. p. 11.

† Withering, "Botanical Arrangement" (1787), 2nd edit., p. 366.

In 1804, this country suffered severely from an outbreak of wheat mildew, in consequence of which Mr. Arthur Young, the Secretary to the Board of Agriculture, issued a circular of questions, so as to obtain a consensus of opinion, from farmers, landowners, and others interested in the subject, as to various points connected with causation of mildew. The ninth question ran thus: "Have you made any observation on the barberry as locally affecting wheat?" The replies to these questions were published,* and from them the following are selected :—

Isaac King, Esq., Wycombe, Bucks, in answer to the question about the barberry, says, "In 1795, a field of about twenty acres had two large barberry bushes growing within twenty yards of it. These appeared to be the focus of destruction to several acres; in front, close to the hedge, the wheat was as black as ink, and further off it was affected to a less degree. . . . In short, I had fifteen acres very good, and five of very little value. You may conclude the barberries were destroyed."

Mr. S. Johnson, Thurning, Norfolk, says,† "My observations on the barberry have been for several years. I have seen the blast from a small stem blown on the wheat in one direction upwards of two furlongs, like smoke from a chimney."

Mr. W. Maxey, Knotting, Bucks, says,‡ "When passing a wheat-field a few years ago on the eve of harvest, I noticed some streaks of a different darker hue across a furlong of wheat from the hedge directly opposite; at the end of each streak was a barberry bush."

Mr. James Sheppard, Chippendale, Newmarket, says,§ "I have never seen an instance of wheat growing near a barberry not being injured more or less."

* A. Young, "Annals of Agriculture" (1805), vol. xliii. p. 457.

† Marshall, *loc. cit.*, p. 469.

‡ *Loc. cit.*, p. 505.

§ *Loc. cit.*, p. 510.

It is quite unnecessary to quote any further from Mr. Young's correspondents upon this point.

The injurious influence of the barberry was, however, a matter of observation only at this time. Various suggestions were made as to the cause of this: some affirmed the barberry bush exhaled a noxious effluvium; others, that the pollen of its flowers poisoned the wheat; others, again, that it appropriated to itself all the nourishment from the soil in its vicinity.

In 1805, however, Sir Joseph Banks, in his paper on "Wheat Mildew," alluding to the subject before us, mentions the belief as being prevalent amongst farmers, but scarcely credited by botanists, and points out the resemblance the yellow fungus on barberry has to rust, although it is larger. He says,* "Is it not more than possible that the parasitic fungus of the barberry and that of wheat are one and the same species, and that the seed is transferred from the barberry to the corn?"

The suggestion of our eminent countryman was soon put to the test of experiment by a totally independent observer.

The honour of being the first to demonstrate the connection between the barberry Æcidium and the wheat mildew belongs to a Danish schoolmaster, who lived in the village of Hammel, near Aarhus, in Jutland, at the beginning of the present century. In 1818, the Royal Agricultural Society of Denmark published a paper by Schoeler, "On the Pernicious Influence which the Barberry Bush exercises on Cereals."† This paper was almost overlooked until Mr. Nielsen brought it under notice in 1874, in his capacity as Consulting Botanist to the Royal Agri-

* Banks, "Annals of Agriculture," vol. xliii. p. 521.

† Om Schoeler, "Berberissens skudelige Indflydelse paa Sœden," *Landvkomminske Tidender* (1818), part viii. p. 289; Nielsen, *Ugeskrift for Landmœda*, 1884.

cultural Society of Denmark. Schoeler began the study of the subject in 1807, when, by closely observing the yellow spots on the under side of barberry leaves, he came to the conclusion that they were due to a microscopical fungus. In 1810, he noticed that the barberry bushes were nearly free from this fungus, and that the rye was that year almost free from rust.* "I then thought," he says, "that there might possibly be a close relationship between the rust on the rye and that upon the barberry; and when, in the following year (1811), I noticed that the rust upon the barberry appeared much earlier in the spring than the rust did upon the other plants—grasses and cereals—I thought I had found out the true origin of the rust in rye. Still, however, this question again and again presents itself to my mind, 'Where does the rust on rye come from in those places in which no barberries are to be found?' In the summer of 1812 I convinced myself that the barberry bushes are indeed able to communicate the rust to the rye, by means of the wind, even to a considerable distance."

For several years prior to 1813 he experimented in his garden by planting different kinds of corn around barberry bushes, and found that rye and oats were especially liable to be destroyed almost every year by the rust, which always appeared first nearest the barberries.

From 1813 to 1817 he planted large and small barberry bushes in his rye-field. He found that the larger bushes did not give rise to the rust in rye, when they lost their foliage in the process of transplanting; but, on the contrary, the smaller bushes, which did not lose their leaves so readily, did give rise to the rust in the rye to a very marked degree.

One of his experiments he thus describes: "I planted

* By the word here translated "rust" is evidently meant, not only the Uredo, but also the mildew.

in May, 1816, in a rye-field, three small barberries; one of them did not thrive, but the other two developed nineteen leaves before the end of May, nearly all of which had the yellow spots (of *Æcidium berberidis*) upon them. When, in the middle of the following June, the rye immediately surrounding these two bushes became rusty, I invited, by a notice in the newspapers, every one interested in the question to come and convince themselves of the pernicious influence which the barberry exerts. On account of this invitation many people came to visit me, and all of them, the learned as well as the unlearned, declared that they could accept no other cause for the rust on the rye than the small barberry plants. They were astonished that so small a cause should have produced so great an effect. On June 22, most of the rye plants for thirty to forty feet (square feet) around these small barberries were more or less rusty, mostly so to the north and north-west, but for a long time afterwards not even a single rust spot could be found elsewhere in the field."

In the same year (1816) Schoeler performed the following experiment:—Some fresh branches of the barberry bush having rusty leaves upon them were cut off, put into a box, and carried to a rye-field, where the rye was still moist with dew. The rusty barberry leaves were applied to some of the rye plants—to the straw as well as to the leaves—by rubbing them with the underside of the affected barberry leaves, until he could see some of the "yellow dust" (spores) of the fungus adhered to the rye plants. The infected rye plants were then marked by tying them to sticks driven into the ground. In five days' time these plants were badly affected with rust, "while at the same time," says Schoeler, "not one rusty plant could be found anywhere else in the whole rye-field."

The question, however, was not even now fully decided;

for although the above facts were in themselves unanswered, yet the so-called scientific botanists urged that the fungus upon the barberry leaves belonged to a totally distinct genus (Æcidium) to that upon the wheat (Puccinia). There is evidence to show that many careful observers, even at this time, suspected that the Puccinia was connected in some way with the uredospores which occur as its pre-cursors. This remained a suspicion only until Tulasne demonstrated that the connection between the uredospore and the teleutospore existed not only in the species in question, but was the general rule amongst the Uredineæ. In 1861, De Bary showed that many of the Uredineæ not only had uredospores and teleutospores, but also that the latter gave rise in many cases (but not in all) to æcidio-spores, and conversely the æcidiospores to uredospores.

De Bary also pointed out that in certain cases the sowing of germinating teleutospores upon the same species of host-plant which bore them was not followed by any result. Amongst these were *Puccinia graminis*. It further occurred to him that, as there were several æcidia unac-companied on their host-plants by any other spore-form, these might belong to Uredines which passed a part of their life upon one plant and the remainder upon another. Familiar with the facts already known to the practical agriculturist concerning the barberry and wheat mildew, he put the matter to the test of actual experiment. In 1864, he sowed *Puccinia graminis* on barberry and produced the Æcidium, and in 1865 he did the converse culture, by sowing the æcidiospores upon rye.

The results obtained by his experiments with *P. graminis* led De Bary to investigate the life-histories of other æcidia, which, like *Æc. berberidis*, are unaccompanied by any other spore-form on the same host-plant. Thus he found that *P. rubigo-vera* has its æcidiospores upon *Lycopsis arvensis*, and *P. coronata* upon *Rhamnus frangula.*[1] In the following year, 1866, Oersted showed that *Gymnosporangium sabinæ* and *juniperinum* were similarly connected with *Ræstelia cancellata* and *cornuta;* and in 1867, that *G. clavariæforme* was connected with *R. lacerata.* In 1869, Fuckel indicated the connection of *Uromyces junci* with *Æc. zonale.* In 1873, Magnus worked out the life-history of *P. caricis* in its relationship to *Æc. urticæ;* and Schröter, in the same year, the heterœcism of *Æc. ranunculi bulbosi* and *Uromyces dactylidis.* Since this time these facts have been repeatedly verified by numerous workers, and our knowledge of the subject has continuously increased, as the subjoined tabular statement will show.

| | | | |
|---|---|---|---|
| * Puccinia graminis | Æcidium berberidis | De Bary | 1864 |
| * Puccinia rubigo-vera | Æcidium asperfolii | De Bary | 1865 |
| * Puccinia coronata | Æcidium rhamni | De Bary | 1865 |
| * Gymnosporangium sabinæ | Rœstelia cancellata | Oersted | 1866 |
| * Gymnosporangium juniperinum | Rœstelia cornuta | Oersted | 1866 |
| * Gymnosporangium clavariæforme | Rœstelia lacerata | Oersted | 1867 |
| * Uromyces junci | Æcidium zonale | Fuckel | 1869 |
| * Puccinia caricis | Æcidium urticæ | Magnus | 1873 |
| * Uromyces dactylidis | Æcidium ranunculi-bulbosi | Schröter | 1873 |
| * Coleosporium senecionis | Peridermium pini | Wolff | 1874 |
| † Puccinia moliniæ | Æcidium orchidearum | Rostrup | 1874 |
| * Puccinia sessilis | Æcidium allii | Winter | 1874 |
| * Puccinia phragmitis | Æcidium rumicis | Winter | 1874 |
| Uromyces pisi | Æcidium cyparissiæ | Schröter | 1875 |
| * Puccinia poarum | Æcidium tussilaginis | Nielsen | 1876 |
| Puccinia limosæ | Æcidium lysimachiæ | Magnus | 1877 |
| Puccinia sesleriæ | Æcidium rhamni-saxatilis | Reichardt | 1877 |
| Puccinia sylvatica | Æcidium taraxaci | Schröter | 1879 |
| * Uromyces poæ | Æcidium ficariæ | Schröter | 1879 |
| Chrysomyxa rhododendri | Æcidium abietinum | De Bary | 1879 |
| Chrysomyxa ledi | Æcidium abietinum | De Bary | 1879 |
| Calyptospora goeppertiana | Æcidium columnare | Hartig | 1880 |
| † Melampsora populina | Æcidium clematidis | Ráthay | 1881 |
| * Puccinia magnusiana | Æcidium ranunculi-repentis | Cornu | 1882 |
| † Melampsora caprearum | Cæoma euonymi | Rostrup | 1883 |
| † Melampsora hartigii | Cæoma ribesii | Rostrup | 1883 |
| † Melampsora tremulæ | Cæoma mercurialis | Rostrup | 1883 |
| Puccinia dioicæ | Æcidium cirsii | Rostrup | 1883 |

[1] De Bary, "Neue Untersuch. über Uredineen," 2nd paper (1886), pp. 208—214.

| | | | |
|---|---|---|---|
| Puccinia eriophori | Æcidium cineraria | Rostrup | 1883 |
| * Puccinia obscura | Æcidium bellidis | Plowright | 1884 |
| * Puccinia schöleriana | Æcidium jacobææ | Plowright | 1884 |
| * Puccinia perplexans | Æcidium ranunculi-acridis | Plowright | 1884 |
| Puccinia vulpinæ | Æcidium tanaceti | Schröter | 1884 |
| * Puccinia arenariicola | Æcidium centaureæ | Plowright | 1885 |
| * Puccinia phalaridis | Æcidium ari | Plowright | 1885 |
| † Melampsora tremulæ | Cæoma laricis | Hartig | 1885 |
| * Gymnosporangium biseptatum | Rœstelia botryapites | Farlow | 1885 |
| * Gymnosporangium confusum | Æcidium mespili | Plowright | 1886 |
| Puccinia polliniæ | Æcidium strobilanthis | Barclay | 1886 |
| Gymnosporangium calvipes | Rœstelia aurantiaca | Thaxter | 1886 |
| Gymnosporangium macropus | Rœstelia pyrata | Thaxter | 1886 |
| Cronartium asclepiadeum | Peridermium acicola | Cornu | 1887 |
| * Puccinia extensicola | Æcidium asteris | Plowright | 1888 |
| * Melampsora æcidioides | Cæoma mercurialis | Plowright | 1888 |
| * Puccinia paludosa | Æcidium pedicularis | Plowright | 1888 |
| * Puccinia persistens | Æcidium thalictri-flavi | Plowright | 1888 |
| * Puccinia trailii | Æcidium acetosæ | Plowright | 1888 |

* is affixed to those species which I have personally investigated.

† to those of which I have repeated the cultures, but am not able to confirm the above statements.

The question naturally presents itself to us, Why are some species heterœcious and others not? One reason is pretty obvious, namely, that those Pucciniæ and Uromyces which are heterœcious occur upon host-plants whose cuticle is, if not silicous, at least very hard and difficult for the germ-tube of the promycelial spore to pierce—namely, on grasses, Carices and Junci. This, however, can hardly be the only reason, since Schröter has produced the Æcidium on *Euphorbia cyparissias* from *Uromyces pisi*; in this case both the host-plants have soft epidermal cells. The Coleosporia and Melampsoræ afford similar instances. Whatever may have been the cause or causes in bygone ages, the fact is that at the present time so completely have these parasites become heterœcismal in habit, that the most profuse application of their promycelial spores to the graminaceous host is always without result.

## CHAPTER VIII.

### MYCELIUM OF THE USTILAGINEÆ.

THE vegetative mycelium of the Ustilagineæ is to the parasite one of its most important organs, for by it, and by it alone, does the fungus derive its nutriment from the host-plant upon which it subsists. Yet the mycelium is of all parts that which is least frequently observed; nor can this be wondered at when one remembers that, like the mycelium of the Uredineæ, it can be seen only by careful search — by cutting and teazing out numerous thin sections of the host-plant. We owe most of the information we possess upon the mycelium of the Ustilagineæ to Fischer von Waldheim,* although, among others, both Kühn † and Hoffmann ‡ had previously figured it. It exists most abundantly in the tissues of the host-plant, in the immediate vicinity of those places in which the spores are developed; but it can also be found in other parts of the affected plant—in the monocotyledons particularly in the stem, but especially in the nodes and in the root-stock. In the dicotyledons it is not so easily found at a distance from the spore-beds; still, in them too it has been seen in the root-stock and

* Fischer von Waldheim, "Pringsheim Jahrbücher" (1869), vol. vii. pp. 1, 2.

† Kühn, "Krankheiten der Kulturgewächse," 2 aufl. 1859.

‡ Hoffmann, "Ueber der Flugbrand." 1866.

stem, especially in the nodes. The mycelium consists of hyaline tubes, frequently septate, and enclosing watery or pellucid, frequently vacuolated contents (Plate V. Figs. 1, 2). Its walls vary in thickness, but very often they have a distinctly double contour. The number and frequency of the septa are subject to much variation; in some instances they are close together, at others they are only found at distant intervals. This is also the case with the mycelial ramifications; sometimes the hyphæ do not extend for more than 2μ without branching, at others they extend for 20μ or more without dividing. These long unbranched hyphæ are found mostly in the internodes; in the nodes themselves not only are the branches more abundant and convoluted, but here too are encountered, more abundantly than elsewhere, the little intercellular haustoria, or suckers, which characterize the mycelia of many of the Ustilagineæ. The mycelium ramifies not only between the cells of the host-plant, but, frequently piercing their walls, grows through them. Its diameter varies from 2 to 5μ. The addition of caustic potash to a section of the host-plant containing mycelium renders the latter more distinct, and otherwise clears up the preparation; so does prolonged treatment in glycerine. Its walls are not composed of cellulose, as they do not show any blue reaction when treated with sulphuric acid and iodine; but in some cases, as with *U. maydis* and *Sorosporium saponariæ*, they do get an external coating of cellulose from the tissues of the host-plant, which, completely investing them, hides them from view (Plate VI. Fig. 3). The mycelium of almost all the Ustilagineæ permeates more or less the whole of the affected plant, and although in the advanced state we can find it only near the spore-beds, yet originally it could be found in all parts of the axis of the young plant. In this it differs from the localized mycelia of most of the

Uredineæ. These permeating mycelia, whether of the Uredineæ or Ustilagineæ, are perennial. When the plant, if perennial, dies down in winter, the mycelium, of course, dies down with it, but remains alive, although quiescent, in the upper part of the root-stock; and when fresh shoots are sent up in spring, the mycelium is sent up in them. One peculiarity of most of the Ustilagineous mycelia is that, although it pervades more or less the whole plant, it produces its spore-formation at certain favoured places only; these are, for the most part, in the flowers or seeds of the plants, but not always, sometimes on the stems or in the leaves. The place of spore-formation, however, is constant with each species. If a plant affected with one of the Ustilagineæ be transplanted into a garden, it will, year after year, be affected with the parasite. In my own garden at King's Lynn I have had growing for the past six years, plants of *Colchicum autumnale* affected with *Urocystis colchici*, *Triticum repens* with *U. hypodytes*, and *Avena elatior* with *U. segetum*. De Bary mentions that a plant of *Saponaria officinalis*, in the Freiburg Botanic Garden, was for more than ten successive years affected with *U. violacea*.

# CHAPTER IX.

## FORMATION OF THE TELEUTOSPORES OF THE USTILAGINEÆ.

WHILE it is true that the spores of Ustilagineæ are formed from the mycelium, yet the process does not take place directly from the vegetative mycelium which has just been described. On the contrary, at those favoured parts of the affected host-plant at which the spores are developed the vegetative mycelium often quite suddenly changes its character. The double-contoured hyphæ with pellucid vacuolate contents lose their double contour, become swollen or distended, and contain, instead of a clear watery fluid, a gelatinous, granular protoplasm in which numerous oleaginous particles may often be seen (Plate V. Figs. 3–6). The gelatinization of these spore-forming hyphæ is a great character of the Ustilagineæ; it does not, however, occur in all species. The first change observable in the mycelium before it becomes a spore-forming hypha, is that its walls increase in thickness at the expense of its calibre, which becomes proportionately diminished; soon, however, the whole hypha becomes dilated, so that its lumen is increased. Its contents can now, by the action of reagents, be shown to consist of protoplasm. Spore-formation takes place, after these changes in the mycelium, so differently in the different genera that it will be necessary to describe the process, as it takes place in each one, separately.

*Ustilago.*—The spore-forming hyphæ enlarge and branch in various ways. The gelatinization of their interior takes place to such an extent as to almost obliterate their lumen, which, however, may frequently be seen as a narrow shining line in the middle of the hyphæ (Plate V. Fig. 7). At certain points the surface of these hyphæ enlarge, so that they appear nodose. The increase in size of the hypha continues, so that adjacent hyphæ become variously tangled and intertwined together, and eventually many of the hyphæ appear glued together, or to have coalesced. The irregularities of the hyphæ become more marked, and it is obvious that each tumefaction will eventually become a spore, inasmuch as they gradually get more and more rounded (Plate V. Fig. 8). It is in the interior of these distended hyphæ that spore-formation takes place. It is, however, always found that the external spores are the most developed, the formation being, in fact, centripetal. The commencement of the differentiation of the protoplasmic contents is at the exterior of the mass, and it gradually proceeds inward towards the centre. The spores when first formed have gelatinous envelopes, and gradually become more or less polygonal from mutual compression. The interior of the spore is now seen to have a distinct contour, and to contain fatty granules. The outer edge of this contour darkens, and even while it is still surrounded by a thin gelatinous envelope the irregularities of the epispore begin to be apparent. As the spores ripen this gelatinous membrane disappears, so that at their maturity they have no remnant of it; nor are any remains of the mycelial hyphæ attached to them, as is often seen in Tilletia.

*Sphacelotheca.*—This genus differs from Ustilago in the spore-mass being developed in a receptacle. De Bary thus describes its development: "The vegetative mycelium, entering the ovary through the flower-stalk, sends its hyphæ through the funiculus into the ovule, which becomes permeated by densely interwoven hyphæ. The micropylar end of the integuments alone escapes and remains, as a cap on the top of the diseased ovule, for some time,

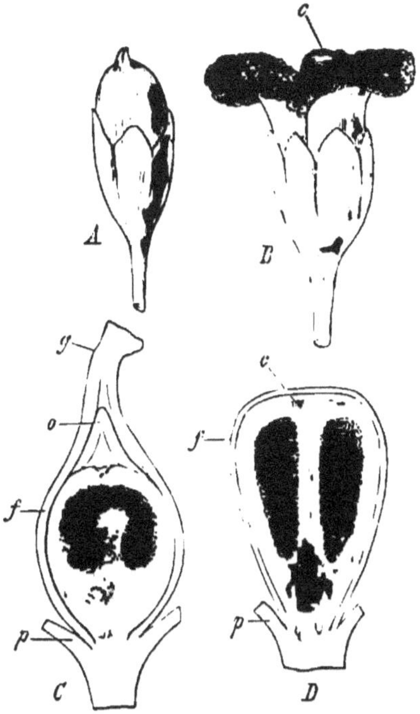


Fig. 4.—*Sphacelotheca hydropiperis* (Schum.) *A*, Ovary and perianth of *Polygonum hydropiper* affected with *S. hydropiperis*. *B*, The same more advanced, showing the micropylar cap (*c*). *C*, Section of ovary and perianth at an early stage; *g*, the style; *o*, the integument of the micropylar end of the ovule; *f*, wall of the ovary; *p*, the perianth. The spore-formation is seen to be commencing above, and the rudimentary columella is visible. *D*, Section of a more advanced ovary, showing the walls of the receptacle and the columella (*c*). Slightly magnified. (De Bary.)

but eventually falls off. The hyphæ develop partly into spores and partly into the receptacle. The latter consists of a thick external case with a central columella. The cells of which it is composed are but loosely compacted, colourless, and about the size of the spores. The least injury fractures this case and allows the escape of the spores." The spore-formation itself is similar to that of Ustilago,* and, commencing above, proceeds downwards.

*Sorosporium.*—The spore-formation in Sorosporium differs considerably from that which has just been described in Ustilago, although it is obviously of the same type. In Ustilago the mature spores are separate and distinct, forming usually a pulverulent mass. In Sorosporium, on the other hand, they are in their perfect state aggregated into spore-balls, which individually often contain fifty or a hundred separate spores. The process of spore-formation has been studied by Von Waldheim with *S. saponariæ*, and is as follows:—The mycelium, which is very abundant in the blossom and ovary, rapidly changes into spore-forming hyphæ, from 4 to 7μ in diameter, which are gelatinous and full of shining protoplasm. The free ends of these hyphæ have a tendency to curve inwards and roll themselves up (Plate V. Fig. 9). The spore-forming hyphæ from several contiguous mycelial branches, incline together, and twist themselves into a ball, as happens in the formation of a lichen thallus. These convoluted and contorted spore-forming hyphæ, being gelatinous, soon become so intertwined and entangled that they cease to be individually recognizable; to all appearances they coalesce together in part, if not entirely, and on the exterior of this gelatinous ball other hyphæ are now seen encircling it (Plate V. Figs. 10, 11). These latter, also being gelatinous, soon lose their individuality, although at times traces of their concentric arrangement can be made out. Spore-formation takes place only in the central gelatinous ball, in the middle of which it commences by the central part darkening in colour and becoming differentiated into spore-like bodies, which vary in number from four to sixteen. Apparently these bodies again subdivide, so

* De Bary, "Vergleichung," p. 187.

that when the spores arrive at their maturity the spore-balls contain sixty to a hundred or more spores. In the young state these developing spores are polygonal from mutual pressure, and they are to be found in spore-balls not more than 50μ in diameter. In the subsequent development of the spores the balls increase in size, and the gelatinous zone swells also. When, however, the spores assume their characteristic deep brown colour this gelatinous zone begins to be absorbed, having been utilized in the development of the spores. In spore-balls of 70μ in diameter the gelatinous zone is only from 4 to 6μ thick. It entirely disappears when the spores have attained their full maturity. In a certain sense it may be said that the spore-formation is centrifugal, inasmuch as it commences in the centre of the gelatinous ball; but the peripheral spores are as in Ustilago, the oldest, having been pushed outward by the continued formation of the younger spores in the centre of the mass. These externally placed spores either continue their development independently, or, what is more probable, their spore-forming hyphæ have become greatly elongated, but still remain in connection with the host-plant (Plate V. Fig. 12). In Sorosporium, however, certain solitary spores occur independent of those aggregated into spore-balls. The development of these isolated spores takes place in single hyphæ. The end of the hypha becomes gelatinized, swells up, and a spore is developed inside. They are at first surrounded by the gelatinized hyphæ, which generally, however, disappear entirely by the time the spores are mature.

*Tubercinia.*—The spore-formation of Tubercinia has been worked out by Woronin * in *Tubercinia trientalis.*

* Woronin, "Beitrag. sur Kenntniss der Ustilagineen" (1882), pp. 4–16, t. ii. figs. 3–10. De Bary and Woronin, "Beiträge zur Morphol. und Physiol. der Pilze," 5 reihe. 1882.

Subjoined is a summary of his observations. The mycelium pervades all parts of the affected plant, excepting the roots. It consists of intercellular, branched, sparsely septated, hyaline hyphæ, having a diameter of from 2 to 3μ. It is most abundantly distributed in the cortical tissues, and sends botryform haustoria (Plate VI. Fig. 1) into the adjacent cells. At certain places it gives off smaller, more richly septate branches, unprovided with haustoria, which are destined to form the spore-beds. This is accomplished in the following manner:—A number of straight and rather larger branches are given off, which soon become curved and interwoven in various ways, generally more or less spirally, so as to form an entangled knot. The spore-bed which this entanglement forms develops spores from within outwards; each spore-ball contains from fifty to a hundred spores. The spores measure from 55 to 75μ in diameter, and consist of an endospore, containing granular protoplasm, and an exospore. No gelatinization of the spore-forming hyphæ, such as takes place in the spore-formation of *Sorosporium saponariæ*, was observed.

*Tilletia.*—The spore-formation in this genus was first indicated by Tulasne,* afterwards by Kühn,† who gave a figure of the process; but it is to Von Waldheim‡ that we are indebted for the most detailed account. He thus describes the process: The vegetative mycelium of *Tilletia tritici* is nearly 2μ in diameter. The swelling and gelatinization of the spore-forming hyphæ is not so marked as in Ustilago, so that the lumen of the hyphæ is never so contracted. Spore-formation begins by the hyphæ giving off pyriform buds, 1·5μ across (Plate VI. Fig. 4), in succession, from their sides; these outgrowths in-

* Tulasne, "1re Mémoire sur les Ustilaginées comparées aux Uredinées," p. 27, *et seq.*

† Kühn, "Krank. der Kulturgewächse," p. 56, t. iv. fig. 5.

‡ F. von Waldheim, *loc. cit.*

crease in length and breadth, so that the pyriform swellings become spheres, from 2 to 3μ across, attached to the hypha by thin stems about 1μ thick. These become differentiated; their contents vacuolated and oleaginous, and the connecting branches speedily wither (Plate VI. Figs. 5, 6). The epispore subsequently darkens and becomes uneven, while to many ripe spores the remains of the spore-forming hyphæ continue attached.

In *T. striæformis* the process is very similar, but the spore-forming hyphæ are larger and more gelatinized, and invest the spores to their maturity, after which they disappear without leaving any trace.

*Doassansia.**—The mycelial hyphæ in the tissues of the host-plant give off branches, which at certain points become interlaced into tangled knots. From these knots are formed the spores; the central portions forming the true spores, while the external develop into a layer of oblong or wedge-shaped sterile cells, which constitute an investing peridium. The peridial cells are darker in colour than the spores (Plate VIII. Fig. 4).

*Entyloma.*—The spore-formation in Entyloma is very similar to that of Tilletia. De Bary † investigated it in *E. microsporum* and *calendulæ.* The much-branched mycelium is principally intracellular. At certain places blister-like swellings appear along the spore-forming hyphæ, and also at their ends. The contents of these swellings become differentiated into spores, so that they are intercalated in the hyphæ (Fig. 5). Often a series of spores, formed one behind the other, may be seen still con-

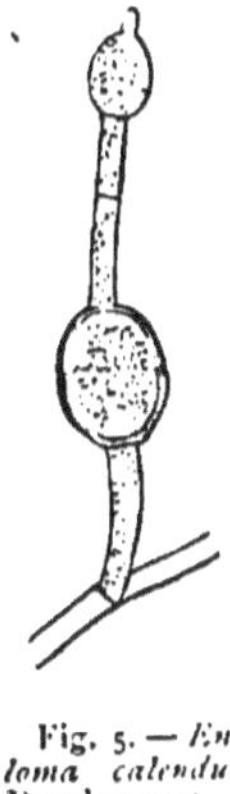

Fig. 5. — *Entyloma calendulæ.* Development of the teleutospores within the spore-forming hypha. (De Bary.)

* Cornu, *Ann. des Science Nat.*, 6e sér., Bot., tome xv. p. 280, *et seq.*
† De Bary, *Bot. Zeitung* (1874), pp. 81–93; pp. 97–108, t. ii.

nected with the remains of the spore-forming hyphæ. At maturity these spores do not break up into dusty masses like Ustilago, but remain as compact colonies embedded in the tissues of the host-plant. Each spore has two coats, and contains coarse protoplasm. The outer coat is sometimes gelatinous at its maturity.

*Urocystis.*—The process has been studied by Kühn,* De Bary,† Wolff,‡ and Winter §. The spore-forming hyphæ become swollen at their ends; two or more of these branch and wind themselves together, generally in a spiral manner, so as to form a glomerulus, and then, becoming highly gelatinized, they are quite indistinguishable from each other. In *U. occulta*, Wolff considers that they have a common investing membrane. At the same time, other branches are given off from the spore-forming hyphæ, which apply themselves to the outside of the glomerulus (Plate V. Figs. 13, 14, 15). As Winter points out, this might be looked upon as a sexual act, the central spiral branches of the glomerulus being the carpogonium, and the external enveloping branches the pollinodium; but there is no proof that this is really a sexual act, especially as similar conditions occur in the spore-formation of various Ustilagineæ, and in Geminella, in which no sexuality occurs. The spores are formed entirely from the gelatinized central glomerulus, and, as De Bary first showed, the pseudospores are formed by the outer branches. Prillieux ‖ considers that the spores of Urocystis are formed in a similar manner to those of Sorosporium. The details of his observations are given on the plate which accompanies his paper.

* Kühn, *loc. cit.*, pp. 78, 79.

† De Bary, "Morph. und Physiol.," p. 125.

‡ Wolff, *Bot. Zeitung*, 1873; "Der Brand des Getreides," 1874.

§ Winter, *Flora*, 1876, "Ustilagineen."

‖ Prillieux, "Sur la formation et la germination des Spores des Urocystis," *Ann. des Science Nat.*, 6e sér., Bot., vol. x. pl. i.

*Localization of Spore-beds.*—The bulk of the Ustilagineæ are characterized by the constancy by which they reproduce their spore-formation, in the same tissues of their respective host-plants. In the majority of the species the spores are localized somewhere in the reproductive organs of the host-plant; very frequently in the ovary (*Tilletia tritici, decipiens, U. caricis, Thecaphora hyalina*), or in the anthers (*U. violacea, scabiosæ*); often the blossoming and fructifying organs are attacked and destroyed (*U. segetum, bromivora, kühneana, Sphacelotheca*); sometimes the leaves are affected (*U. longissima, T. striæformis, Uroc. occulta*), or the stems (*U. hypodytes, grandis*), or even the subterranean organs (*U. hypogæa, Uroc. gladioli*). With Entyloma and its allies, however, this selection of tissue by the parasite does not obtain; with these species the spore-formation seems to occur in almost any part of the plant.

*Structure of the Teleutospores.*—The teleutospores of the Ustilagineæ consist of two membranes: an outer, which is thick and generally dark-coloured—the exospore; and an inner, which is thin and hyaline—the endospore. The exospore is subject to considerable variety; it may be quite smooth, or covered with extensive reticulations. Between these various intermediate conditions occur; thus in *U. segetum* it has generally been regarded as smooth, but it is rather to be described as granular, although Winter speaks of it as being "generally very minutely verrucose, and seldom quite smooth." The degree of roughness of the epispore varies from extremely minute elevated points to marked tubercles. These elevations may be evenly distributed over the whole spore, or they may be confluent in lines or ridges. Sometimes these ridges anastomose, and so form a reticulate or alveolate spore (*T. tritici, Sphacelotheca*). The colour of the spore depends upon that of the exospore; as a general rule the spores, as seen *en masse*, are blackish, but frequently, however, with an olive-brown or yellowish lustre, especially when viewed in an oblique light. *U. segetum*, when it occurs in wheat, has a distinctly golden lustre, but when on *Avena elatior* it is sooty black. Physiological research will possibly show that these two forms are specifically distinct. With *U. scabiosæ* the spores in bulk are flesh-coloured, and in *U. succisæ* they are quite white. Individually, the spores of the various species, as seen under the microscope, afford a considerable range of colour—black, dark violet, brown, olive-brown, and yellowish, or quite colourless. In some species a germ-pore is said to exist, through which the promycelium is protruded in germination; but these germ-pores of the Ustilagineæ are by no means so marked a formation as in the Uredineæ. In *U. tragopogi* the germ-pore is said to occupy from one-quarter to one-half the epispore.* Much more commonly do we find, as in *T. tritici*,† a small opening which splits into a rift as the promycelium grows out. In *Thecaphora hyalina* the germ-pore is round and paler in colour than the rest of the epispore; moreover, it is smooth, while the epispore is verrucose. Upon the whole, although germ-pores probably exist in all species, they are inconspicuous, and are very easily overlooked in the smaller spores. The reticulations on the epispore of Sphacelotheca are shown by the action of sulphuric acid, when examined by a high magnifying power, to consist of a series of distinct palisades, placed vertically.‡ As a general rule, the spores are globose, but in most species this is subject to a certain amount of variation; they often have one diameter rather longer than the other, but more frequently they show the result

* De Bary, "Morph. und Physiol.," p. 128.

† Brefeld, "Hefenpilze," p. 48.

‡ F. von Waldheim, "Sur la structure des spores des Ustilaginées" (1867), pp. 243–245.

of mutual compression by being flattened in one or more directions. In *U. caricis* they are often flattened on one side, so as to be subhemispherical. In size they vary from 4$\mu$ in *U. hypodytes* to 30$\mu$ or more in *Urocystis fischeri.* They may be simple or collected into spore-balls (Sorosporium), in which case they appear, when seen separately, as in *T. hyalina*, to be segments of a sphere, being convex externally, but internally more or less wedge-shaped. In Urocystis the spore-balls are surrounded by a variable number of barren spores or pseudospores, which are paler in colour, often almost hyaline ; these do not germinate.

Hartsen * found that the spores of the *U. maydis* would not yield their colouring matter to any of the ordinary solvents, but that strong sulphuric and nitric acids decolourized the epispore, rendering it more transparent without at once destroying it, so that the contour of the exterior remains unaffected. Although the spores sometimes burst, yet by these reagents the structure of the epispore can be conveniently examined. In nitric acid the spores of the *U. maydis* swell up, and after a time they dissolve, giving off an odour of bitter almonds. Sulphuric acid is the better reagent to employ in the examination of the spores, as, although it decolourizes and renders the epispore transparent, yet it does not so rapidly destroy the latter.

* Hartsen, "Compt. rendus" (1874), pp. 441, 442.

## CHAPTER X.

### GERMINATION OF THE TELEUTOSPORES OF THE USTILAGINEÆ.

At the beginning of the present century, Prevost* discovered the fact that the spores of *U. segetum* and *T. tritici*, when placed in water, would germinate. He observed the process in Tilletia to consist of the protrusion of a germ-tube and the development upon it of primary spores, which became united in pairs below, and which bore above the secondary spores. His observations were confirmed by De Candolle,† by Caron and Vandenhecke,‡ and by Mr. Berkeley,§ and they have been accurately described and delineated by almost all the more recent observers. With *U. segetum* Prevost observed the germ-tube, and that it gave off small secondary branches. His observations were confirmed by Tulasne,|| Bonorden,¶ Kühn,** and others. Little additional light had been thrown upon the subject,

* Prevost, B., "Mém. sur la cause immédiate de la Carie." Montauban: 1807.

† De Candolle, "Physiol. Veget." (1832), vol. iii. p. 1436.

‡ In 1835 referred to by Tulasne, "1re Mémoire," p. 38.

§ Berkeley, "Propagation of Bunt," *Trans. Hort. Soc. London* (1847), vol. ii. p. 113.

|| Tulasne, "Mém. sur les Ured. et Ust.," *Ann. de Sci. Nat.*, 4e sér., tome ii. p. 113.

¶ Bonorden, "Handb. d. Allg. Mykologie" (1851), p. 39.

** Kühn, "Krank. Kult.," 2 aufl. 1859.

until Brefeld,* in 1882, published his investigations, by which it appears that the Ustilagineæ are capable of perpetuating themselves for long periods, outside and independently of their host-plant, in the excreta of herbivorous animals. It will be more convenient to describe the germination of the spores of the various Ustilagineæ in detail, as was done with their spore-formation.

Much confusion exists in the works of various authors who have written upon the subject, from the diverse applications to which the word "spore" has been made. For instance, Cooke † speaks of the teleutospores of the Ustilagineæ as pseudospores, which term he applies to the æcidiospores, uredospores, and teleutospores of the Uredineæ. Again, these bodies are sometimes called conidia, which term is applied by Brefeld to the secondary spores of the Ustilagineæ. In order to avoid confusion, it may as well be stated at once that while the term "spore" may be correctly applied to all the reproductive bodies possessed by these fungi, in a general sense, yet it becomes necessary to affix to it some qualifying word, such as teleuto-spore, resting-spore, uredo-spore, promycelial spore, æcidio-spore, and so forth. In speaking of the Ustilagineæ, the word "spore" has generally been applied to the perfect, last-formed bodies—the teleutospores, analogous to the teleutospores of the Uredineæ; those bodies, in fact, with which we are all familiar as the black dust of bunt and smut. These black teleutospores, when they germinate, protrude a germ-tube—the promycelium. This promycelium bears certain very small hyaline bodies, which are spores, and may be fairly enough designated promycelial spores, inasmuch as they have been produced by the promycelium; as, however, the term "sporidia" is very commonly applied

* Brefeld, "Hefenpilze." 1883.

† Cooke and Berkeley, "Fungi," International Scientific Series, vol. xiv. 1875.

to them by Continental botanists, we may use it when writing or speaking in English.

The promycelial spores very commonly produce secondary promycelial spores by budding. Hence we may speak of primary and secondary promycelial spores. When these promycelium spores continue many times to multiply themselves by budding, after the manner of Saccharomyces, Uredine spermatia, etc., they may very well be called, as Brefeld suggests, "yeast-spores" and "yeast-colonies." The term "conidia" will be confined to that form of fruit in Tubercinia and Entyloma which is produced in the air from the mycelium in the living host-plant.

*Ustilago.*—The germination of the spores of Ustilago varies somewhat in different species. The commonest type is that of *U. segetum*, *violacea*, *maydis*, *kühneana*, *scabiosæ*, etc.

*U. segetum.*—If a few spores be placed in a drop of water, they will begin to evince signs of vitality in six or eight hours. Germination occurs more rapidly in summer than in winter, and in fresh spores than in those which have been kept some months. At one point of its surface the spore emits a germ-tube, which grows straight outwards, until it is from three to four times as long as the spore is wide; and under certain circumstances this tube may, according to Kühn, have the functions of a germ-tube (Plate VII. Fig. 6), entering by its pointed extremity the tissues of the host-plant. Normally, however, it becomes divided by septa into from three to five compartments, generally into four. This germ-tube is an outgrowth of the endospore, which is pushed upward through the exospore. It is from 30 to 40$\mu$ long, and from 4 to 5$\mu$ broad at its maturity. At first it is in direct communication with the endospore, and the protoplasm therein contained passes into the germ-tube and fills it. Adopting the phraseology of De Bary, this germ-tube constitutes the promycelium. In the course of a few hours the promycelium has received into itself all the protoplasm originally contained in the spore. Transverse septa now make their appearance in it, and it thus becomes divided into four or five equal compartments. From the outer walls of the now septate promycelial tube little offshoots or buds arise, into which the protoplasm of the tube passes (Plate VII. Fig. 1). These buds continue to increase in size until they become elongated, ovate, or elliptical promycelial spores; they then fall off. Generally they are produced from the side wall of the promycelium, near the septa, and almost always one is produced from the apex of this structure. If the protoplasm in each segment be exhausted by the production of promycelial spores, then spore-formation from it ceases; but if all the protoplasm be not used up in the formation of the first promycelial spore, a second but usually a smaller one is budded off. In fact, spore-development from the promycelium goes on until its protoplasmic contents are exhausted. It is not at all uncommon for one of these primary promycelial spores to remain attached to the promycelium instead of falling off, and at its free end to give off a small bud, which gradually grows into a secondary promycelial spore, the latter being of smaller dimensions than the one from which it sprang (Plate VII. Fig. 2). These primary and secondary promycelial spores, after they have fallen away from the promycelium, show still further developmental changes in water. (1) They may, as Tulasne has figured, emit a germ-tube (Fig. 12)—a very narrow tube pointed at its extremity, into which the contents of the promycelial spore are passed.* (2) Two promycelial spores, being near one another, may become joined by a transverse branch, through which the contents of one of

* Tulasne, "2e Mémoire," pl. 12, figs. 22 24.

them passes into the other (Fig. 11). When the first promycelial spore has become emptied of its contents, the second emits a tube which may remain as a germ-tube, or, as it becomes full of protoplasm, its end may swell out and form a third spore. (3) At a variable distance from the spore from which it arose a detached spore may form a connection with one of the segments of another promycelium ; sometimes as many as three spores may thus become united.*

Germination, however, does not always occur in the above typical manner, namely, by the development of promycelium and promycelial spores. From some of the largest teleutospores two promycelia are occasionally given off (Fig. 7). More commonly we find that, instead of promycelial spores being produced in the regular manner above indicated, only one or two segments give rise to them. The others send off branches, into which their contents are emptied in the same manner as occurred when spores were formed. The free ends of the promycelial branches often come into contact with one another. When this happens they fuse together and become one continuous tube. (1) Thus a tube given off from one of the upper segments may form a connection with one of the lower segments of the same promycelium in the form of a bow (Fig. 4). (2) Two continuous segments may become united by forming what Brefeld calls a buckle-joint. This consists of the unequal growth of one side of the promycelium at the level of one of the septa ; as this growing-out continues the promycelium itself becomes bent at an angle, at first obtuse, but eventually acute. A reference to the figures (Plate VII. Figs. 3 and 9) will render this obvious. (3) Promycelia from two different spores may unite by branches in various ways, either by their ends or at any

* Brefeld, *loc. cit.*, pp. 54–67, t. ii., iii., figs. 1–17.

part of their length; or the end of one may become united with the central segment of another, and so forth (Fig. 5). These various fusions take place only in spores grown in water. Brefeld has shown, however, that when cultivated in a medium which is capable of supplying suitable nutriment to the spores, these various fusions do not take place. He discovered that an aqueous extract of the excreta of herbivorous animals, sterilized by discontinuous boiling, afforded such a medium (*nährlösung*). Spores of *U. segetum* placed in Brefeld's nutrient fluid germinate sooner than in water. Not only so, but the promycelia and the promycelial spores are larger than those produced in water. No fusion of spores or buckling of promycelia occur. The most remarkable fact is that the promycelial spores multiply themselves by budding, very much after the manner in which yeast-cells multiply themselves in saccharine fluids (Plate VII. Figs. 8–10). Brefeld has kept them thus reproducing themselves for more than a year, by replacing the nutrient fluid as it became exhausted. The promycelial spores continue to reproduce themselves by budding, as long as the nutrient fluid remains unexhausted; when this occurs they cease to bud, and fuse in various ways, as is seen when germination takes place in water. The yeast-spores of this species produced in his nutrient fluid are rather larger than the primary spores produced from the promycelium in water; they are, however, of the same shape, and measure from 9 to 30$\mu$ in length, and from 3 to 5$\mu$ in breadth.* Kept moist, they retain their vitality for about two months, but if allowed to dry, none germinated after the sixth week. The teleutospores, on the contrary, if kept dry, retain their germinative faculties for a very long period. Brefeld found that at the end of two years they germinated as freely as when fresh; other

* Brefeld, *loc. cit.*, p. 13.

observers have found that they even germinate after seven and a half years.*

*U. Cardui*, F. v. W.—Kühn found that teleutospores in water produced promycelia and small ovate promycelial spores.† In nährlösung yeast-colonies are produced very abundantly, being from 5 to 8$\mu$ long, and from 3 to 5$\mu$ wide; they very much resemble the cells of beer-yeast, but of course they do not bud in saccharine media.‡

*U. flosculorum*, D.C. (*U. intermedia*, Schröter).—The germination has been studied by Schröter,§ who found the spores, when placed in water, germinated very quickly, for in twelve hours all had produced promycelia. These attained a length of from 16 to 20$\mu$, and a width of from 5 to 6$\mu$; they generally become triseptate, and bear sporidia both laterally and terminally. These are shortly ovate in form, and measure 6$\mu$ in length, and from 4 to 5$\mu$ in width. In the course of thirty-six hours the sporidia tend to become spherical. The promycelia have a great tendency to fall away from the teleutospores and subsequently to produce promycelial spores, so that they become sporophores (*fruchtträger*). The budding in nährlösung Brefeld ‖ found to be so profuse as to be quite phenomenal. The yeast-spores are from 4 to 5$\mu$ long, and 4$\mu$ wide.

— *U. violacea*, Pers. (*U. antherarum*, Fries).—The teleutospores of this species are even better suited for observation than those of the one just described, inasmuch as, being rather larger, the promycelia are proportionately bigger, and the septation and spore-formation more easily observed. The teleutospores germinate after a very short

* Von Liebenberg, "Oesterr landw. Wochenblatt" (1879), Nos. 43, 44.

† Kühn in Rabenhorst's "Fungi Europæi," No. 1798.

‡ Brefeld, *loc. cit.*, pp. 86-88, t. vi. figs. 1-16.

§ Schröter, "Beitrage zur Biol.," bd. ii. heft. iii. pp. 352, 353.

‖ Brefeld, *loc. cit.*, pp. 89, 90, t. vi. figs. 17-27.

immersion in water. The process has been described and figured by Tulasne * and Von Waldheim.† Brefeld ‡ found the production of the yeast-cell colonies was very prolific in nährlösung, where they multiplied themselves through endless generations, whereas in water no budding took place after the second day. These yeast-spores are ovate, but somewhat elongated; from 5 to 7μ long, and from 3 to 4μ wide. The germinative faculty lasts for about six weeks. The spores in water or in exhausted nährlösung frequently unite in the same manner as those of *U. segetum*.

When this fungus attacks the anthers of *Lychnis diurna*, a plant which is usually unisexual, the styles which would normally be short, acute, and erect, become long and recurved, as they are in the female flower. This has been pointed out by M. Cornu, as well as by other observers.

*U. maydis.*—The germination of these spores was observed by Kühn § and Wolff,‖ and does not materially differ from the above: the yeast-spores,¶ however, are elongated and fusiform, being from 10 to 36μ long, and from 3 to 5μ wide.

*U. scabiosæ* (Sow.) (*U. flosculorum*).—The spores germinate very freely and very soon in water. According to Schröter ** the promycelia are three or four-partite; about from 20 to 22μ long, and 4μ wide; the promycelial spores are about 4μ long. In nährlösung Brefeld †† found the yeast-spores to be from 4 to 8μ long, and from 1·5 to 2μ wide. They were produced continuously and

* Tulasne, "1re Mémoire," t. iv. fig. 18.
† F. von Waldheim, *loc. cit.*, t. xii.
‡ Brefeld, *loc. cit.*, pp. 36–54, t. i. figs. 1–27.
§ Kühn, "Krankh d. Kulturgew." p. 260, t. iii. figs. 22, 23.
‖ Wolff, "Brand des Getreides," p. 11 t. i. fig. *c*.
¶ Brefeld, *loc. cit.*, pp. 67–75, t. iv. figs. 1–17.
** Schröter, "Cohn Beiträge," vol. ii. 1877.
†† Brefeld, *loc. cit.*, pp. 78–81, t. v. figs. 1–6.

abundantly. As the teleutospores are of pretty large size, their germination is easily observed. I have found them germinate much more freely in summer than in late autumn (Plate VII. Fig. 18).

*U. tragopogi* (Pers.) (*U. receptaculorum*, Fries).—The germination was observed by Tulasne,* and by Von Waldheim,† to consist in the protrusion of a promycelium through a very marked germ-pore in the epispore. It becomes three or four-septate, and produces subcylindrical promycelial spores, which are rounded, especially at their distal ends. These often grow nearly parallel to the promycelium, as figured by Von Waldheim; sometimes they are produced terminally upon branches given off from the segments of the promycelium. Conjugation often takes place between them after they have fallen off. In nährlösung Brefeld ‡ found them to be abundantly reproduced by budding, but they were larger, measuring from 5 to 20$\mu$ in length, and from 5 to 7$\mu$ in breadth.

Fig. 6.—*Ustilago tragopogi.* Teleutospore germinating, and promycelial spores conjugating. (Tulasne.)

*U. kühneana*, Wolff.—The promycelium is three to four-septate, the lowermost compartment being the longest and empty. It produces numerous promycelial spores in whorls at each septum (Plate VII. Fig. 17). In nährlösung § they budded very profusely, forming very characteristic small yeast-spores, from 3 to 5$\mu$ long, and from 3 to 5$\mu$ wide. Wolff ‖ also investigated the germination in water.

*U. hypodytes* (Schlecht.).—The germination of this plant

* Tulasne, "2e Mémoire," pp. 159–160, t. xii. figs. 34–40.

† F. von Waldheim, *loc. cit.*, t. xi. figs. 27–37.

‡ Brefeld, *loc. cit.*, pp. 81, 82, t. v. figs. 7–11.

§ Brefeld, *loc. cit.*, pp. 83–88, t. v. figs. 12–20.

‖ Wolff, *Bot. Zeitung* (1874), pp. 814, 815.

is markedly different from that of any of the former species. Winter * figures an elongated promycelium with a promycelial spore, borne laterally upon a long pedicel. He found that the spores germinate freely in water, and that the endospore sends out a process—the promycelium—from the interior of the spore, which grows to 30 or 50$\mu$ in length, but is only about 3$\mu$ wide. It becomes septate, and gives off short lateral branches, which become spores. They are slightly clavate, and from 6 to 7$\mu$ long.† Brefeld ‡ found that in nährlösung they only produced mycelial hyphæ without spores. Although I have tried many times, I have never succeeded in getting the teleutospores of this species to germinate.

*U. longissima* (Sow.).—When the spores of this species are placed in water they very soon begin to germinate. The process, as carried on in this species, differs very materially from that which obtains with the previously mentioned species. This, as was first pointed out by Von Waldheim,§ consists in the protrusion of a very narrow straight tube through a small opening in the epispore. This acquires a length of about 10 or 12$\mu$, when it becomes divided below by a septum into two unequal parts (Plate VII. Fig. 14), the upper of which is about 6 or 8$\mu$, and the lower 3 or 4$\mu$ long. The lower portion is, moreover, narrower than the upper, and is the true promycelium, the upper being the promycelial spore. The promycelial spore soon falls off, and the promycelium produces, in about an hour, Von Waldheim says (but I have personally made no observation as to time), a second promycelial spore, which in like manner falls off, and is followed by a third (Plate VII. Figs. 15, 16).

* Winter in Rabenhorst, "Kryptogam, Flora," vol. i. p. 81, fig. 4.
† "Flora" (1876), Nos. 10, 11. ‡ Brefeld, *loc. cit.*, p. 103.
§ F. von Waldheim, *loc. cit.*, t. v. figs. 42–46.

According to Brefeld,* it is seldom that more than three promycelial spores are produced from one teleutospore before it becomes emptied of protoplasm and exhausted. Be this as it may, if a few teleutospores be placed in a drop of water on a glass slide, and examined at intervals for two or three days, one can see with the naked eye that there has fallen to the bottom of the drop a whitish cloud. Upon microscopic observation, this cloud is found to consist of an immense assemblage of promycelial spores. They are cylindrical bodies, with somewhat attenuated extremities, and often measure from 8 to 10$\mu$ in length, and from 1·5 to 2$\mu$ in breadth. Hence it appears that they have increased in size since they fell off the promycelium. After a time this increase in size ceases, but not before some few odd ones here and there have attained a length of from 20 to 30$\mu$.

Brefeld found, by the culture of isolated promycelial spores in nährlösung, that after these bodies had fallen away from the teleutospore which produced them they not only multiplied themselves, but increased enormously in length and thickness. This they did with great rapidity. They more resembled hyphæ than promycelial spores, and each soon became more or less septate. They multiplied by giving off a small bud-like projection laterally, and at a short distance from one or other of their extremities. This bud rapidly grew into a second spore, but before it attained the dimensions of its parent the latter had given off a similar bud towards its opposite extremity; and so the process of multiplication goes on until the nährlösung is exhausted. When this takes place, instead of multiplying in the manner above described, the promycelial spores give off hyphæ of considerable length, which become septate at intervals from below upwards, and the protoplasm is passed

* Brefeld, *loc. cit.*, pp. 104–116, t. viii., ix., figs. 1–16.

along to its growing end. Frequently two promycelial spores become united by a transverse bridge (conjugation); one of them then gives off a germ-tube, as in the case of single promycelial spores, only it is longer.

*U. grandis.*—In 1876, Kühn* investigated the germination of this species. He found that the promycelia, which are about from 50 to 60$\mu$ long, and from 5 to 8$\mu$ wide, had a great tendency to fall off from the teleutospores before they produced promycelial spores, although this was not by any means always the case. Brefeld † found in nährlösung that the promycelia were not only larger, but produced more and also larger promycelial spores. These sporidia not only reproduced themselves, but also grew into sporophores (*Fruchtträger*), which were indistinguishable from the original promycelia, inasmuch as they were cylindrical septate tubes, which in their turn budded off spores. Thus colonies of yeast-cells do not occur in the life-cycle of *U. grandis* any more than they do in that of *U. longissima.* With *U. grandis* the promycelium produces spores which grow out into sporophores, and they in their turn produce spores again. These sporophores are multicellular.

*U. bromivora*, Tul.‡—In water each spore produces a small promycelium through a minute opening in the epispore, very much after the manner of *U. longissima.* This bears terminally a spore which soon falls away. Between the fallen-off spores conjugations are frequently to be seen; sometimes they become uniseptate and buckle-jointed. In nährlösung the typical germination takes place in the production of single promycelial spores from a short promycelium. These spores increase in size and become uniseptate, or, as Brefeld terms them,§ bicellular sporo-

* Kuhn in Rabenhorst "Fungi Europæi," cent. xxiii. No. 2299, fig.

† Brefeld, *loc. cit.*, pp. 116–123, t. ix. figs. 17–26.

‡ *Cf.* Kühn, "Vortrages über Getreidebrand." 1874.

§ Brefeld, *loc. cit.*, pp. 123–129, t. x. figs. 2–8.

phores. There is no observable difference between the sporophores produced from the teleutospore and those produced from a promycelial spore. These developments go on until the nährlösung becomes exhausted, when the promycelial spores and sporophores alike give off mycelium-like tubes and fuse in various ways. In other words, *U. bromivora* is characterized by its promycelial spores growing into bicellular sporophores, which sprout directly into new promycelial spores. True yeast-cell colonies do not occur.

In the teleutospores of this species which Mr. Soppitt sent me I found that promycelia were freely produced in water, and that they developed elongate, elliptical promycelial spores (10—12 × 3—4$\mu$), which tended to become vacuolate after they had fallen off, and afterwards emitted pointed germ-tubes. I found that teleutospores gathered in June germinated freely in September.

*U. olivacea* (D.C.).—The teleutospores germinate,* after a few hours in water, very much like those of *U. longissima*. The promycelium is, however, so curtailed as practically not to exist, and the promycelial spores are really produced at once out of the teleutospores without any promycelium. These promycelial spores are variable in size; each is subfusiform, and measures from 5 to 20$\mu$ in length, and from 2 to 3$\mu$ in breadth. In nährlösung they form yeast-colonies.

*U. major*, Schröter.—I gathered some specimens of this fungus near Paris in the middle of October, 1887. The spores germinated very readily when placed in water. In twenty-four hours they had developed cylindrico-fusiform promycelia (10—12 × 2$\mu$), which fell off from the teleutospores (Plate VII. Figs. 19—22) much after the manner of *U. longissima*. In forty hours these had attained a

* Brefeld, *loc. cit.*, pp. 129–133, t. x. figs. 9–26.

size of 15—18 × 3·5—4·5μ. They formed an opalescent cloud at the bottom of the culture-drop. Each promycelium was cylindrical in form, vacuolate, and eventually became triseptate, and produced promycelial spores both laterally and also at either end (Plate VII. Figs. 22, 24, 25). The latter were elliptical in form, and measured from 3 to 4μ in length by 2μ in width. Many of the promycelia were observed to be slightly curved.

*U. utriculosa* (Nees).—The teleutospores of this species are by no means easy to germinate, and although I have made a great many attempts to do so, I have always failed. Schröter* states that they emit a cylindrical promycelium, which becomes triseptate, and produces elliptical promycelial spores in pairs, which conjugate in couples.

*Sphacelotheca.*—The spores do not germinate at all readily, and I have been unsuccessful in observing the process. Schröter † states that the teleutospores emit a cylindrical promycelium, which becomes triseptate, and bears elliptical promycelial spores laterally, and that these conjugate in pairs at their bases.‡

*Sorosporium.*—Woronin § succeeded in getting the spores of *S. saponariæ*, Rud. on *Lychnis dioica* and *Saponaria officinalis* to emit a germ-tube, but no spore-formation was

* Schroter, Cohn's "Krypt. Flora von Schlesien," vol. iii. p. 273.

† *Loc. cit.*, p. 275.

‡ The two following species of Ustilago can in no sense be regarded as being British, yet they occur in this country sufficiently commonly to render them objects of interest.

*U. ficuum*, Richdt., is often met with on the cheaper kinds of figs, known in the trade as "natural figs." The spores are formed in the interior of the fruit, and are black or dark violet, smooth, globose, from 6 to 8μ across. I found no germination took place below 10° C., but between 10° and 13° C., when placed in water, they emitted promycelia 20—150 × 4 5μ, but I was unable to observe any further development.

*U. phœnicis*, Corda, is a closely allied species, which is frequently to be met with on cheap dates. The spores are globose, smooth, dark violet, from 4 to 5μ in diameter.

§ Woronin, *loc. cit.*, pp. 18, 19, t. iii. figs. 13-18.

observed. The spores collected in June germinated in December, after being placed in water from three to five days.

*Thecaphora.*—The germination of *T. hyalina* has been investigated by Woronin.* He found that spores collected in August germinated in October and November, after fourteen to eighteen days' maceration in water, but older spores did not germinate at all. Each spore is provided with a germ-pore in the epispore, which is pale in colour and free from any of the verrucosities which occur upon the other parts of the epispore. Every germ-tube becomes filled with protoplasm, and generally contains four nuclei. It becomes septate, and each compartment contains one of the nuclei. From each segment of the promycelium narrower lateral branches are given off. Those from the upper compartments tend to grow downwards, while those from the lower, on the contrary, grow upwards. If one of the upper branches comes in contact with one of the lower, they unite at their ends and form a bow-like conjugation. From this a long germ-tube is given off, into the end of which the protoplasm is passed. No spore-formation was observed.

I have made many attempts, but have always been unsuccessful in getting the teleutospores of this species to germinate.

Brefeld,† in an allied species (*T. lathyri*, Kühn), found promycelia, at the end of which spherical promycelial spores were formed. These promycelial spores in nährlösung germinated and produced a mycelial mass, which in turn also produced spores upon those of its branches which came in contact with the air.

*Tilletia.*—The germination of *Tilletia tritici* has been

* Woronin, *loc. cit.*, pp. 21, 22, t. iii. figs. 19–28.
† Brefeld, *loc. cit.*, pp. 134–138, t. xi. figs. 2-12.

known since 1807, when Prevost * figured not only the promycelium, but the primary and secondary promycelial spores. Mr. Berkeley,† in 1847, discovered the conjugation of the primary spores, which was again more fully investigated by Tulasne.‡ Kühn § gives a full account of the process. Since then nothing has been added to our knowledge of the subject, till Wolff ‖ showed the method by which the germ-tube enters the host-plant, and Brefeld ¶ investigated the further development of the spores in nährlösung.

The spores do not germinate until they have been placed in water for some considerable time, not before forty-eight or fifty hours; but often I have found them to take a much longer period. They retain their germinative power for two or three years, and one author says as long as eight and a half years.** The process differs materially from that previously described in the other genera. The promycelial tube is emitted from a small germ-pore, but very soon, as the tube increases in diameter, it causes the epispore to split. Its length varies according to circumstances, its diameter being about 8$\mu$. If it be given out from a spore under water, at the bottom of the culture-drop, it grows upwards until its apex reaches the air. As soon as the promycelium has reached the air several tuberculations appear upon its summit. The protoplasmic contents of the spore are passed along the promycelium to its extremity. If the promycelium happen to be a very long one, then numerous septa occur from below upwards; but,

* Prevost, "Mémoire sur la cause immédiate de la Carie." 1807.

† Berkeley, "Propag. of Bunt," *Trans. Roy. Hort. Soc.* (1847), vol. ii. p. 113.

‡ Tulasne, "1re Mém. sur les Ured. et les Ustilag." 1854.

§ Kuhn, "Krank. der Kulturgew." 1859.

‖ Wolff, "Der Brand des Getreides." 1874.

¶ Brefeld, *loc. cit.*, pp. 146–163, t. xii., xiii. figs. 25–52.

** Liebenburg, *loc. cit.*

however short it may be, there is always one septum developed near its upper end. The tubercles above mentioned increase rapidly in length, and become the primary spores (Plate VI. Fig. 7). They are filiform bodies, curved in various ways, and measure from 80 to 100$\mu$ in length; in number they vary from four to twelve or more, according to the size of the spores from which they are developed. When all the protoplasm from the promycelium has been absorbed into these primary spores, they become cut off from it by septa at their attached ends. If the promycelium be so situated that it cannot reach the air, no primary spores are produced. Shortly after their maturity these primary spores conjugate (Plate VI. Fig. 8), or become united by transverse bridges, usually in pairs. The primary spore is possibly a wind-carried spore, but, as in artificial cultures they are not exposed to this force, they frequently germinate *in situ*. This they do by a repetition of the oft-described process of protoplasmic migration, with septation of the emptied parts; the protrusion of a bud-like process into which the protoplasm is emptied, and which becomes a secondary spore. The end of the promycelium, after the primary spores have fallen off, remains tuberculated, showing the points of their attachment. The conjugation of two primary spores cannot be considered a sexual act, inasmuch as the single spores, which have not been subjected to it in any way, germinate as freely, and produce

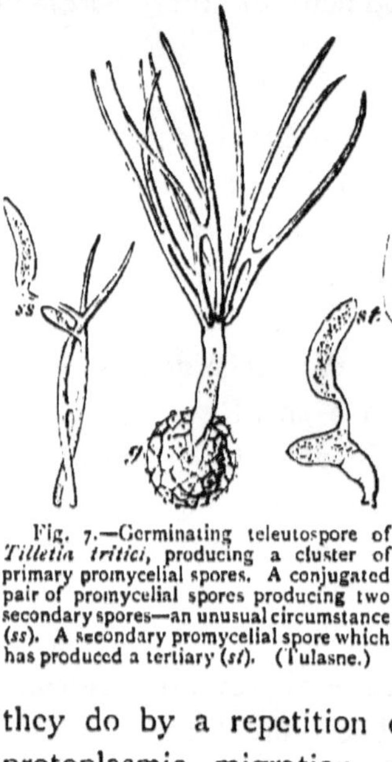


Fig. 7.—Germinating teleutospore of *Tilletia tritici*, producing a cluster of primary promycelial spores. A conjugated pair of promycelial spores producing two secondary spores—an unusual circumstance (*ss*). A secondary promycelial spore which has produced a tertiary (*st*). (Tulasne.)

secondary spores as effectually, as when two or three have become connected, the only difference being that with fused or conjugated spores larger germ-tubes and larger secondary spores are produced. The secondary spores (Plate VI. Figs. 9, 10) are at first cylindrical, but they soon become reniform, and at length, by the attenuation of their ends, more or less crescent-shaped. They may be produced from any part of the primary spores, even from the connecting bridge, and at almost any distance; very seldom is more than one produced from each fused pair of primary spores. Three or more primary spores have been seen connected,* and double fusion between two has been also observed.† Not only do the primary produce secondary spores, but they may emit a pointed germ-tube, for direct penetration of the host-plant. In like manner, between the secondary spores all sorts of connections and conjugations or fusions occur, with the same emptying of the contents from one spore into the other (Figs. 11, 12). The secondary spores are, however, essentially the spores the germ-tubes of which enter the host-plant.

In nährlösung, according to Brefeld, the primary and secondary spores are larger, and are produced in greater abundance. No conjugations or fusions occur, except when it becomes exhausted. The primary spores, however, comport themselves very differently in nährlösung; they send out germ-tubes which are narrower than those given off by them in water. These tubes are unseptate, but branch and inosculate with each other so as to form a mycelium (Plate VI. Fig. 14), which, as it grows out into the air, forms a white floccose mass. For five or six days no secondary spores are found in well-nourished mycelia, whereas spore-formation occurs soon in badly nourished ones. The spores are produced only on those

* Brefeld, *loc. cit.*, t. xiii. fig. 30. † Ibid., fig. 38.

branches of the mycelium which are given off into the air. Upon these numerous short, lateral branches are given off, which swell up at the ends and become crescentic spores (Fig. 15). No conjugation takes place between these bodies. The branches of the mycelium which are given off in the fluid do not produce spores, but grow outwards until at length they reach the air, when they produce terminal spores or they remain sterile. In the latter case the hyphæ are empty and septate. Brefeld has further observed that by long-continued culture the hyphæ, under certain circumstances, become nodose, and apparently develop certain globose bodies which closely resemble the original teleutospore.*

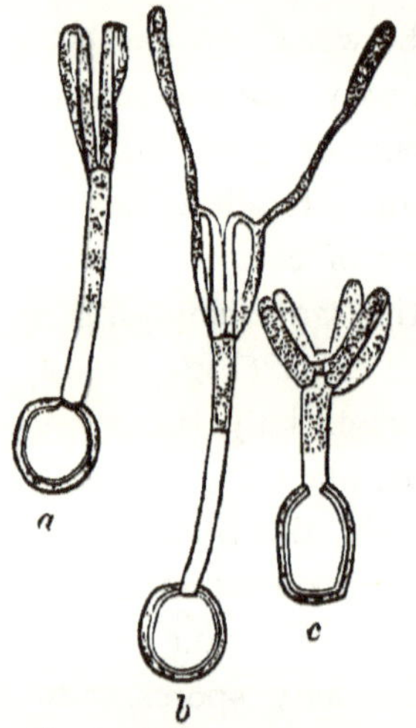


Fig. 8.—*Entyloma microsporum.* *a*, Teleutospore germinating (the promycelial spores have conjugated at their upper ends); *b*, two secondary spores produced from the conjugated pairs of primary promycelial spores; *c*, teleutospore of *E. calendulæ*, which has produced five promycelial spores, four of which have conjugated below. (De Bary.)

*Entyloma.*—The germination of Entyloma, though similar to that of Tilletia, is far less complex.

*E. microsporum.*—De Bary † found that if the spores were wholly immersed in water, they, in the course of twenty-four hours, would give out a germ-tube from four to ten times the length of the spore. At its rounded summit this promycelium gave off six or seven branches, each of which was dilated upwards; and when they attained a length measuring 30 or 40$\mu$, each became cut off by a basal septum. They conjugate in pairs, by a transverse bridge,

* Brefeld, *loc. cit.*, t. xiii. figs. 46–52.

† De Bary, *Bot. Zeitung* (1874), pp. 81–92, 97–108, t. ii.

either at their lower or upper ends; after which one of the branches grows onwards in its original direction, and at its end develops a long, fusiform, secondary spore (Fig. 8). This falls off, and in its turn germinates by the protrusion of a long, very narrow germ-tube. If the number of the primary branches be odd, the odd one does not conjugate, but grows more slowly, and ultimately shows no further change. Various deviations from the typical germination take place; thus the promycelium itself can grow into a germ-tube, or the primary whorl of branches may send a branch downwards. In *E. calendulæ* the process is similar.

*E. ranunculi.*—In this species Brefeld * found that conidia were produced upon the host-plant, and that in the interior of the leaves a richly developed mycelium existed, sending up conidiophores through the stomata. The fresh conidia in nährlösung produced a mycelium less extended, but little different from that produced by the spores of Tilletia under similar conditions. This very soon becomes covered with conidia, which in their turn produce mycelia and conidia again. Marshall Ward † found that the conidia which are produced in spring are clavate or elongate-oval in form; that they germinate in from twenty-four to thirty hours by emitting a delicate germ-tube from both extremities, one of which grows, the other becoming empty of protoplasm and septate. The growing germ-tube generally becomes swollen into a secondary conidium when the culture is made in water. From this secondary conidium a branched germ-tube is emitted. If the conidia germinate on a leaf, the formation of the secondary conidia is rare. The germ-tubes enter the stomata, and

* Brefeld, *loc. cit.*, pp. 163, 164.

† Marshall Ward, *Phil. Trans. of the Roy. Soc.*, vol. 178 (1887), B., pp. 173-185, plates 10-13.

the conidia reproduce themselves in from fifteen to twenty days. The teleutospores are developed in spherical dilatations, in the continuity of the mycelial tubes, rarely at the ends of branches. Conjugation between conidia sometimes takes place. The conidiophores emerge either through the stomata or between the epidermal cells. The conidia germinate more rapidly and throw out larger germ-tubes when this process takes place on the living leaves of the host-plant than when it does so in water. This may be in part due to the more abundant supply of oxygen which they would receive in the former situation.

*E. canescens.*—Schröter * finds the spores germinate as soon as they are ripe by protruding a germ-tube, from 20 to 30$\mu$ long, and 4$\mu$ wide, on the end of which a tuft of cylindrico-fusiform spores are produced. They measure from 25 to 40$\mu$ in length, and from 2·5 to 3$\mu$ in thickness.

*Doassansia.*—The germination of this genus is identical with that of its ally, Entyloma, consisting in the protrusion of a promycelium of limited growth, which develops apically a tuft of promycelial spores.† In *D. alismatis* the promycelial spores are long and cylindrical, and they are produced in great numbers (Plate VIII. Fig. 5).

The process is also similar in *D. sagittariæ*, as observed by Fisch,‡ who observed conjugation to take place between the fallen-off promycelial spores. He found that the germ-tubes of these spores entered the sides of the cells of the host-plant, having insinuated themselves between the epidermal cells. The teleutospores of *D. alismatis* germinated as soon as they were ripe, but those of *D. sagittariæ* did not do so until the ensuing spring.

* Schröter, "Cohn Beiträge," vol. ii. (1877), p. 372.

† Cornu, *Ann. des Scienc. Nat. Bot.*, 6e sér., tome xv. p. 281.

‡ Fisch, "Entwickelungsgeschichte von Doassansia Sagittariæ," *Berichte der deutschen botan. Gesellschaft*," September, 1882, bd. ii. t. x.

*Urocystis.*—The germination of Urocystis was first observed by Kühn in *U. occulta,** and is also described by Wolff.† *U. colchici* was studied by Winter; ‡ *U. anemones*, by Von Waldheim; § *U. violæ*, by Prillieux; ‖ and *U. primulicola*, by Pirotta.¶

The process consists in the protrusion of a promycelium from the inner coloured spores (the paler peripheral pseudo-spores do not germinate), into which the protoplasm passes to the upper end, where it gives rise to a variable number of primary spores. If the promycelium be produced under water, no spore-formation occurs until its point comes into the air. The primary spores fall off, and occasionally conjugate in various ways, but not so constantly as in Tilletia; they also frequently germinate whilst attached to the promycelium.

*U. occulta.*—The central, dark-coloured spores (as first described by Kühn **) emit a promycelium, at the apex of which from two to six primary spores are borne. These sometimes conjugate by a transverse bridge at their upper ends, and often germinate—as Wolff has more recently shown—while still attached to the promycelium, from their lower ends, sending out a long, narrow germ-tube, which receives the protoplasm from the interior of the spore, so that the upper part of the spore is first emptied of its contents.††

*U. fischeri.*—The spores of this species, which Mr. Soppitt was kind enough to send me, germinated only after a con-

* Kühn, *loc. cit.*, pp. 78–80, t. ii. figs. 13–34.

† Wolff, "Der Brand des Getreides," pp. 16, 17, t. ii. figs. 1–10.

‡ Winter, "Ustilagineen Flora" (1876), Nos. 10, 11.

§ F. von. Waldheim, *loc. cit.*, t. vi. figs. 38–43.

‖ Prillieux, *Ann. des Scienc. Nat. Bot.*, 6e sér., tome x. (1880), p. 49, t. i.

¶ Pirotta, "R. Nuovo Giornale Bot. Ital.," vol. xiii., 12 Luglio (1881), No. 3.

** Kühn, *loc. cit.*, t. ii. fig. 20*a*.

†† Wolff, *loc. cit.*, t. ii. B. figs. 7, 8.

siderable period of soaking in water. The germ-tube was larger than in any of the other species of Urocystis the germination of which I have watched; the promycelial spores were also not only larger, but more numerous. I counted as many as eight on some of the promycelia (Plate VII. Figs. 34, 35).

*U. anemones.*—Von Waldheim * points out that unless the promycelium grows in the air no spore-formation takes place. At its end it divides into three or four branches, which become spores. They are elongated, oval, and generally wider at their upper end. In length, they measure from 10 to 14μ; and in breadth, from 3 to 3·5μ. They become vacuolate, and enlarge in size till they often measure 22 by 4μ. After several hours' (forty-eight and more) immersion in water, in November and December, I found the teleutospores germinated. The promycelial spores were of the same size and form as described by Von Waldheim (Plate VII. Fig. 31); I also observed they became vacuolate when old (Figs. 32, 33). The promycelial spores were applied to the foliage of *Ranunculus repens*, in two experimental cultures, on December 12, 1884. No change was observed in the plants until February, when it was noted that they showed signs of the formation of spore-beds. On February 11 in one experiment, and on the 22nd in the second, spores were developed. This is one of the few species in which mycelium is localized, and the infection of the host-plant occurs at the same place at which the teleutospores are subsequently formed.

*U. violæ.*—The spore-balls generally produce only one promycelium, which bears at the end a cluster of five or six fusiform spores. If the promycelium remain short, spores are produced; but if it grow to any great length, either no spores at all are formed or only small ones. Of

* Waldheim, *loc. cit.*

the six spores generally only three germinate, and produce at their distal extremities secondary spores similar in size and shape to themselves.*

*U. primulicola*, Magnus.—Pirotta † found that the fresh ripe spores germinated in water in about ten hours, by emitting a short cylindrical promycelium, which at its extremity gave off three or four branches that became spores, measuring from 9 to 18$\mu$ in length, and from 4 to 9$\mu$ in width. These, while still attached, produced secondary spores from their ends. The secondary spores germinated by the protrusion of a germ-tube (about 3$\mu$ wide, and from 10 to 20 times as long as the spore), into which the protoplasm migrated. Lateral conjugation was occasionally observed.

This species occurred in 1884, in Rev. C. Wolley Dod's garden, on *P. farinosa*. In August of that year I received some specimens from Mr. Dod. The spores germinated readily in water, and emitted short promycelia, which bore a cluster of promycelial spores as figured by Pirotta. I found that no spore-formation took place unless the end of the promycelium grew in the air. If a spore germinated at the bottom of a drop of water, the promycelium grew upwards through the water until it reached the air. In these cases the lower part of the promycelium became emptied of its protoplasm and septate, just as one sees in Tilletia (Plate VII. Fig. 26, 27). The promycelial spores varied from 12 to 20$\mu$ in length, and were 4 or 5$\mu$ in width (Figs. 28, 29, 30). After keeping the promycelial spores in nährlösung for two hundred and sixty-four hours, no further spore-formation was observed; but they became septate and nucleate (Fig. 29).

*Melanotænium.*—The mycelium is principally inter-

* Prillieux, *loc. cit.*

† Pirotta, "Nuovo Giornale Bot. Ital.," vol. xii. (1881), pp. 235–239, t. vi.

cellular, and pervades all parts of the affected plants, especially the cortical parts of the stem, and also to some extent the pith. It also occurs in the upper part of the root-stock. It is hyaline, about 4·5$\mu$ wide, richly branched, and septate. Its contents are colourless and vacuolated. The spores are formed inside the mycelial hyphæ, where it becomes coarsely granular, much after the manner of Entyloma.* Woronin † finds that the mycelium is abundantly provided with very marked haustoria, which enter the cells. These botryform prolongations enter and occupy a third or a half of their interior. Germination takes place in autumn. Specimens gathered in June germinated in October and November. The epispore splits, and the endospore grows out as a blunt cylindrical promycelium. At its extremity it emits a cluster of from four to seven apical branches. The outgrowth of the endospore is at first often in the form of two equal branches, one of which develops into the promycelium, while the other ceases to grow and has become emptied of its protoplasm, which passes into the developed branch. Towards the upper half of the promycelium a septum appears, cutting off the protoplasm above from the empty tube below; but true spore-formation was not observed.

*Tubercinia.*—The germination of the spores of *T. trientalis* has been worked out by Woronin.‡ Teleutospores collected at the end of September and the beginning of October were found often to have already germinated upon the plant. Placed in a damp atmosphere, they germinated freely after the manner of Tilletia, each spore producing a promycelium surmounted by a cluster of spores. All the teleutospores of one spore-ball do not germinate at the

* De Bary, *loc. cit.*

† Woronin, *loc. cit.*, pp. 27, 28, t. iv. figs. 27–35.

‡ Woronin, "De Bary und Woronin Beiträge," 5 reihe (1882), pp. 4–16, t. i., ii., iii. figs. 1–12.

same time. The promycelium emerges through a small round opening in the epispore, its length corresponding to the size of the spore. On the upper, free, blunt end of the promycelium from four to eight protuberances appear, which elongate themselves into branches and become the cylindrico-fusiform promycelial spores. After all the protoplasm from the interior of the teleutospores has been passed into the upper end of the promycelium and into the developing promycelial spores, a septum is formed close to its upper end. If the promycelium happen to be a very long one, two or more septa occur. The promycelial spores, while still attached to the promycelium, become united in pairs by a bridge-like connection. This conjugation takes place at the bases of the promycelial spores, and but rarely at their summits. One of the conjugated spores then buds out a secondary spore, which in its turn sometimes produces a tertiary; sometimes all these may be observed in a chain. If there be an odd spore on the promycelium which has not conjugated, it does not bud.

*Conidia.**—These are produced from a mycelial mesh that exists for the most part just beneath the epidermal structures, and is provided with very numerous botryform haustoriæ (Plate VI. Fig. 1). The conidiophores emerge through the stomata, or between the epidermal cells (Plate VIII. Fig. 1). The conidia are borne almost horizontally; they are from 11 to 15$\mu$ long, and consist of subpyriform cells attached by their larger end. A thin hyaline membrane encloses the granular protoplasm, in which a small vacuole may be observed. When placed in a damp atmosphere the vacuole enlarges and a germ-tube is produced, generally from the larger end of the conidium; into this germ-tube the protoplasmic contents of the conidium are received and passed onwards as it elongates

* Woronin, *loc. cit.*

(Plate VIII. Fig. 2). If the conidium germinate upon a leaf, the germ-tube squeezes its point between the two epidermal cells (Fig. 3), and soon produces in the leaf a mycelium with haustoria. In from twelve to twenty days after infection this mycelium produces the black teleutospores, but not the conidia.

The life-history of this species is peculiar: the teleutospores germinating in autumn produce promycelial spores, which, entering the young subterranean shoots of the host-plant, develop a mycelium, which remains quiescent during the winter, and in the spring produces, first the conidia on the leaves, and afterwards teleutospores mostly in the stem.

The entrance of the germ-tube in this species is (as already stated above) between the epidermal cells. It grows downwards in the partition wall, splitting it into two laminæ, and so makes its way through the epidermis.

# CHAPTER XI.

## INFECTION OF THE HOST-PLANTS BY THE USTILAGINEÆ.

THE manner in which the Ustilagineæ gain admission into their respective host-plants has been studied very carefully by many botanists, but is not yet fully understood. With those species which affect the flowering parts of annual graminaceous plants, such as *Tilletia tritici* and *U. segetum* on wheat, it is noteworthy that not only are all the blossoms or fruits upon an ear affected, but also all the ears which arise from one plant. It is very exceptional ever to find one sound ear upon a plant of which the others are diseased; in like manner, it is very unusual to find a sound kernel upon an ear in which the other kernels are affected. Coupled with the fact that in diseased plants the mycelium of the fungus can be found in all parts of the axis, it is obvious that the parasite gained admission into the plant at an early stage of its growth. Kühn * specially investigated this point with *T. tritici*, and found, in very young wheat seedlings, that the mycelium was present in them. Hoffmann † came to the conclusion that the spores entered between the split in the young sheath and the rootlet. He also figures the spores forming a mycelium which enters the stomata of the young

* Kühn, *loc. cit.*, pp. 48, 49.

† Hoffman, "Flugbrand," pp. 202–206, t. xiv. figs. 14–18.

plant, but this is probably incorrect. To Wolff,* however, we owe the first accurate explanation of this process. He investigated it with *U. segetum* and *maydis*, *Urocystis occulta* and *T. tritici.* The outcome of his observations is that the germ-tube of the promycelial spores of the species is capable of piercing the embryonic plant at any time before the primary enveloping sheath of the young plant is ruptured. The germ-tubes of *T. tritici* squeeze their points through the epidermal cells of the young plant, at first piercing through the outer epidermis of the primary sheath; they then grow through the cells of the sheath itself, then through the inner epidermal cells of the sheath, across the interspace to the outer epidermal cells of the embryo, and so into the embryo itself (Plate VI. Fig. 2). With certain species the entering germ-tube acquires for itself an investing sheath from the cells through which it passes (Plate VI. Fig. 3)—a sort of invagination of the outer wall of the outer epidermal cell, which is continued over the young mycelium as it grows through one cell after another. With *Urocystis occulta* the investing sheath exists only where the mycelium passes through the first epidermal cell. Kühn † subsequently repeated and confirmed Wolff's observations as far as they went, but he also found that the germ-tubes could enter, not only through the primary sheath-leaf, but also into the true root-node at the base of the sheath, and, in point of fact, into almost any part of the embryo. While it has long been known that, by merely dusting wheat with the teleutospores of *T. tritici* and planting it, it became affected with bunt, yet with *U. segetum* such dusting rarely, if ever, succeeds in producing the disease. Hoffmann was able to produce only a few smutted plants in many hundred

* Wolf, "Roggenstengelbrand," *Bot. Zeitung* (1873), t. viii.; "Der Brand des Getreides" (1874), pp. 18–24, t. iii., iv.

† Kühn, *Bot. Zeitung* (1874), pp. 121–124; *Fahlingsladw. Zeitung* (1879) p. 84.

experiments in which he applied the teleutospores to the young plant. F. von Waldheim was equally unsuccessful. Kühn asserts that he found that, if too many germ-tubes entered an embryo plant, they developed into spore-forming hyphæ and formed a "brand-knot" in the sheath-node, and killed the young plant. All the experiments which I have conducted with a view of infecting the young plants of wheat, barley, and oats with *U. segetum* have uniformly failed. I have attempted the infection in various ways: dusted the spores on the dry grain and planted it; soaked the grain in water, and then dusted it with the dry spores; planted the grain in flower-pots, and dusted the spores thickly on the surface of the soil; the grain allowed to germinate, and applied the dry spores to the embryos just as they emerged from the seed-corn; placed the teleutospores of *U. segetum* in water for twelve hours, and, when an abundant development of promycelial spores had taken place, applied the spore-charged water to the emerging embryos; germinated the teleutospores of *U. segetum* in nährlösung and dipped the young embryos in it; watered the grain, after it was planted, and before it came up, with nährlösung, containing spores, but the result was uniform failure. Wolff has stated that the infection will not be successful if the infected plants be kept too moist at first. This point was attended to, but the result was the same. Mr. A. S. Wilson, however, has been more successful, for he showed me some oat-plants which he had artificially infected by removing the glumellæ and applying the spores to that part of the kernel from which the embryo emerges; but he also informed me that he had many failures.

But more than this remains to be considered. *T. tritici* matures its teleutospores at the same time that the wheat-plant matures its fruit; but with *U. segetum* the case is altogether different, for the smut is formed at the time, or soon after, the cereals are in blossom, and long before harvest it has been scattered by the winds, so that in the harvest field one never finds a smutted ear. We know, moreover, that when once the teleutospores fall on the ground, or in any way become damp, they forthwith germinate, and although they are capable of retaining their power of germination for some years, it is only when they are perfectly dry—a condition which never obtains with them in a state of nature. There must, therefore, be some means by which the interval is bridged over between the ripening of the teleutospores of *U. segetum*, which takes place in early summer, and the time when the grain itself germinates, for this, under any circumstances, can only be one or two months later. This may be by a metœcism, but there is no proof whatever that any such occurs; or it may be by the continued reproduction of yeast-spores, as Brefeld suggests taking place in manure heaps. My own experiments, however, with nährlösung containing *U. segetum* spores have all been negative.

There is a certain point in connection with the reproduction of smut (*U. segetum*) wherein it differs essentially from bunt (*T. tritici*); it is this—that however carefully wheat may be dressed with cupric sulphate, arsenic, brine, lime, etc., while such dressing almost absolutely protects the crop from bunt, yet it has no appreciable affect upon smut. This fact is obvious to any one residing in an agricultural district. The wheats are dressed for bunt on every well-managed farm, but they are as much affected with smut as the barley and oat crops, which latter, never being affected with bunt, are never subjected to protective dressing.

In 1883, I made a series of experiments by applying the teleutospores of *U. segetum* to the wheat and oat plants while they were in flower, but the plants were subsequently destroyed by an accident.

Recently Mr. J. L. Jensen has published the results of his experiments and observations conducted on the experimental farm of the Royal Agricultural School near Copenhagen. He believes the spores of *U. segetum* effect their entrance into the host-plant at the time it is flowering, and either infect the ovum or remain quiescent, enclosed within the grain until the ensuing spring, when they germinate when the grain does, and so cause its infection.

He found that barley grown for twenty-five years consecutively upon one experimental plot was not more affected with *U. segetum* than when grown in the ordinary rotation with other crops, which clearly shows the teleutospores do not remain effective in the soil. He further found that manuring with farmyard manure does not produce more *U. segetum* in the crop than occurred when artificial manures were employed. But he did find that seed obtained from a field in which the fungus had been abundant produced a more severely diseased crop than when the seed was taken from a healthy field; but that, if the seed oats were dipped in water at a temperature of 57° C. (134° F.), and allowed to remain there for five minutes, the disease was prevented, and, moreover, the vitality of the seed was unimpaired.*

With regard to *Tilletia tritici*, the important question of the protective dressing of the seed corn has long ago engaged the attention of agriculturists. Many have been employed, but that most generally used (and probably the best) is a ·5% solution of cupric sulphate in water. Alum, ferrous sulphate, and even sulphuric acid, unless used sufficiently concentrated to injure the seed corn, were

* Some additional observations of Mr. Jensen on this subject, made during the year 1888, will be found under *Ustilago segetum*. See Descriptions.

proved by Wolff to be quite useless. With regard to temperature, a series of experiments made by Schindler * shows that the teleutospores of *T. tritici* will withstand a dry heat of 65° C. before altogether losing their power of germination; but with moist heat they were sterilized between 45° and 50° C. Cold, on the other hand, had little effect upon them, even after exposure to – 20° C. for a prolonged period.

* Schindler, "Ueber den Einfluss verschiedenea Temperaturen auf die Keimfahigkeit den Steinbrandsporen," "Forschungen auf Geb. der Agrikultur-physik. bd. iii." (1880), pp. 288-293.

# CHAPTER XII.

## SPORE-CULTURE.

THE microscopic examination of the Uredineæ and Ustilagineæ is a very simple matter. The æcidia viewed as opaque objects with a low power are always very attractive objects. To examine the various spore-forms, all that is necessary is to remove a small quantity with the point of a penknife, place them in a drop of water on a glass slide, and, having covered them with a circle of thin glass, view the preparation as a transparent object with a quarter-inch objective. The various markings on the exterior of the æcidiospores and uredospores are more readily seen if they be examined dry. In order to obtain more accurate information of the structure of the spore-beds, thin sections must be cut with a sharp knife, including both the spore-bed and a small portion of the host-plant. With a little patience, and by teazing out the cells of the host-plant, the mycelium can be observed. This is often rendered more conspicuous by the application of a drop of caustic potash.

To observe the germination of the spores is not difficult, and can be accomplished without the aid of expensive or elaborate apparatus. All that is necessary is to place the spores in a sufficiently humid atmosphere, or in a sufficiently moist place. This may be conveniently accomplished by placing a drop of pure water upon an ordinary glass slide, and putting into this drop of water the spores whose germination it is desired to watch; as, however, the spores do not germinate for several hours, the drop of water would evaporate, unless means be taken to prevent it, before the germination takes place. This is easily done by placing the glass slide under a bell-glass, inverted over a plate of water, so that it is kept continuously in a saturated atmosphere. Any suitable object will do to place the slide on; but the most convenient appliance will be found to be a simple stand or rack (Fig. 9), which can be constructed

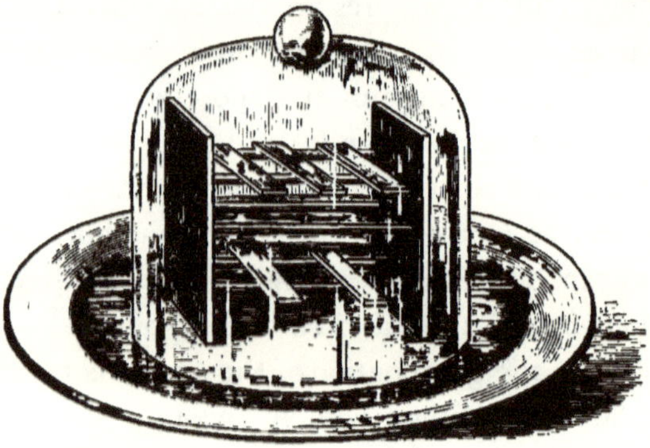

Fig. 9.—Stand with slides for the culture of Uredine spores, placed on a plate of water, and covered by a bell-glass.

in a few minutes out of two flat pieces of gutta-percha and four glass rods. The gutta-percha which is used for soling boots answers well enough. If two oblong pieces of equal size, say three and a half by four and a half inches (9 × 12 cm.), be taken, and a hole bored at each of the four corners, and through each pair of holes let a glass rod be passed, a very simple stand is made at the cost of a few pence. For the glass rods an old barometer tube, cut into suitable lengths with a file, does well enough. The great advantage of such a stand is, that when the slides are laid across the rods, they touch only at two points, so that when placed under the microscope there is no necessity to wipe the lower side of the slide. Of course, the length and height of the stand must be proportionate to the size of the bell-glass. The best form of bell-glass is that known by gardeners as a propagating glass, which is rather flatter than the ordinary bell-glasses ; the advantage being that the slides are not too far above the surface of the water in the plate, so that the drops of water on them do not evaporate so rapidly.

Supposing we wish to observe the germination of the æcidiospores of *Puccinia graminis*, having obtained a perfectly fresh-gathered leaf of barberry with the Æcidium on it, we proceed as follows. The spores can either be brushed upon the slide with a camel-hair pencil, or what will often be found more convenient, the æcidium can be gently struck upon the dry slide, and a drop of water let fall upon the tiny heap of golden spores that have been displaced. The ripe spores will most of them float on the top of the water and can be readily observed with a quarter-inch objective. The preparation must, of course, not be covered with a cover-glass, and it takes a little patience to examine these uncovered objects, because the front of the object-glass is apt to become bedewed and misty. The only plan is to raise it by the coarse adjustment, wipe it dry, and try again.

Germination will be well advanced in the course of ten or twelve hours, and the migration of the yellow endochrome along the germ-tube will by that time have taken place. This will be followed by their circumnutatory movements and ultimate branching. Earlier examination of the preparation will show the germ-tubes emerging from the germ-pores. It is useless to attempt to get æcidiospores to germinate unless they are perfectly fresh and perfectly ripe. For instance, the spores dug out from the bottom of an æcidial cup with a needle will not germinate ;

nor will they if they have once become thoroughly dry. The same method is to be adopted with the uredospores, and the same precautions observed. With regard to the teleutospores, certain modifications of the above are necessary. If one of the Leptopucciniæ is to be examined, all that is requisite is to cut up one of the sori and place the fragments in a drop of water, and in a few hours the promycelia will be developed. Those species which have a prolonged period of rest in their life-history, it is, of course, useless to attempt to germinate except at the proper season of the year. Suppose it is desired to observe the germination of *P. graminis*, in the autumn some specimens of mildewed straw must be procured, and preserved through the winter. I have always found the best plan is to tie them up in a bundle and keep them out-of-doors, so that they are exposed to the same vicissitudes of temperature and moisture as would happen to them in a state of nature; for it is obvious that if they be kept throughout the winter indoors, they will not only be maintained at a higher temperature, but also will become more completely dried than is natural to them. Under such circumstances they neither germinate so freely nor so uniformly as they do when they have passed the winter in the open air. Specimens of *P. graminis* may be obtained on *Triticum repens*, in February or March, from the immediate vicinity of any barberry bush, and these will be found to germinate very readily. Having obtained the material, in March or April, however it may have been preserved, in order to get it to germinate all that is required is to place it in water. If some of the spore-beds be cut into pieces about one-eighth of an inch (2 or 3 mm.) across, and placed in water in a watch-glass, under the bell-glass, the process of germination soon commences, perhaps in twelve hours, perhaps longer, according to the manner in which the material has been preserved, and according to the temperature of the atmosphere at the time the experiment is made. I never remember having seen any teleutospore germinate if the temperature within the bell-glass was below 5° C. Germination is very partial and very slow at 8° C., but at from 10° to 15° C. it is both vigorous and rapid. The germination can be recognized to have taken place, if it be at all free, by the naked eye; the clusters of spores will then be seen to be surrounded by an opalescent, hazy cloud, which, when placed under the microscope, will be found to consist of myriads of promycelia. A very convenient method of preserving material on grasses—such, for instance, as *Uromyces poæ*, which occurs on the leaves of the grass that in the ordinary course of events become disintegrated by decay during the winter—is to gather a small bundle of affected leaves, attached to the stems, place this bundle in a flower-pot just as if it were a living plant, to cover the flower-pot with a bell-glass, and keep it out-of-doors in a shady place all winter. There will be enough moisture in the atmosphere to prevent the material from being injured by desiccation, but not enough to allow the teleutospores to germinate until they are purposely placed in water. The bell-glass will protect the grass from injury by wind and weather; so that when spring comes you will have abundance of material ready to hand, in excellent condition for germination, which you can induce at pleasure, by merely soaking it in water. The same method may be conveniently adopted with those species which occur on leaves, and in which the spores are lost by their decay in the ordinary course of events. The Melampsoræ on willow and poplar may thus be kept out-of-doors under a bell-glass with great facility.

With those species which occur on leaves of plants which have a very perishable foliage, it is necessary to collect specimens which are perfectly mature, if possible, upon leaves that are beginning to fade from age, and dry them in the ordinary way between blotting-paper. When the spring comes, the affected leaves must be soaked for twenty-four hours in water, and the spores examined to see if any attempt at germination is observable. If not, the soaked leaves may be wrapped in an old, well-washed piece of calico, and buried for a day or two in the ground; after which treatment, a few spores must be tried in a drop of water on a glass slide. Should they fail to evince any signs of vitality, the leaves must again be buried for a day or two longer, and re-examined.

The entrance of the germ-tube into the host-plant can be observed in various ways. The promycelial spores can be applied to the surface of a leaf, and sections made a few hours afterwards. This is, however, an exceedingly delicate process, and requires not only patience, but considerable manipulative skill. A piece of the epidermis may be stripped off and laid flat upon a moistened slide, and the spores placed on it; if the preparation be kept in a moist atmosphere for a few hours, the germ-tubes can be seen boring through the cells. Another plan is to place a mass of teleutospores, which has first been seen by the microscope to be in active germination, on a leaf, and to keep it for a few hours under a bell-glass in a moist atmosphere. The teleutospore mass can be seen by the naked eye, and is a guide to the exact part of the leaf to be examined. By a little deft manœuvring, pieces of the epidermis at this spot can be ripped off with the point of a penknife, and examined either on their external or internal surfaces. By similar methods, the entrance of the uredospore and æcidiospore germ-tubes can be observed, only, being larger and containing yellow endochrome, the process is less difficult. With most of the Ustilagineæ, all that is necessary is to place the spores in water, and they will germinate at once—for example, *U. segetum*, *longissima*, etc.; but some require a longer period of immersion, *e.g.* *T. tritici*, which will not germinate till after being several days in water. The germination of the Tilletia spores can also be conducted in a hanging-drop culture in the following manner:—A piece of glass tube, about half an inch (12 mm.) in diameter and about three-quarters of an inch (18 mm.) long, is cemented on an ordinary glass slide, so as to form a deep cell (Fig. 10).

Fig. 10.—Deep cell for hanging-drop cultures, made by cementing a piece of glass or lead tube upon an ordinary glass slide.

Into this is placed a small quantity of water. The drop containing the spores to be germinated is placed on the centre of a circular cover-glass, which will fit the top of the cell (Fig. 11). If the upper edge of the tube which forms

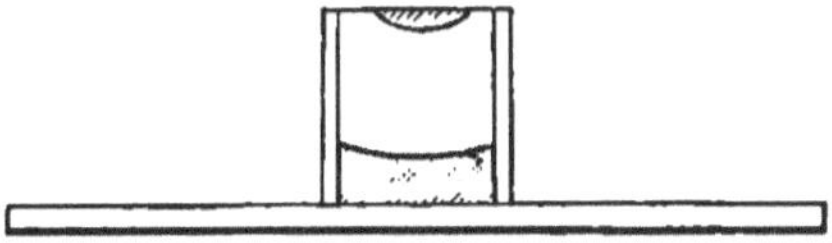

Fig. 11.—Hanging-drop culture, seen in section.

the cell be oiled, no air can get into the culture, and the germination can be watched for months, as the hanging-drop will not evaporate, because the water placed in the bottom of the cell keeps the atmosphere saturated, and any loss which it may sustain from evaporation is made up from the water at the bottom of the cell.

There is no need to keep these sealed hanging-drop cultures in a moist atmosphere, because the necessary moisture is contained in the cell itself; all that is required is to place them under a bell-glass to keep them free from dust. The disadvantage of this mode of culture is, that all spores will not germinate normally unless they get a supply of free air. To obviate this hindrance, the glass cover may be fixed to the cell with three tiny fragments of wax. These cells may be readily constructed with lead tubing; an ordinary piece of gas-piping, cut into suitable lengths with a knife, and the ends smoothed on a whetstone, will answer all the requirements of the case.

The germination of the spores in *nährlösung*, however, requires more care. The nährlösung is prepared by boiling fresh horse-dung in pure water, and filtering first through

Fig. 12.—One of Brefeld's "*kammer*" for nährlösung cultures. It consists of a glass tube open at both ends; in the middle a bulb has been blown which has been compressed laterally, so that its sides are parallel to each other, or nearly so. When in use the two open ends of the tube are closed with cotton wool.

coarse filtering-paper, then through fine; then boiling again for a short time in a flask, the mouth of which is closed, while the steam is issuing from it, by a firm plug of cotton wool. After three or four hours the flask is again boiled for a short time. To sterilize the fluid this boiling requires to be repeated several times, carefully closing the mouth of the flask each time with a plug of cotton wool. The spores to be germinated are put in a small quantity of sterilized nährlösung, and the process watched *in camerâ*. The little apparatus necessary consists of a glass tube about eight or ten inches long, open at both ends; in the middle of this tube a bulb has been blown, the sides of which have been compressed laterally (Fig. 12), so that they are flat and parallel to each other. Before using, the apparatus (*kammer*) must be sterilized by heat; and after the nährlösung has been introduced, the two ends of the tube are closed by cotton wool; this allows free access of air, but filters out any germs that may be floating in it.

Or the culture may be made in a hanging-drop cell made of lead tubing, in the sides of which two holes have been cut (Fig. 13). I have found it most convenient to

Fig. 13.—Hanging-drop culture cell, made of lead tubing, with two holes for the admission of air; when in use with nährlösung these openings are closed with cotton wool.

wrap cotton wool round the cell, and to hold it in its place by a small elastic band. These cells require to be sterilized by washing with a weak solution of corrosive sublimate.

X March 13,

# CHAPTER XIII.

## THE ARTIFICIAL INFECTION OF PLANTS.

IN order to ensure success in the artificial infection of plants, attention to several little details is absolutely necessary. Of course, if you simply wish to produce *Æcidium urticæ* on a cluster of nettles, you may throw a handful of *Carex hirta* affected with Puccinia upon the ground where the nettles grow in autumn, and, trusting to chance, you will probably find them bearing the Æcidium the following spring. But such a procedure is open to many objections; the wind may blow away your Carex during the long winter and spring months either before the Puccinia has germinated or before the nettles have appeared above ground. A still greater objection is, that even if a few clusters of æcidia happen to be produced on the nettles, you have no proof that they arose from the Puccinia you threw down. Still more important is it to avoid this clumsy method of "laying on," if you are investigating the life-history of any particular species of Uredine, for it often happens that more than one species attacks the same host-plant; *P. magnusiana, trailii*, and *phragmitis*, on the reed, for instance.

The first thing to be done is to provide suitable plants for infection. These should, it is hardly necessary to state, be healthy, and have had time to become established before they are infected. It is a good plan to establish a number of plants, say half a dozen, in the autumn; they will then be ready for use in the following spring. It is often convenient to infect every alternate plant, so that the remaining plants may be kept as control specimens. The reason for using established plants is that the young foliage is so much more easily infected by the Uredineæ than the older; in fact, it is by no means uncommon for an old leaf to die off before the Uredine has had time to complete its development.

Let us suppose we wish to perform the classical infection of the barberry with *P. graminis*. In the autumn, six young barberries, small enough to be covered with a bell-glass, having been planted, as soon as their leaves are fully developed in the spring they may be infected in the following manner. A quantity of *P. graminis* having also been provided in the autumn and kept during the winter in the mode before explained, as soon as the barberry foliage is ready, test the germinative power of the *P. graminis* by placing a few fragments in water in a watch-glass. If it germinate freely and produce a good crop of promycelial spores, as proved by microscopic examination, the contents of the watch-glass may be at once employed. It is best to do your infection experiments in the evening. Water one of the barberries freely through the rose of a watering-can and then cover it with a bell-glass; then water the outside of the bell-glass. By so doing the temperature of the enclosed air is reduced, and the inside of the bell-glass as well as the leaves of the barberry become bedewed with condensed vapour. After leaving it a few minutes, remove the bell-glass, and apply the germinating spores with a camel-hair pencil. As the promycelial spores easily become diffused in the water in the watch-glass, by stirring it with the camel-hair pencil the water becomes equally charged with them; then, by simply brushing the water on the leaves, you may be pretty sure of successfully infecting the plant. Replace the bell-glass, and give it another douching outside with the watering-can. If sufficient material has been prepared, each alternate barberry may be infected in the same manner. The bell-glass need not be kept over the infected plants more than two or three days. If the weather be very bright, the bell-glasses should be shaded by putting a piece of matting or carpet over them to prevent the foliage being scorched by the sun. In the course of eight or ten days the yellow spots on which the spermogonia are produced will appear, and in two or three weeks the perfect æcidiospores will be developed. It will then be seen that only those barberries to which the spores were applied have the æcidiospores on them, while the alternate plants remain free. If an attempt be made to infect a plant in the day-time, when the sun's rays are full upon it, it will be found that the water all runs off the leaves; but by operating in the evening in the manner directed, the leaves are bedewed with a thin layer of moisture, and no difficulty will be found in applying the spore-charged water.

Should it be desired to perform the converse culture—the infection of wheat plants with the barberry æcidiospores—greater care is necessary to prevent the control plants from becoming infected, because the æcidiospores will not diffuse themselves in water, and are very readily carried away by currents of air. The simplest way is to plant some wheat in two flower-pots, and at once to place the pots on two plates of water and cover each with a bell-glass. As soon as the young wheat plants have made a green leaf, those in one of the flower-pots may be infected, using the same precautions as are given above. The æcidiospores may be collected by brushing them into a watch-glass of water with a dry camel-hair pencil, taking care to use only those which will readily brush off. The contents of the watch-glass may then be applied to the wheat plants. By this method the accidental infection of the control plants is avoided, because the bell-glass is never removed from them.

Personally, I have found infection with promycelial spores more certain than with the æcidiospores, because we can see that they are actually germinating at the time they are used, while with the æcidiospores this cannot be done so certainly; moreover, the æcidiospores being generally brought from a distance, they are apt to lose this germinative power, unless used immediately, from becoming too dry on the one hand, or on the other, if kept in too moist an atmosphere during their transit, from many of them having already germinated before they are employed. It is only by attention to these minute details that we can ensure uniform success.

The Gymnosporangia are very easy to cultivate. A few seedling hawthorns can be obtained anywhere, and it is necessary only to soak the *Gymnosporangium clavariæforme* for twelve hours, when the golden promycelial spores will be visible to the naked eye.

In producing *R. cancellata* on pear, it is necessary to infect two-year old plants, because if seedlings be infected the spermogonia alone will be produced, because seedling pears lose their foliage before the Rœstelia has had time to develop; with these plants it is essential to success that they should be thoroughly established before they are made the subject of experiment.

In working out the life-history of the allied species duplicated cultures are very valuable. Suppose we wish to produce the æcidia of *P. magnusiana* and *P. phragmitis*. Having provided the proper material and a number of growing plants of *Rumex obtusifoliius* and *Ranunculus repens*, germinate a quantity of the *P. magnusiana* in a watch-glass, and then put half on a Rumex and the other half on a Ranunculus ; in a week or ten days we shall find the Ranunculus affected and the Rumex free. Care, of course, must be taken that there is no mixture of teleutospores in the watch-glass. In like manner the *P. phragmitis* may, on a subsequent occasion, be applied to the other two plants, when we shall find the Rumex become affected and the Ranunculus will remain free.

The main points to be attended to in order to ensure success in performing these cultures are, first and foremost, to have ocular demonstration that your infecting material is actually germinating at the time you use it; and, secondly, to infect the young growing foliage of established plants.

# DESCRIPTIONS OF THE BRITISH UREDINEÆ.

## UREDINEÆ. Tulasne.

MYCELIUM parasitic in living plants. Spores formed from the ends of erect, crowded hyphæ, usually of more than one kind. Teleutospores germinating by a short promycelium.

## UROMYCES. Link.

Teleutospores separate, unicellular, pedicellate, produced in flat sori (spore-beds), apex perforated by a single germ-pore.

### I. EUUROMYCES. Schröter.

Having spermogonia, æcidiospores, uredospores, and teleutospores, the latter germinating only after a period of rest.

#### A. AUTEUUROMYCES.

Having all spore-forms on the same host-plant.

### Uromyces fabæ. (Pers.)

*Æcidiospores*—Pseudoperidia generally crowded upon whitish spots, which are more or less circular, short, slightly prominent, flat, with torn white edges. Spores subglobose, orange-yellow, finely echinulate, 15–25$\mu$ in diameter.

*Uredospores*—Sori chestnut-brown, roundish, amphigenous, scattered, often confluent, soon naked. Spores subglobose or ovate, with three germ-pores, shortly echinulate, yellowish brown, 20–30 × 17–20$\mu$.

*Teleutospores*—Sori rounded on the leaves, more abundant and elongate on the stems, persistent, black. Spores variable in form, ovate or broadly clavate, dark brown, smooth, apex much thickened (8–10$\mu$), rounded or conical with one infundibuliform germ-pore, 25–40 × 20–25μ. Pedicels long, persistent, pale brown.

*Synonyms.*

*Uromyces orobi*, Pers. Winter in Rabh., "Krypt. Flor.," 2nd edit., vol. i. p. 158, in part.

*Uromyces fabæ*. Cooke, "Grevillea," vol. vii. p. 135.

*Uromyces appendiculata*, Lév. Cooke, "Hdbk.," p. 518; "Micro. Fungi," 4th edit., p. 212.

*Uredo fabæ*, Pers. Romer, "New Mag.," vol. i. p. 92. Grev., "Scot. Crypt. Flor.," t. 95; "Flor. Edin.," p. 436, in part.

*Uredo fusca*. Purton, "Midl. Flor.," vols. ii. and iii., No. 1130.

*Uredo leguminosarum*, Link. Berk., "Eng. Flor.," p. 383.

*Uredo appendiculosa*. Berk., "Eng. Flor.," p. 383.

*Trichobasis fabæ*, Lév. Cooke, "Micro. Fungi," 4th edit., p. 225.

*Puccinia globosa*. Grev., "Flor. Edin.," p. 368; "Scot. Crypt. Flor.," t. 29.

*Puccinia fabæ*, Link. Berk., "Eng. Flor.," p. 434. Cooke, "Hdbk.," p. 508; "Micro. Fungi," 4th edit., p. 211.

*Exsiccati.*

Cooke, i. 71; ii. 52. Vize, "Micro. Fungi," 44; "Micro. Fungi Brit.," 63, 223.

On *Faba vulgaris*, *Vicia cracca*, *sepium*, *sativa*, *Lathyrus pratensis*, *Pisum sativum*.

Æcidiospores, April and May; uredospores, May to July; teleutospores, July to November, and lasting through the winter on the dead stems.

Biology.—There are several species of Uromyces parasitical upon the Leguminosæ. By most botanists the above is considered identical with *U. ervi*. As early in the year as February 8 (1884), I produced the Æcidium upon a bean plant (*V. fabæ*) from teleutospores on bean straw which had been grown for agricultural purposes. The latent period between placing the teleutospores on the host-plant and the appearance of the spermogonia was twenty-three days. The culture was repeated on March 20 in two experiments, both of which were successful, the spermogonia in each case showing on April 16. In 1886, four cultures were made with *U. ervi* by placing the germinating teleutospores on bean, pea (*Pisum sativum*), and two vetch seedlings. Although the teleutospores were germinating very freely, yet they produced no result. In 1888 two further series of cultures were made with *U. fabæ* on bean, pea, *Vicia sativa*, *cracca*, *Lathyrus pratensis*, and *Ervum hirsutum*. No æcidiospores were produced except upon the bean and the pea. The æcidium on the bean occurs on white spots, which are thickened and very conspicuous. On pea the spots are pale dirty yellow, and the pseudoperidia few and scattered, while on bean they are numerous and crowded together. There can be no doubt that *U. fabæ* and *ervi* are biologically distinct, but the 1888 cultures show that continued investigation will probably lead to further subdivision of the forms now grouped under *U. fabæ*.

De Bary says that in some instances the same mycelium which produces the æcidia of *U. fabæ* gives rise to a few isolated uredospores.

### Uromyces orobi. (Pers.)

*Æcidiospores*—Spots scattered or circinate. Pseudoperidia flat or slightly prominent, with torn whitish edges. Spores subglobose or polygonal from mutual pressure, orange, echinulate, 16–27$\mu$ in diameter.

*Uredospores*—Sori small, crowded or scattered, soon naked, elongated on the stems. Spores subglobose or ovate, echinulate, yellowish, 15–28 × 16–22$\mu$.

*Teleutospores*—Sori roundish or elongate, very dark brown, at first covered with epidermis. Spores oblong or subpyriform, apiculate above, often obliquely, from a thickening of the epispore, dark brown, smooth, 25–40 × 18–28$\mu$. Pedicels very long, persistent, pale brown.

*Synonyms.*

*Uromyces orobi* (Pers.). Winter in Rabh., "Krypt. Flor.," vol. i. p. 158, in part.

*Æcidium orobi*, Pers. Romer, "New Mag.," vol. i. p. 92.

*Æcidium orobi*, D. C. Cooke, "Hdbk.," p. 542; "Micro. Fungi," 4th edit., p. 197. Berk., "Eng. Flor.," vol. v. p. 378.

*Puccinia fabæ.* Johnst., "Flor. Berw.," vol. ii. p. 197.

*Exsiccati.*

Vize, "Micro. Fungi Brit.," p. 327.

On *Lathyrus macrorrhizus* (*Orobus tuberosus*).

July to October.

### Uromyces phaseoli. (Pers.)

*Æcidiospores*—On circular spots about 2 mm. across. Pseudoperidia cup shaped, crowded, with everted, whitish, deeply toothed edges. Spores polygonal, finely verrucose, colourless, 20–25 × 16–18μ.

*Uredospores*—Sori scattered, pale cinnamon brown. Spores rounded or shortly elliptical, pale brown, echinulate, 25–34 × 15–18μ.

*Teleutospores*—Sori black-brown, soon naked. Spores spherical or shortly elliptical, apex thickened, with a wide germ-pore and often a colourless papilla, smooth, dark brown, 26–35 × 22–36μ. Pedicels short, deciduous.

*Synonyms.*

*Uromyces phaseoli* (Pers.). Winter in Rabh., "Krypt. Flor.," vol. i. p. 157.

*Uredo appendiculata*, var. *phaseoli*, Pers. Observat. in "Usteri ann. d. Botan.," vol. xv. p. 17.

*Uromyces phaseolorum*, De Bary. Cooke, "Grevillea," vol. vii. p. 135.

On *Phaseolus vulgaris.*

May and October.

BIOLOGY.—The spots on which the æcidia occur are at first pale, becoming yellowish.

I have never seen British specimens of this, which is inserted on the faith of Dr. Cooke's paper upon Uromyces in "Grevillea."

### Uromyces limonii. (D. C.)

*Æcidiospores*—On purplish spots. Pseudoperidia hypophyllous, white, cylindrical, with much-torn white edges. Spores subglobose or ovate, yellow, minutely verrucose, 16–25 × 15–20μ.

*Uredospores*—Sori brown, roundish, bullate, then naked, scattered. Spores globose or ovate, pale brown, finely verrucose, 30–35 × 25–30μ.

*Teleutospores*—Sori small, rounded, black. Spores ovate or subpyriform, darker and thickened at the apex, often attenuated below, smooth, rich brown, 25–50 × 15–25μ. Pedicels very long.

*Synonyms.*

*Uromyces limonii* (D. C.). Winter in Rabh., "Krypt. Flor.," vol. i. p. 156.

*Puccinia limonii.* D. C., "Flore franç.," vol. ii. p. 595.

*Æcidium statices*, Desm. Cooke, "Micro. Fungi," 4th edit., p. 197.

*Uredo statices*, Desm. Cooke, "Hdbk.," p. 528; "Micro. Fungi," 4th edit., p. 217.

*Uredo armeriæ*, Duby. Berk., "Eng. Flor.," p. 377.

*Uromyces limonii*, Lév. Cooke, "Hdbk.," p. 518; "Micro. Fungi," 4th edit., p. 212.

*Exsiccati.*

Cooke, i. 632, 444, 591; ii. 83, 324. Vize, "Fungi Brit.," 60, 71; "Micro. Fungi Brit.," 128.

On *Statice limonium, Armeria vulgaris.*

Æcidiospores, May and June; uredospores, June and July; teleutospores, July to October.

BIOLOGY.—The æcidia occur on pallid spots, which are usually tinged with purple round their circumference; on the stems they cause considerable distortion. As the fungus occurs in situations which are often covered by the spring tides, it is obvious that the presence of salt is not inimical to it.

## **Uromyces polygoni.** (Pers.)

*Æcidiospores*—Spots generally hypophyllous, rarely cauline. Pseudoperidia in small irregular clusters, rather flat, with broad, whitish torn edges. Spores subglobose, finely verrucose, pale yellow, 15–23$\mu$ in diameter.

*Uredospores*—Sori brown, scattered, rarely circinating, amphigenous, pulverulent. Spores globose or ovate, pale brown, finely echinulate, 20–25 × 15–20$\mu$.

*Teleutospores*—Sori blackish, on the leaves roundish, on the stems elongated. Spores globose or elliptical, smooth, chestnut brown, apices thickened, rounded or conical, 23–35 × 15–20$\mu$. Pedicels yellowish, long, persistent.

*Synonyms.*

*Uromyces polygoni* (Pers.). Winter in Rabh., "Krypt. Flor.," vol. i. p. 154.

*Puccinia polygoni.* Pers., "Disp. Meth.," p. 39.

*Æcidium aviculariæ*, Kze. Cooke, "Micro. Fungi," 4th edit., p. 199; "Hdbk.," p. 545.

*Uredo polygonorum.* Grev., "Scot. Crypt. Flor.," t. 80; "Flor. Edin.," p. 434. Johnst., "Flor. Berw.," vol. ii. p. 201. Berk., "Eng. Flor.," p. 377.

*Trichobasis polygonorum*, B. Cooke, "Micro. Fungi," 4th edit., p. 226.

*Puccinia vaginalium*, Link. Cooke, "Hdbk.," p. 519. Berk., "Eng. Flor.," vol. v. p. 363. Cooke, "Micro. Fungi," 4th edit., p. 204.

*Puccinia aviculariæ.* Grev., "Flor. Edin.," p. 429. Johnst., "Flor. Berw.," vol. ii. p. 195.

*Uromyces polygoni*, Fckl. Cooke, "Hdbk.," p. 519; "Micro. Fungi," 4th edit., p. 213.

*Uromyces aviculariæ*, Schröt. Cooke, "Grevillea," vol. vii. p. 136.

*Exsiccati.*

Cooke, i. 123; ii. 144, 312. Vize, "Micro. Fungi Brit.," 30; "Micro. Fungi," 170.

On *Polygonum aviculare.*

Æcidiospores, May; uredospores, May to July; teleutospores, July to November.

Biology.—The æcidium usually occurs on yellowish spots, which are often surrounded by a reddish margin. On the stems it causes considerable distortion.

## **Uromyces trifolii.** (Alb. and Schw.)

*Æcidiospores* in circular clusters, on pallid spots. Pseudoperidia shortly cylindrical, flattish, on the stems in elongated groups; edges whitish, torn. Spores subglobose or irregular, finely verrucose, pale orange, 14–23$\mu$ in diameter.

*Uredospores*—Sori pale brown, rounded, scattered, surrounded by the torn epidermis. Spores round or ovate, with three or four germ-pores, echinulate, brown, 20–26 × 18–20μ.

*Teleutospores*—Sori small, rounded, almost black, long covered by the epidermis. Spores globose, elliptical or subpyriform, with wart-like incrassations on their summits, smooth, dark brown, 22–30 × 15–20μ. Pedicels long, deciduous.

*Synonyms.*

*Uromyces trifolii* (Alb. and Schw.). Winter in Rabh., "Krypt. Flor.," vol. i. p. 159.

*Uredo fabæ, β. trifolii.* Alb. and Schw., "Consp.," p. 127.

*Puccinia fallens.* Cooke, "Hdbk.," p. 508; "Micro. Fungi," 4th edit., p. 212, in part.

*Trichobasis fallens.* Cooke, "Micro. Fungi," 4th edit., p. 226.

*Uromyces apiculatus,* Lév. Cooke, "Grevillea," vol. vii. p. 136.

*Exsiccati.*

Cooke, i. 116; "L. F.," p. 40. Vize, "Fungi Brit.," 18.

On *Trifolium pratense,* and *repens.*

Æcidiospores, May; uredospores, May; and teleutospores, May to November.

BIOLOGY.—The Uredo and Uromyces frequently attack the petioles, where they cause elongated swellings and distortions. Schröter has found the æcidiospores only on *Trifolium repens*, and states that this stage is of very short duration. A plant of *T. repens*, with the Uromyces upon it, was in October brought indoors and kept there until the following summer. During all this time it produced only teleutospores, no æcidiospores. In the open air the foliage would probably have been destroyed by the cold, so that the fungus would, therefore, have been unable to have kept itself alive, and would of necessity have been compelled to develop æcidiospores in spring from the last year's teleutospores (Schröter, "Cohn's Beitrage," vol. ii. p. 78). Dr. Cooke has observed bicellular teleutospores (*Seem. Jour.*, vol. iv., 1866) on *Vicia sepium;* they were few in number, and mixed with the uredospores.

**Uromyces geranii.** (D. C.)

*Æcidiospores*—Pseudoperidia crowded in irregular or roundish patches on reddish spots, shortly cylindrical, edges white, at first adpressed, afterwards recurved, toothed. Spores roundish, finely verrucose, orange, 20–30 × 15–20$\mu$.

*Uredospores*—Sori dark chestnut-brown, in rounded groups, small rounded, soon naked. Spores spherical, shortly elliptical, echinulate, pale brown, 20–25 × 19–23$\mu$.

*Teleutospores*—Sori blackish, rounded, often circinate. Spores elliptical or pyriform, with a colourless wart-like papilla on the summit, smooth, brown, 20–25 × 17–23$\mu$. Pedicels short, deciduous.

*Synonyms.*

*Uromyces geranii* (D. C.). Winter in Rabh., "Krypt. Flor.," vol. i. p. 160.

*Uredo geranii.* D. C., "Synop. Plant.," p. 47. Berk., "Eng. Flor.," vol. v. p. 380. Sow., t. 398, fig. 5. Grev., "Scot. Crypt. Flor.," t. 8; "Flor. Edin.," p. 434. Johnst., "Flor. Berw.," vol. ii. p. 201.

*Trichobasis geranii*, Berk. Cooke, "Hdbk.," p. 530.

*Æcidium geranii*, D. C. Cooke, "Micro. Fungi," 4th edit., p. 199; "Hdbk.," p. 543. Berk., "Eng. Flor.," vol. v. p. 371. Johnst., "Flor. Berw.," vol. ii. p. 205.

*Uromyces geranii.* Cooke, "Micro. Fungi," 4th edit., p. 213; "Grevillea," vol. vii. p. 134.

*Exsiccati.*

Cooke, i. 107, 440; ii. 50.

On *Geranium sylvaticum*, *pratense*, *dissectum*, and *molle*.
May to October.

BIOLOGY.—At the Mycological Conference in Paris, in 1887, Dr. Richon exhibited a figure of the uredospores accompanied by large clavate, hyaline paraphyses, but I have not observed these in any British specimen I have examined. The æcidiospores often cause great distortion when they occur on the stems. On the leaves they usually occur on reddish spots. Professor Trail finds this fungus near Aberdeen, on the two first-named host-plants, but not upon *G. dissectum* or *molle*.

### Uromyces betæ. (Pers.)

*Æcidiospores*—On yellowish rounded or elongated spots. Pseudoperidia irregularly scattered or circinate, cup-shaped, with whitish torn edges. Spores polygonal, isodiametric, orange-yellow, smooth, 15–25μ in diameter.

*Uredospores*—Sori brown, irregularly roundish, surrounded by the ruptured epidermis, scattered or circinate. Spores ovate or elliptical, pale yellowish brown, echinulate, 25–30 × 16–25μ.

*Teleutospores*—Sori black brown. Spores roundish, elliptical, ovate or obovate, dark brown, smooth, with a colourless papilla on their summits, 25–35 × 20–25μ. Pedicels long, deciduous.

*Synonyms.*

*Uromyces betæ* (Pers.). Winter in Rabh., "Krypt. Flor.," vol. i. p. 155.

*Uredo betæ.* Pers., "Syn.," p. 220. Berk., "Eng. Flor.," vol. v. p. 377.

*Uromyces betæ*, Kühn. Cooke, "Grevillea," vol. vii. p. 136; "Micro. Fungi," 4th edit., p. 213.

*Trichobasis betæ*, Lév. Cooke, "Hdbk.," No. 1587; "Micro. Fungi," 4th edit., p. 225.

*Exsiccati.*

Cooke, i. 70; ii. 31; "L. F.," 39. Vize, "Micro. Fungi Brit.," 553.

On *Beta vulgaris* and *maritima*.

Æcidiospores, April and May; uredospores, June and July; teleutospores, August to October.

BIOLOGY.—The æcidiospores are very rarely found in this country in a state of nature. In 1885, I produced the æcidia on two plants of mangold from the teleutospores from wild plants of *Beta maritima* grown on the banks of the River Ouse at West Lynn. Some dead stems of the Beta were laid on the mangold plants on March 20, and on April 21 the æcidiospores were found, which in due course were followed by the uredospores. The æcidium was found in April, on the wild *Beta maritima*.

### Uromyces valerianæ. (Schum.)

*Æcidiospores*—Pseudoperidia hypophyllous, rarely cauline, circinate, scattered, cup-shaped, on the stem often elongate, slightly prominent, with torn, erect, white edges. Spores polygonal, finely echinulate, orange-yellow, 17–24μ in diameter.

*Uredospores*—Sori small, reddish brown, rounded, scattered or clustered, amphigenous. Spores spherical or elliptical, echinulate, pale brown, 20–30 × 18–20μ.

*Teleutospores*—Spots irregular, dark brown, sometimes forming dendritic figures. Sori long covered by the epidermis, amphigenous, slightly elevated. Spores elliptical or ovate, summits thickened, smooth, chestnut brown, 20–25 × 15–20μ. Pedicels short, deciduous.

*Synonyms.*

*Uromyces valerianæ* (Schum.). Winter in Rabh., "Krypt. Flor.," vol. i. p. 157. Cooke, "Grevillea," vol. vii. p. 137.

*Uredo valerianæ*, Schum. "Enum. Plant. Sæll.," vol. ii. p. 233.

*Lecythea valerianæ*. Berk., "Outl.," p. 334. Cooke, "Hdbk.," p. 523; "Micro. Fungi," 4th edit., p. 222.

*Æcidium valerianaccarum*, Duby. Cooke, "Hdbk.," p. 540; "Micro. Fungi," 4th edit., p. 196; "Eng. Flor.," vol. v. p. 370. Johnst., "Flor. Berw.," vol. ii. p. 206.

*Exsiccati.*

Berk., 349. Cooke, i. 63, 103; ii. 64, 88; "L. F.," 32, 56. Vize, "Micro. Fungi Brit.," 132, 448; "Fungi Brit.," 68.

On *Valeriana officinalis* and *dioica*.

Æcidiospores, May and June; uredospores, June and July; teleutospores, July to September.

Biology.—The presence of the mycelium in the leaves and stems causes the æcidiospore-sori to be seated on the thickened spots.

### Uromyces parnassiæ. (D. C.)

*Æcidiospores*—Hypophyllous on pallid spots, in rounded patches. Pseudoperidia, tawny yellow, between urceolate and concave, with thick edges. Spores pallid.

*Uredospores*—Spores spherical, rough, 20–25μ in diameter.

*Teleutospores*—Sori amphigenous, at first bullate, then rupturing the epidermis, scattered, often confluent. Spores subglobose, ovoid, brown, smooth, 25–30 × 20–22μ.

*Synonyms.*

*Uredo parnassiæ.* D. C., "Flore franç.," vol. vi. p. 68.

*Æcidium parnassiæ*, Grev. Cooke, "Micro. Fungi, 4th edit., p. 198.

*Trichobasis parnassiæ.* Cooke, *Seem. Jour. Bot.*, vol. ii. p. 344; "Hdbk.," p. 531.

*Uromyces parnassiæ*, Schröt. Cooke, "Grevillea," vol. vii. p. 134.

*Exsiccati.*

Cooke, i. 74. Vize, "Micro. Fungi," 4, 226.

On *Parnassia palustris.*

The æcidiospores were found by Dr. Greville near Glasgow, and by Professor Trail near Aberdeen; the teleutospores by Dr. Cooke at Irstead, Norfolk, in 1864.

### Uromyces salicorniæ. (D. C.)

*Æcidiospores*—Pseudoperidia scattered or in small clusters, at first hemispherical, then shortly cylindrical, with erect, white torn edges. Spores polygonal, isodiametric, finely verrucose, orange-yellow, 17–35μ in diameter.

*Uredospores*—Sori rounded, small, long surrounded by the ruptured epidermis. Spores oblong or subpyriform, echinulate 20–35 × 18–20μ.

*Teleutospores*—Sori generally larger than those of the uredospores, pulverulent, dark brown, soon naked. Spores rounded, subpyriform, apex often thickened, smooth, dark brown, 24–36 × 15–26μ. Pedicels long, persistent.

*Synonyms.*

*Uromyces salicorniæ* (D. C.). Winter in Rabh., "Krypt. Flor.," vol. i. p. 156. Cooke, "Grevillea," vol. vii. p. 137.

*Æcidium salicorniæ.* D. C., "Flore franç.," vol. vi p. 92.

*Exsiccati.*

Cooke, i. 538; ii. 143. Vize, "Micro. Fungi," 139.

On *Salicornia herbacea.*

Biology.—The æcidiospores often occur on the young cotyledonary leaves, on yellowish spots.

B. Heteruromyces. Schröt.

Having the spermogonia and æcidiospores on one host-plant, and the uredospores and teleutospores upon another of a different genus.

## Uromyces dactylidis. Otth.

*Æcidiospores*—Pseudoperidia on rounded or elongated spots, often in confluent clusters, cup-shaped, with everted torn white edges. Spores polygonal, subglobose, or isodiametric, 15–25$\mu$ in diameter.

*Uredospores*—Sori small, elliptical or oblong, scattered, long covered by the epidermis. Spores almost spherical, rarely ovate, echinulate, orange-yellow, 18–30 × 15–20$\mu$, without paraphyses.

*Teleutospores*—Sori small, elongated or roundish, long covered by the epidermis. Spores irregularly rounded or oblong, somewhat thickened and darker above, smooth, brown, 18–20 × 14–17$\mu$. Pedicels short, persistent.

*Synonyms.*

*Æcidium ranunculacearum*, D. C. in part. Cooke, "Hdbk.," p. 539; "Micro. Fungi," 4th edit., p. 196.

*Uromyces dactylidis.* Otth., "Nat. Ges. in Bern." (1861), p. 85. Winter in Rabh., "Krypt. Flor.," vol. i. p. 161.

*Uromyces graminum.* Cooke, "Hdbk.," p. 520; "Grevillea," vol. vii. p. 138; "Micro. Fungi," 4th edit., p. 214.

*Exsiccati.*

Cooke, i. 537. Vize, "Fungi Brit.," 37; "Micro. Fungi Brit.," 139.

Æcidiospores on *Ranunculus bulbosus*, May and June.

Teleutospores on *Dactylis glomerata*, July to October, and continuing on the dead stems until the following spring.

BIOLOGY.—This species has been the subject of many cultures by me. It has been stated that the uredospores are provided with paraphyses; in this country they certainly are not. It has been affirmed, too, that *Ranunculus repens* and *acris* bear the æcidiospores; but in numerous cultures, many of which were serial (*i.e.* the same infecting material was simultaneously applied to a series of plants), no result was obtained on the above-named plants (*R. acris* and *repens*). The series included *R. bulbosus*, and on it, and on it alone, the æcidium developed, the other species named above (*R. repens* and *acris*), as well as *R. ficaria* and *auricomus* remaining free from the parasite; conversely, the æcidiospores from *R. bulbosus* applied to *Poa pratensis* and *trivialis* produced no result.

For a detailed account of these cultures, see *Quart. Jour. of Micro. Science*, vol. xxv., new series, pp. 152–156.

## **Uromyces poæ.** Rabh.

*Æcidiospores*—Similar to the preceding. Spores 15–20μ in diameter.

*Uredospores*—Sori orange, rounded, elliptical, or linear; at first covered by the epidermis, which splits longitudinally. Spores rounded, elliptical, or ovate, finely echinulate, orange-yellow 16–26μ in diameter, without paraphyses.

*Teleutospores*—Sori brown, small, punctiform or elongate, covered by the epidermis. Spores generally irregular in form, often elliptical or ovate, pale brown, with a smooth epispore, 17–25 × 25–40μ. Pedicels long, narrow, rather persistent.

*Synonyms.*

*Æcidium ficariæ.* Pers., "Obs. Myc.," vol. ii. p. 23. Purton, "Midl. Flor.," vol. iii. p. 333. Sow, t. 397, fig. 4.

*Æcidium ranunculacearum*, D. C. in part. Cooke, "Hdbk.," p. 539; "Micro. Fungi," 4th edit., p. 196, plate ii. figs. 12–14. Johnst., "Flor. Berw.," vol. ii. p. 206. Berk., "Eng. Flor.," vol. v. p. 370.

*Æcidium confertum.* Grev, "Flor. Edin.," p. 446. Johnst., "Flor. Berw.," vol. ii. p. 205.

*Uromyces poæ.* Rabh., "Unio. Itin." (1866), No. xxxviii. Winter in Rabh., "Krypt. Flor.," vol. i. p. 162.

*Exsiccati.*

Cooke, i. 8; ii. 87; "L. F.," 55. Vize, "Fungi Brit.," 72.

Æcidiospores on *Ranunculus ficaria, repens,* and *bulbosus*, March to May.

Teleutospores on *Poa trivialis, pratensis,* and *annua*, April to July.

BIOLOGY.—This species is said to occur on *P. nemoralis* as well as on *P. pratensis* and *trivialis*, but I have been unable to produce it on the first mentioned, although I have done so several times on the two latter. In one serial culture in which the Uromyces from *P. trivialis* was applied to *R. ficaria*, (2) *repens*, and (3) *bulbosa*, the æcidium was produced on the two latter, but not on the first named. No result was obtained on *R. auricomus* and *acris*. At present it is safer to say that *Ur. poæ* has its æcidiospores on *R. ficaria, repens*, and apparently on *R. bulbosus* (because I do not like to be too confident about one culture), and that it has its teleutospores on *Poa trivialis* and *pratensis*. I have failed more than once in producing the æcidium on *R. ficaria* from the Uromyces on *P. trivialis*, although I have always succeeded with teleutospores from *P. trivialis* on *R. repens*, and I think it quite possible that there may be two species, the one having its teleutospores on *P. trivialis*, and its æcidiospores on *R. repens*, the other with its teleutospores on *P. pratensis*, and its æcidiospores on *R. ficaria*. Further cultures can alone determine the truth of this surmise.

## **Uromyces junci.** (Desm.)

*Æcidiospores*—Pseudoperidia circinating, cup-shaped, with whitish torn edges. Spores polygonal, irregular, globose or elongate, pale orange, smooth, $15-23\mu$ in diameter.

*Uredospores*—Sori on brown or yellowish elongated spots, scattered, irregular, rounded or elongated, confluent. Spores rounded or elliptical, echinulate, pale brown, $17-28 \times 15-17\mu$.

*Teleutospores*—Sori round or elongate. Spores dark brown, usually elliptical or cuneiform, with much thickened and often attenuated summits, deep brown, smooth, $20-40 \times 15-20\mu$. Pedicels long, thick, pale brown.

*Synonyms.*

*Uromyces junci* (Desm.). Winter in Rabh., "Krypt. Flor.," vol. i. p. 162.

*Puccinia junci*, Desm. "Plant. Crypt. Edin.," 2nd edit., No. 170.

*Æcidium zonale*, Duby. "Bot. Gall.," vol. ii. p. 906.

*Uromyces junci*, Tul. Cooke, "Grevillea," vol. vii. p. 139; "Micro. Fungi," 4th edit., p. 213.

*Exsiccati.*

Vize, "Micro. Fungi Brit.," 445.

Æcidiospores on *Inula dysenterica*, May to July.

Teleutospores on *Juncus obtusiflorus*, July to October, and lasting through the winter on the dead stems.

BIOLOGY.—The presence of the mycelium of the æcidiospores in the leaves causes round spots, which are yellowish in colour, and surrounded by a purple line. Fuckel suggested the probability of the connection between the æcidiospores and the teleutospores, but I believe the actual demonstration of their relationship by culture was never made until I did so in 1882.

## Uromyces pisi. (Pers.)

*Æcidiospores*—Scattered over the whole leaf surface. Pseudoperidia cup-shaped, with whitish edges. Spores subglobose or polygonal, orange, finely verrucose, 17–26$\mu$ in diameter.

*Uredospores*—Sori roundish, scattered or crowded, cinnamon-brown. Spores subglobose or elongate, yellowish brown, echinulate, 17–20 × 20–25$\mu$.

*Teleutospores*—Sori roundish or elliptical, blackish. Spores subglobose or shortly elliptical, finely but closely punctate when recent, apex only slightly thickened, 20–30 × 17–20$\mu$ Pedicels long, colourless, fragile.

*Synonyms.*

*Uromyces pisi* (Pers.). Winter in Rabh., "Krypt. Flor.," vol. i. p. 163.

*Uredo appendiculata, β. pisi.* Pers., "Observ. myc." in Usteri, *Annal. d. Botan.*, vol. xv. p. 17.

*Uromyces pisi*, De Bary. Cooke, "Grevillea," vol. vii. p. 135.

*Æcidium cyparissiæ.* D. C., "Flore franç.," vol. ii. p. 240.

Æcidiospores on *Euphorbia cyparissias.*

Uredospores and teleutospores, on *Pisum sativum.*

BIOLOGY.—I have no acquaintance with this as a British species, and the host-plant of its æcidiospores is not native in this country. Its life-history was worked out by Schröter, who states that the teleutospores occur on *Vicia cracca*, *Pisum sativum*, *Lathyrus pratensis* and *sylvestris*. He further finds that another æcidium on *E. cyparissias* produces another Uromyces—*U. striatus*, with brown teleutospores; which, when recent, are delicately striate with wavy lines, and which have a flat conchiform brown cap over the germ-pore. The latter occurs on *Lotus corniculatus*, *Trifolium arvense*, *Medicago sativa*, etc.

### II. BRACHYUROMYCES. Schröt.

Having spermogonia, uredospores, and teleutospores.

### III. HEMIUROMYCES. Schröt.

Having only uredospores and teleutospores, which occur on the same host-plant.

### **Uromyces scutellatus.** (Schrank.)

*Uredospores*—Sori small, roundish, at first covered by the epidermis. Spores scanty, mixed with the teleutospores, roundish or subpyriform, with a thick colourless or yellowish-brown spore-membrane, smooth, verrucose or finely echinulate, 20–35 × 15–25$\mu$.

*Teleutospores*—Irregular, rounded, ovate, or oblong, apices of the spores often surmounted with a broad, flat, pale or colourless papilla. Spores brown, smooth, tuberculate or reticulate, 20–40 × 15–25$\mu$. Pedicels rather long, deciduous.

*Synonyms.*

*Lycoperdon scutellatum*, Schrank. "Baiersch Flor.," vol. ii. p. 631.

*Uredo excavata*, D. C. "Synop. Plant.," p. 47.

*Uromyces scutellatus*, Lév. Cooke, "Grevillea," vol. vii. p. 137, in part.

*Uromyces excavatus*, D. C. Cooke, "Grevillea," vol. vii. p. 138; "Micro. Fungi," 4th edit., p. 213.

*Uromyces scutellatus* (Schrank). Winter in Rabh., "Crypt. Flor.," vol. i. p. 144.

On various Euphorbiæ (?).

BIOLOGY.—The mycelium of the teleutospores permeates the whole plant. The foliage of the affected plants is altered by its presence, being shorter, broader, and thicker; moreover, they seldom blossom. The sori are scattered over the whole plant. Winter considers this species to belong to Hemiuromyces, Schröter that it is without uredospores. It is doubtfully British.

### **Uromyces anthyllidis.** (Grev.)

*Uredospores*—Sori roundish, chestnut-brown. Spores subglobose, 22–24$\mu$ in diameter, echinulate, chestnut-brown, with four or five germ-pores, contents orange-red.

*Teleutospores*—Sori brownish black. Spores short, elliptical or globose, 19–22 × 17–20$\mu$, dark chestnut-brown, markedly verrucose, apex rounded. Pedicels short, deciduous.

*Synonyms.*

*Uredo anthyllidis.* Grev. in Hook. Herb. Berk., "Eng. Flor.," vol. v. p. 383.

*Uromyces anthyllidis.* Schröt., "Krypt. Flor. Schl.," vol. iii. p. 308.

On *Anthyllis vulneraria.* June to October.

### **Uromyces rumicis.** (Schum.)

*Uredospores*—Sori amphigenous, brown, small, round, scattered. Spores elliptical or subrotund, echinulate, pale brown, 20–25 × 20–30$\mu$.

*Teleutospores*—Sori dark brown, roundish, scattered. Spores roundish, elliptical or subpyriform, chestnut-brown, smooth, 25–35 × 15–25$\mu$, apex of the spores having a pale, rounded, wart-like point. Pedicels short, deciduous.

*Synonyms.*

*Uromyces rumicis* (Schum.). Winter in Rabh., "Krypt. Flor.," vol. i. p. 145.

*Uredo rumicis*, Schum. "Enum. Plant. Saell.," vol. ii. p. 231. Purton, "Midl. Flor.," vol. iii. No. 1544.

*Uredo bifrons.* Grev., "Flor. Edin.," p. 435. Berk., "Eng. Flor.," vol. v. p. 382. Johnst., "Flor. Berw.," vol. ii. p. 201. Cooke, "Hdbk.," p. 528; "Micro. Fungi," 4th edit., p. 217, t. vii. figs. 137–139.

*Uredo apiculosa*, Link. Berk., "Eng. Flor.," vol. v. p. 382. Purton, "Midl. Flor.," vol. iii. p. 297. Grev., "Flor. Edin.," p. 436.

*Trichobasis rumicum*, D. C. Cooke, "Micro. Fungi," 4th edit., p. 225.

*Uromyces apiculosa*, Lév. Cooke, "Hdbk.," p. 518; "Micro. Fungi," 4th edit., p. 212, t. vii. figs. 154–155.

*Uromyces rumicum*, Lév. Cooke, "Grevillea," vol. vii. p. 136.

*Exsiccati.*

Cooke, i. 318, 322; "L. F.," p. 26. Vize, "Fungi Brit.," 63, 55; "Micro. Fungi Brit.," 225.

On *Rumex conglomeratus, obtusifolius, crispus, hydrolapathum, acetosa.* May to September.

BIOLOGY.—The sori are often accompanied by very little discoloration of the foliage, but the presence of the mycelium often causes those parts of the leaf which are adjacent to the sori to retain their original green colour long after the unaffected portions of the leaves have become yellow from age. Sometimes, however, on *R. acetosa* there is considerable red discoloration.

**Uromyces sparsus.** (Kze. and Schm.)

*Uredospores*—Sori on pale spots, often convex, round or elliptical, amphigenous or cauline, long covered, at length surrounded by the ruptured epidermis. Spores round or oblong, 20–23$\mu$ in diameter, to 30$\mu$ in length.

*Teleutospores*—Round or ovoid, frequently attenuated below, brown, smooth, epispore thickened above, sometimes hooded, 30–40 × 15–24$\mu$. Pedicels long and persistent.

*Synonyms.*

*Uromyces sparsus* (Kze. and Schm.). Winter in Rabh., "Krypt. Flor.," vol. i. p. 148.

*Uredo sparsus*, Kze. and Schm. "Deutsch. Schwamme," 170.

*Uromyces sparsa*, Lév. Cooke, "Hdbk.," p. 519; "Grevillea," vol. vii. p. 137; "Micro. Fungi," 4th edit., p. 214.

On *Spergularia rubra.* May to July.

**Uromyces alchemillæ.** (Pers.)

*Uredospores*—Sori golden-scarlet, hypogenous, rounded or elliptical, frequently arranged radially, following venation of the leaves, becoming confluent. Spores globose, elliptical, or oval, epispore thickly covered with short sharp points, orange-yellow, 15–25 × 15–20μ.

*Teleutospores*—Sori chestnut-brown, hypogenous, scattered, roundish, discrete, rarely confluent. Spores elliptical or ovate, verrucose, brown, 30–40 × 20–35μ. Pedicels rather long, deciduous.

*Synonyms.*

*Uromyces alchemillæ* (Pers.). Winter in Rabh., "Krypt. Flor.," vol. i. p. 146.

*Uredo alchemillæ.* Pers., "Syn.," p. 215. Grev., "Flor. Edin.," p. 439. Johnst., "Flor. Berw.," vol. ii. p. 199.

*Uredo intrusa.* Grev., "Flor. Edin.," p. 436. Berk., "Eng. Flor.," vol. ii. p. 382. Johnst., "Flor. Berw.," vol. ii. p. 201.

*Uromyces intrusa*, Lév. Cooke, "Hdbk.," p. 519; "Micro. Fungi," 4th edit., p. 213.

*Uromyces alchemillæ*, Fckl. Cooke, "Grevillea," vol. vii. p. 136.

*Trachyspora alchemillæ*, Fckl. *Bot. Zeit.*, 1861, p. 250. Schröter, "Krypt. Flor. Schlesien," vol. iii. p. 350.

*Exsiccati.*

Cooke, i. 121; "L. F.," 27. Vize, "Fungi Brit.," 40; "Micro. Fungi Brit.," 43.

On *Alchemilla vulgaris.* May to September.

BIOLOGY.—Those leaves which are attacked by the mycelium of the uredospores do not properly develop, and are usually more elongated in the stalk than the healthy ones. The teleutospores are much less conspicuous than the uredospores.

**Uromyces alliorum.** (D. C.)

*Uredospores*—Subglobose, pale, 22 × 25μ. Epispore thin.

*Teleutospores*—Elliptical, brown, 30–35 × 15–18μ. Pedicels very evanescent.

*Synonyms.*

*Uromyces alliorum*, D. C. Cooke, "Grevillea," vol. vii. p. 138; "Micro. Fungi," 4th edit., p. 212.

*Uredo alliorum*, D. C. Cooke, "Micro. Fungi," 4th edit., p. 217, in part.

I am unacquainted with this species.

IV. UROMYCOPSIS. Schröt.

Having spermogonia, æcidiospores, and teleutospores. Uredospores absent.

### Uromyces behenis. (D. C.)

*Æcidiospores*—Pseudoperidia in round, often confluent, circinate clusters, sometimes almost covering the affected leaf, slightly prominent, rather short, with broad, torn, whitish-yellow edges. Spores round or elongate, finely echinulate, orange-yellow, 15–20$\mu$ in diameter.

*Teleutospores*—Sori dark brown, often occurring with the æcidia in small, roundish clusters, sometimes confluent, long covered by the epidermis. Spores elliptical or ovate, smooth, with rather markedly thickened summits, 25–40 × 17–25$\mu$. Pedicels very long, stout, persistent, hyaline or yellowish.

*Synonyms.*

*Uromyces behenis*, D. C. Winter in Rabh., "Krypt. Flor.," vol. i. p. 153.

*Æcidium behenis*, D. C. "Encycl.," vol. viii. p. 239. Berk., "Eng. Flor.," vol. v. p. 372. Cooke, "Hdbk.," p. 541; "Micro. Fungi," 4th edit., p. 197.

*Uromyces behenis*, Lév. Cooke, "Grevillea," vol. vii. p. 134; "Micro. Fungi," 4th edit., p. 213.

*Exsiccati.*

Baxt., 90. Cooke, i. 442. Vize, "Fungi Brit.," 167; "Micro. Fungi Brit.," 134.

On *Silene inflata, maritima.* July to September.

Biology.—The mycelium causes considerable discoloration of the host-plant, the affected spots being sometimes yellow, sometimes brown, and generally having a purple margin. Both æcidiospores and teleutospores are produced from the same mycelium, as is the case with the Chilian species, *Puccinia berberidis.*

### Uromyces scrophulariæ. (D. C.)

*Æcidiospores*—Pseudoperidia on yellowish spots, in roundish circinate clusters, generally hypophyllous, with rather prominent, yellowish-white, erect or sometimes inverted, entire edges. Spores rounded, polygonal, finely verrucose, 17–30$\mu$ in diameter.

*Teleutospores*—Sori frequently intermingled with the accompanying æcidia, circinate, confluent or following the venation, small, round or elliptical. Spores round, obovate or oblong, or subpyriform, with much-thickened, often conically attenuated summits, chestnut-brown, smooth, 20–35 × 10–20$\mu$. Pedicels rather long, deciduous.

*Synonyms.*

*Uromyces scrophulariæ* (D. C.). Winter in Rabh., "Krypt. Flor.," vol. i. p. 151.

*Æcidium scrophulariæ.* D. C., "Flore franç.," vol. vi. p. 91. Cooke, "Hdbk.," p. 544; "Micro. Fungi," 4th edit., p. 199.

*Uromyces concomitans*, B. and Br. Cooke, "Micro. Fungi," 4th edit., p. 213.

*Uromyces scrophulariæ*, Lév. Cooke, "Grevillea," vol. vii. p 136; "Micro. Fungi," 4th edit., p. 213.

*Puccinia scrophulariæ*, Lév. Cooke, "Hdbk.," p. 497.

*Exsiccati.*

Cooke, i. 209, ii. 82, 637. Vize, "Micro. Fungi Brit.," 41.

On *Scrophularia nodosa.*
May to October.

BIOLOGY.—The mycelium causes considerable distortions of the stem and petioles; the spots are yellowish, usually surrounded by a purplish margin. In this species the same mycelium which produces the æcidiospores also gives rise to the teleutospores.

## Uromyces ervi. (Wallr.)

*Æcidiospores*—Scattered over both surfaces of the leaves and on the stems. Pseudoperidia cylindrical or abbreviated, with torn white edges. Spores subglobose, orange, often polygonal, 10–12$\mu$ in diameter.

*Teleutospores*—Sori oval, elliptical or linear, shining black, persistent, erumpent, mostly cauline. Spores ovate or subglobose, smooth, brown, apex thickened and darker, rounded or flattened, often with a blunt conical incrassation, 10–12$\mu$ high 30–35 × 15–25$\mu$. Pedicels brown, long, firmly attached, stout.

*Synonym.*

*Æcidium ervi.* Wallr., "Flor. Crypt. Germ.," vol. ii. p. 247.

On *Vicia hirsuta* (*Ervum hirsutum*, L.).

Æcidiospores, May to June and August to October; teleutospores, from July, lasting on the dead stems through the winter.

BIOLOGY.—Intermixed with the teleutospores occasionally are found a few ovate, pale brown uredospores, 20 × 10$\mu$. This species morphologically closely resembles *U. orobi*, *fabæ*, and *pisi;* but I found, in 1886, that the germinating teleutospores, when placed on young plants of *Pisum sativum*, *Faba vulgaris*, and a vetch seedling, produced no effect (Exp. 623, 645, 646, 647). In 1888, no result was obtained on *Vicia cracca*, *sativa*, and *Lathyrus pratensis*, but only upon *Ervum hirsutum* (Exp. 813 to 827).

### V. MICRUROMYCES. Schröt.

Having only teleutospores, which germinate after a period of rest. Rarely there are found hidden between the teleutospores a few solitary uredospores.

## Uromyces ficariæ. (Schum.)

*Teleutospores*—Sori numerous, amphigenous, in more or less rounded clusters. Spores mostly ovoid, obovate, or pyriform, smooth, brown; the apex of each has usually a wart-like, pale brown papilla, 25–44 × 16–26$\mu$. Pedicels short, colourless.

*Synonyms.*

*Uromyces ficariæ* (Schum.). Winter in Rabh., "Krypt. Flor.," vol. i. p. 141.

*Uredo ficariæ*, Schum. "Enum. Plant. Sæll.," pt. ii. p. 232. Grev., "Flor. Edin.," p. 434. Johnst., "Flor. Berw.," vol. ii. p. 203. Berk., "Eng. Flor.," vol. v. p. 380.

*Uromyces ficariæ*, Lév. Cooke, "Hdbk.," p. 518; "Micro. Fungi," 4th edit., p. 212, t. vii. figs. 156, 157; "Grevillea," vol. vii. p. 134.

*Exsiccati.*

Cooke, i. 122; ii. 145; "L. F.," 24. Vize, "Fungi Brit.," 41.

On *Ranunculus ficaria*, L.
April to June.

BIOLOGY.—The sori occur upon pale spots on the leaves and stems; on the latter, they cause elongated distortions. This species was at one time considered to be the teleutospores of *Æcidium ficariæ*, until Schröter (Cohn's "Beitrage," vol. ii. pt. iii. p. 63) worked out the life-history of the *Ranunculus Æcidia*.

**Uromyces scillarum.** (Grev.)

*Teleutospores*—Sori amphigenous, brown, arranged more or less concentrically, sometimes irregularly, becoming confluent, especially in the centre of the larger groups, at first covered with the cuticle. Spores roundish, ovate, or elliptical, with a uniformly thick, smooth, brown epispore, apex rounded or flattened, $20–30 \times 15–20\mu$. Pedicels short, slender, deciduous.

*Synonyms.*

*Uromyces scillarum* (Grev.). Winter in Rabh., "Krypt. Flor.," vol. i. p. 142.

*Uredo scillarum*. Grev., "Flor. Edin.," p. 376. Berk., "Eng. Flor.," vol. v. p. 376.

*Uromyces concentrica*, Lév. Cooke, "Hdbk.," p. 519.

*Uromyces concentricus*, Lév. Cooke, "Grevillea," vol. vii. p. 138; "Micro. Fungi," 4th edit., p. 213.

*Puccinia scillarum*. Baxt., Exs. 40.

On *Scilla bifolia* and *nutans.*

May and June.

BIOLOGY.—The presence of the mycelium causes the sori to be produced on paler discoloured spots on the leaves and stems.

### Uromyces ornithogali. (Wallr.)

*Teleutospores*—Sori dark brown, scattered, elliptical or oblong, on the smaller leaves causing various distortions, often confluent, at first covered by the epidermis, then pulverulent. Spores elliptical or subpyriform, sometimes smooth, sometimes rough, pale or chestnut-brown, having a colourless wart-like point above, rounded or attenuated below, $25–40 \times 17–22\mu$. Pedicels long, slender, deciduous.

*Synonyms.*

*Uromyces ornithogali* (Wallr.). Winter in Rabh., "Krypt. Flor.," vol. i. p. 141.

*Erysibe rostellata*, var. *ornithogali.* Wallr., "Flor. Crypt. Germ.," vol. ii. p. 209.

*Uromyces ornithogali*, Lév. Cooke, "Grevillea," vol. vii. p. 138.

On *Gagea lutea*, L.

BIOLOGY.—The presence of the mycelium causes variously shaped pale spots on the affected leaves.

### Uromyces urticæ. Cooke.

*Teleutospores*—Subpyriform, apiculate, pale, $30 \times 18\mu$. Epispore thickened, on hyaline pedicels.

*Synonym.*

*Uromyces urticæ.* Cooke, "Grevillea," vol. vii. p. 137.

On *Urtica dioica*. Shere, Surrey.

This species has never been found but once. It is remarkable that no one except Dr. Cooke should have met with it, seeing how common the host-plant is.

### VI. LEPTUROMYCES. Schröt.

Having only teleutospores, which germinate as soon as they are mature upon the living host-plant.

## PUCCINIA. Pers.

Teleutospores separate, pedicellate, produced in flat sori, consisting of two superimposed cells, each of which is provided with a germ-pore. The superior cell has its germ-pore, as a rule, piercing its apex; in the inferior the germ-pore is placed laterally, immediately below the septum.

### I. EUPUCCINIA. Schröt.

Having spermogonia, æcidiospores, uredospores, and teleutospores; the latter germinating only after a period of rest.

### A. AUTEUPUCCINIA. De Bary.

Spermogonia, æcidiospores, uredospores, and teleutospores on the same host-plant.

### Puccinia galii. (Pers.)

*Æcidiospores*—Pseudoperidia scattered or clustered in irregular groups, edges torn, whitish. Spores roundish or shortly elliptical, orange-yellow, smooth, 16–23$\mu$ in diameter.

*Uredospores*—Sori reddish brown, round or oval, often confluent. Spores globose, oval, or ovate, echinulate, pale brown, 20–30 × 17–22$\mu$.

*Teleutospores*—Sori black, roundish, persistent. Spores elliptical, oblong, or clavate, base attenuated, apex much thickened (9–10$\mu$), often obliquely conical, constriction slight, brown, smooth, 30–55 × 15–25$\mu$. Pedicels rather long, brown.

*Synonyms.*

*Puccinia galii* (Pers.). Winter in Rabh., "Krypt. Flor.," vol. i. p. 210.

*Æcidium galii.* Pers., "Syn.," p. 207. Cooke, "Hdbk.," p. 540; "Micro. Fungi," 4th edit., p. 196, t. ii. figs. 15–17.

*Trichobasis galii,* Lév. Berk., "Outl.," p. 332. Cooke, "Hdbk.," p. 501; "Micro. Fungi," 4th edit., p. 226.

*Puccinia galiorum*, Link. Berk., "Eng. Flor.," vol. v. p. 366. Cooke, "Hdbk.," p. 501; "Micro. Fungi," 4th edit., p. 208, t. viii. figs. 172, 173.

*Puccinia difformis*, Fckl. Cooke, "Hdbk.," p. 501; "Micro. Fungi," 4th edit., p. 208.

*Exsiccati.*

Cooke, i. 9, 72, 113; ii. 318, 325, 575. Vize, "Fungi Brit.," 82, 229; "Micro. Fungi Brit.," 110.

On *Asperula odorata*, *Galium cruciata*, *aparine*, *uliginosum*, *palustre*, *verum*, *mollugo*.

BIOLOGY.—The presence of the mycelium in the stems, especially in *G. aparine*, causes considerable swellings and distortions.

## **Puccinia asparagi.** D. C.

*Æcidiospores*—Pseudoperidia in elongated patches upon the stems and larger branches, short, edges erect, toothed. Spores orange-yellow, round, very finely echinulate, 15–26$\mu$ in diameter.

*Uredospores*—Sori brown, flat, small, long covered by the epidermis. Spores irregularly round or oval, clear brown, echinulate, 17–25 × 20–30$\mu$.

*Teleutospores*—Sori black-brown, compact, pulvinate, elongate or rounded, scattered. Spores oblong or clavate, base rounded, apex thickened, darker, central constriction slight or absent, deep chestnut-brown, 35–50 × 15–25$\mu$. Pedicels persistent, colourless or brownish, as long as or longer than the spores.

*Synonym.*

*Puccinia asparagi.* D. C., "Flore franç.," vol. ii. p. 595. Winter in Rabh., "Krypt. Flor.," vol. i. p. 201. Grev., "Flor. Edin.," p. 429. Berk., "Eng. Flor.," vol. v. p. 363. Cooke, "Hdbk.," p. 494; "Micro. Fungi," 4th edit., p. 203.

*Exsiccati.*

Cooke, i. 111. Vize, "Fungi Brit.," 113.

On *Asparagus officinalis*.
April to October.

**Puccinia thesii.** (Desv.)

*Æcidiospores*—Pseudoperidia scattered and thickly crowded over the whole plant, cylindrical, whitish, with torn edges. Spores nearly spherical, orange-yellow, smooth, 17–26 × 12–17μ.

*Uredospores*—Sori reddish brown, round, subconfluent. Spores round or shortly elliptical, brown, echinulate, 20–28μ in diameter.

*Teleutospores*—Sori compact, round or elongate, brownish black. Spores elliptical or ovate, central constriction slight or absent, base rounded, apex thickened and darker, smooth, brown, 30–40 × 14–28μ. Pedicels brown, persistent, very long.

*Synonyms.*

*Puccinia thesii* (Desv.). Winter in Rabh., "Krypt. Flor.," vol. i. p. 202.

*Æcidium thesii.* Desv., *Jour. de Bot.*, vol. ii. p. 311. Cooke, "Hdbk.," p. 537; "Micro. Fungi," 4th edit., p. 195, t. iii. figs. 50–51. B. and Br., *Ann. Nat. Hist.*, No. 1048.

*Puccinia thesii*, Chail. Cooke, "Hdbk.," p. 495; "Micro. Fungi," 4th edit., p. 204.

*Exsiccati.*

Berk., 318. Cooke, ii. 311. Vize, "Fungi Brit.," 12, 81; "Micro. Fungi Brit.," 214, 457.

On *Thesium humifusum.*

April to October.

**Puccinia calthæ.** Link.

*Æcidiospores*—Pseudoperidia on roundish spots, or when cauline on long swellings, flat, with white torn edges. Spores subglobose, finely verrucose, orange, 20–30μ in diameter.

*Uredospores*—Sori small, round, chestnut-brown, soon scattered. Spores globose or elliptical, echinulate, brown, 22–30 × 20–25μ.

*Teleutospores*—Sori small, black, pulverulent, but persistent. Spores oblong, attenuated towards both extremities, rarely rounded, central constriction little or none, apex conical or with a paler wart-like papilla, smooth, brown, 30–44 × 13–22μ. Pedicels persistent, rather long.

*Synonyms.*

*Puccinia calthæ.* Link., "Sp. Plant," vol. vi. pt. ii. p. 79. Winter in Rabh., "Krypt. Flor.," vol. i. p. 216. Berk., "Eng. Flor.," vol. v. p. 367. Cooke, "Hdbk.," p. 504; "Micro. Fungi," 4th edit., p. 210. Grev., "Flor. Edin.," p. 367. Johnst., "Flor. Berw.," vol. ii. p. 196.

*Æcidium calthæ.* Grev., "Flor. Edin.," p. 446. Berk., "Eng. Flor.," vol. v. p. 371. Cooke, "Hdbk.," p. 539; "Micro. Fungi," 4th edit., p. 196.

*Exsiccati.*

Cooke, i. 114. Vize, "Micro. Fungi," 159; "Micro. Fungi Brit.," 219.

On *Caltha palustris.*
May to September.

**Puccinia convolvuli.** (Pers.)

*Æcidiospores*—In circular clusters on the leaves and on elongated swellings on the stems. Pseudoperidia cup-shaped, with broad, recurved, torn white edges. Spores polygonal, finely verrucose, pale yellow, 17–26 × 25–30μ.

*Uredospores*—Sori scattered or circinate, often confluent, brown, soon naked. Spores subglobose, rarely ovate, echinulate, pale brown, 22–26 × 25–30μ.

*Teleutospores*—Sori scattered or circinate, dark brown, long covered by the epidermis, sometimes confluent. Spores of two kinds: (1) Teleutospores proper, oblong, oval, or subclavate, apex truncate, slightly thickened, or attenuated, or much thickened, base rounded or slightly attenuated, deep brown, 38–66 × 20–30μ. Pedicels stout, brownish. (2) Mesospores, generally ovate, apex much thickened, truncate, or attenuated, brown, 25–35 × 20–25μ. Pedicels pale brown, persistent.

*Synonyms.*

*Puccinia convolvuli* (Pers.). Winter in Rabh., "Krypt. Flor.," vol. i. p. 204.

*Uredo betæ*, var. *convolvuli.* Pers., "Syn.," p. 221.

On *Convolvulus sepium;* Miss Jelly.
June to October

**Puccinia gentianæ.** (Strauss.)

*Æcidiospores*—Pseudoperidia on circular or elongated brownish spots, flat, with white torn edges. Spores globose, orange, finely verrucose, 16–23 × 14–17μ.

*Uredospores* — Sori roundish, scattered or circinating, at first covered by the epidermis, chestnut-brown. Spores elliptical, ovate, or obovate, brown, echinulate, 20–30 × 19–24μ.

*Teleutospores*—Sori small, blackish, pulverulent. Spores elliptical or ovate, rounded at both ends, median constriction slight or absent, smooth, thickened above, 28–38 × 20–25μ. Pedicels colourless, delicate, very deciduous.

*Synonyms.*

*Puccinia gentianæ* (Strauss). Winter in Rabh., "Krypt. Flor.," vol. i. p. 205.

*Uredo gentianæ*, Strauss. "Wetter Ann.," vol. ii. p. 102. W. G. Smith, *Gard. Chron.*, Sept. 19, 1885, p. 372, fig. 82.

On *Gentiana acaulis.* Kew Gardens.

**Puccinia silenes.** Schröt.

*Æcidiospores*—Pseudoperidia on pale yellowish spots, in small clusters, small, white, shortly cylindrical, edges torn. Spores subglobose, granular, orange-yellow, 17–26 × 14–20μ.

*Uredospores*—Sori brown, roundish, scattered or circinating, often confluent. Spores roundish or elliptical, pale brown, echinulate, 20–26 × 17–20μ.

*Teleutospores*—Sori small, pulverulent. Spores elliptical or oblong, very slightly constricted, rounded at both ends, apex slightly thickened, smooth, chestnut-brown, 25–40 × 16–26μ. Pedicels short, deciduous.

*Synonyms.*

*Puccinia silenes*, Schröt. Winter in Rabh., "Krypt. Flor.," vol. i. p. 215.

*Puccinia lychnidearum.* Fckl., "Symb.," p. 50, *pro parte.*

*Puccinia silenes*, Rabh. Cooke, "Micro. Fungi," 4th edit., p. 211.

On *Silene inflata.*

### **Puccinia porri.** (Sow.)

*Æcidiospores*—Pseudoperidia in linear or circinate clusters, shortly cylindrical, with everted torn edges. Spores polygonal, finely verrucose, membrane colourless, contents orange, 19–28$\mu$ in diameter.

*Uredospores*—Sori reddish brown, scattered or in elongate clusters, linear or oblong. Spores roundish, shortly elliptical, very finely echinulate, orange-yellow, 20–27 × 25–30$\mu$.

*Teleutospores*—Sori small, bluish grey from the dark spores being covered by the semitransparent epidermis. Spores clavate or oblong, central constriction slight, generally attenuated towards the stem, apex rounded or truncate, smooth, brown, 30–45 × 20–26$\mu$. Pedicels long, but deciduous.

*Mesospores* numerous, unicellular on long deciduous pedicels, often irregular in form, sometimes thickened above, 22–36 × 17–23$\mu$.

*Synonyms.*

*Puccinia porri* (Sow.). Winter in Rabh., "Krypt. Flor.," vol. i. p. 200.

*Uredo porri.* Sow., "Eng. Fungi," t. 411.

*Uredo alliorum*, D. C. Berk., "Eng. Flor.," vol. v. p. 376, in part. Cooke, "Micro. Fungi," 4th edit., p. 217, in part; "Hdbk.," p. 528.

*Uromyces alliorum.* Cooke, "Hdbk.," p. 518 (?).

*Puccinia mixta*, Fckl. Vize, Exs.

*Exsiccati.*

Cooke, ii. 425. Vize, "Fungi Brit.," 38; "Micro. Fungi Brit.," 430.

On *Allium cepa, schœnoprasum.*

May to August.

### **Puccinia prenanthis.** (Pers.)

*Æcidiospores*—Pseudoperidia in circular or elongate patches, hemispherical or shortly conical, opening above by small irregular clefts without typical pseudoperidium. Spores rounded, pale orange, densely verrucose, 15–25 × 12-20$\mu$.

*Uredospores*—Sori reddish brown, scattered or circinate, small, rounded. Spores spherical, pale yellow, finely echinulate, with very remarkably thickened margins to the three germ-pores, as seen in the moist state, 16–23$\mu$ in diameter.

*Teleutospores*—Sori round, blackish, dusty. Spores elliptical or ovate, central constriction slight, extremities rounded, apex scarcely thickened, finely punctate, brown, 25–35 × 20–25$\mu$. Pedicels deciduous, colourless, short.

*Synonyms.*

*Puccinia prenanthis* (Pers.) Winter in Rabh., "Krypt. Flor.," vol. i. p. 208. Schröt., "Krypt. Flor. Schl.," vol. iii. p. 318.

*Æcidium prenanthis.* Pers., "Symb.," p. 208. Cooke, "Hdbk.," p. 542; "Micro. Fungi," 4th edit., p. 198. Grev., Flor. Edin.," p. 444. Johnst., "Flor. Berw.," vol. ii. p. 205.

*Æcidium compositarum*, Mart. in part. Berk., "Eng. Flor.," vol. v. p. 370.

*Puccinia chondrillæ*, Corda. Vize, Exs.

*Exsiccati.*

Vize, "Micro. Fungi Brit.," 211.

On *Lactuca muralis.*

April to September.

Some of the teleutospores are as much as 40$\mu$ in length.

**Puccinia lapsanæ.** (Schultz.)

*Æcidiospores*—Pseudoperidia flat, amphigenous, with torn, white reflexed edges, on purple, irregular spots. Spores ovate or subrotund, smooth, yellow, about 18$\mu$ in diameter.

*Uredospores*—Sori orbicular, pulverulent, very numerous, often confluent, cinnamon-brown. Spores subglobose or ovate, brown, finely echinulate, 18–21 × 15–17$\mu$.

*Teleutospores*—Sori small, flattish, black, pulverulent, numerous. Spores ovate or roundish, obtuse at both extremities, slightly (if at all) constricted, punctate, dark brown, 23–28 × 18–21$\mu$ Pedicels hyaline, short, often oblique.

*Synonyms.*

*Æcidium lapsanæ.* Schultz, "Prod. Flor. Starg.," p. 454.

*Puccinia lapsanæ,* Fckl. Schröt., "Krypt. Flor. Schl.," p. 318. Cooke, "Micro. Fungi," 4th edit., p. 207.

*Æcidium lapsanæ,* Purton. Cooke, "Hdbk.," p. 543; "Micro. Fungi," 4th edit., p. 198.

*Æcidium prenanthis,* Pers. Cooke, "Hdbk.," p. 542; "Micro. Fungi," 4th edit., p. 198.

*Trichobasis lapsanæ,* Fckl. Cooke, "Micro. Fungi," 4th edit., p. 224.

*Exsiccati.*

Cooke, i. 13; ii. 91, 92. Vize, "Micro. Fungi," 162, 169.

On *Lapsana communis, Crepis paludosa.*

Æcidiospores, March to April; uredospores, April to June; teleutospores, June to August.

BIOLOGY.—This species is said by Schröter to occur on *Crepis paludosa.* It is certainly distinct from *Puccinia variabilis* and *taraxaci* on Taraxacum, as neither the æcidiospores, uredospores, nor teleutospores of *P. lapsanæ*, when applied to Taraxacum in duplicated experiments made in 1885, had any effect (Exp. 481, 497, 499, 500), although they readily enough infected plants of *Lapsana communis.* The species which occurs so frequently on the Compositæ is without æcidiospores.

### **Puccinia variabilis.** (Grev.)

*Æcidiospores*—Spots very small, purple, scattered over the under surface of the leaves, from 2 to 5 mm. across. Pseudoperidia often single, cup-shaped, with torn whitish edges. Spores subglobose or oval, hyaline, with orange contents, echinulate, 20–25 × 15–20$\mu$.

*Uredospores* } Sori small, round or elongate, dark brown, soon
*Teleutospores* } naked, containing both spore-forms.

Uredospores scanty, subglobose, often irregular, brown, echinulate, 20–25$\mu$ in diameter. Teleutospores oval, oblong, often subglobose, frequently distorted in various ways, constriction almost none, dark brown, finely verrucose, 28–30 × 18–20$\mu$. Pedicels hyaline, deciduous, of variable length.

*Synonyms.*

*Puccinia variabilis.* Grev., "Scot. Crypt. Flor.," t. 75; "Flor. Edin.," p. 431. Johnst., "Flor. Berw.," vol. ii. p. 196. Berk., "Eng. Flor.," vol. v. p. 365. Cooke, "Hdbk.," p. 500; "Micro. Fungi," 4th edit., p. 207, t. 4, figs. 82, 83.

*Æcidium taraxaci.* Grev., "Flor. Edin.," p. 444.

*Æcidium grevillei.* Grove, *Jour. of Bot.*, May, 1886.

On *Taraxacum officinale.*
July to October.

BIOLOGY.—Mr. H. T. Soppitt met with this Æcidium at Grassington, in Yorkshire, and found, by applying the æcidiospores to healthy plants of *Taraxacum officinale*, the uredospores and teleutospores were produced in about fourteen days. Greville's description of the Æcidium is clear: "Spreading over the whole leaf, and generally collected into numerous little clusters, with single ones scattered between them." The Æcidium to *Puccinia sylvatica*, which also occurs on Taraxacum, has its pseudoperidia not scattered but collected into large clusters. In July, 1888, I found the æcidiospores of *P. variabilis* at Watlington, near King's Lynn, and successfully repeated Mr. Soppitt's experiment.

**Puccinia pulverulenta.** Grev.

*Æcidiospores*—Pseudoperidia flat, scattered over the whole surface of the leaves. Spores subglobose, finely verrucose, orange-yellow, 16–26μ in diameter.

*Uredospores*—Sori scattered or circinating, often confluent, pulverulent. Spores elliptical or ovate, echinulate, brown, 20–28 × 15–25μ.

*Teleutospores*—Sori dark brown, often concentric, pulverulent. Spores elliptical or oblong, centrally constricted, apex thickened, hooded, base generally rounded, smooth, brown, 24–35 × 16–20μ. Pedicels colourless, deciduous.

*Synonyms.*

*Puccinia epilobii tetragoni* (D. C.). Winter in Rabh., "Krypt. Flor.," vol. i. p. 214.

*Uredo vagans*, var. *epilobii.* D. C., "Flore franç.," vol. ii. p. 228.

*Æcidium epilobii*, D. C. Berk., "Eng. Flor.," vol. v. p. 372. Cooke, "Hdbk.," p. 536; "Micro. Fungi," 4th edit., p. 195. Grev., "Flor. Edin.," p. 444. Johnst., "Flor. Berw.," vol. i. p. 204.

*Uredo epilobii*, D. C. Purton, "Midl. Flor.," vol. iii., No. 1604. Johnst., "Flor. Berw.," vol. ii. p. 200. Berk., "Eng. Flor.," vol. v. p. 381.

*Trichobasis epilobii.* Berk., "Outl.," p. 333. Cooke, "Micro. Fungi," 4th edit., p. 226.

*Puccinia pulverulenta.* Grev., "Flor. Edin.," p. 432. Berk., "Eng. Flor.," vol. v. p. 368. Cooke, "Hdbk.," p. 507; "Micro. Fungi," 4th edit., p. 211, t. iv. figs. 78, 79.

*Exsiccati.*

Berk., 108, 349. Cooke, i. 4, 49; ii. 80; "L. F.," 52. Vize, "Fungi Brit.," 19, 79; "Micro. Fungi Brit.," 36, 215.

On *Epilobium hirsutum, montanum, tetragonum.*

BIOLOGY.—The mycelium of the æcidiospores is diffused through the greater part of the affected plant, but whether it be truly perennial I cannot say. The mycelium of the uredospores and teleutospores is strictly localized. I found in June, 1882, that the æcidiospores sown on seedlings of *E. hirsutum* gave rise to æcidiospores in seventeen days.

## **Puccinia violæ.** (Schum.)

*Æcidiospores*—Pseudoperidia on the leaves in circular concave patches, often causing much distortion on the stems, flat, with white torn edges. Spores subglobose, finely verrucose, orange-yellow, 16–24 × 10–18$\mu$.

*Uredospores*—Sori brown, small, roundish, scattered, soon naked. Spores roundish or elliptical, brown, echinulate, 20–26$\mu$ in diameter.

*Teleutospores*—Sori black, roundish, small, pulverulent. Spores elliptical or oblong, slightly attenuated at the base, with an apical thickening, constriction almost absent, brown, 20–35 × 15–20$\mu$. Pedicels long, deciduous.

*Synonyms.*

*Puccinia violæ* (Schum.). Winter in Rabh., "Krypt. Flor.," vol. i. p. 215. Grev., "Flor. Edin.," p. 432. Johnst., "Flor. Berw.," vol. vi. p. 196.

*Æcidium violæ*, Schum. "Enum. Plant Sæll.," vol. ii. p. 224. Grev., "Edin. Flor.," p. 444. Berk., "Eng. Flor.," vol. v. p. 372. Johnst., "Flor. Berw.," vol. ii. p. 205. Cooke, "Hdbk.," p. 543; "Micro. Fungi," 4th edit., p. 198.

*Uredo violarum*, D. C. Berk., "Eng. Flor.," vol. v. p. 380. Sow., t. 440. Johnst., "Flor. Berw.," vol. ii. p. 202.

*Trichobasis violarum*, Berk. Cooke, "Micro. Fungi," 4th edit., p. 226.

*Puccinia violarum*, Link. Berk., "Eng. Flor.," vol. v. p. 367. Cooke, "Hdbk.," p. 505; "Micro. Fungi," 4th edit., p. 210.

*Exsiccati.*

Berk., 223, 228. Cooke, i. 46, 49, 104; ii. 135; "L. F.," 43. Vize, "Micro. Fungi Brit.," 57–112; "Fungi Brit.," 77.

On *Viola canina, odorata, sylvatica, hirta, tricolor.*
May to October.

**Puccinia albescens.** (Grev.)

*Æcidiospores*—Pseudoperidia short, whitish, with toothed edges, on bleached places on the stems, petioles, leaves, and blossoms. Spores roundish, colourless, 15–22$\mu$ in diameter. Mycelium diffused, perennial.

*Uredospores*—Sori brown, small, round or oblong, long covered by the epidermis. Spores oval or elliptical or subglobose, nucleate, echinulate, clear brown, 20–25 × 15–20$\mu$.

*Teleutospores*—Sori small, round, scattered. Spores often in the same sori as the uredospores, rarely confluent, long covered by the epidermis, fusiform or elliptical, attenuated at both ends, often surmounted by a colourless papilla, constriction slight, smooth, brown, 34–40 × 18–25$\mu$. Pedicels short, hyaline, deciduous.

*Synonym.*

*Æcidium albescens.* Grev., "Flor. Edin.," p. 444. Berk., "Eng. Flor.," vol. v. p. 372. Johnst., "Flor. Berw.," vol. ii.

p. 205. Cooke, "Hdbk.," p. 536; "Micro. Fungi," 4th edit., p. 194.

*Exsiccati.*

Cooke, i. 636; ii. 78. Vize, "Fungi Brit.," 164. "Micro. Fungi Brit.," 561.

On *Adoxa moschatellina.*

March to May.

BIOLOGY.—The mycelium of the æcidiospores is perennial, that of the uredospores and teleutospores annual. Schröter found that when the æcidiospores were sown on healthy plants, a small quantity of the uredospores were produced. I have grown plants affected with the æcidiospore mycelium in a flower-pot, which for three successive years came up affected with the Æcidium. Mr. Soppitt produced the uredospores and teleutospores in June, 1888, from the æcidiospores. He states that the teleutospore sori remains longer covered by the epidermis than those of *P. adoxæ* do.

## **Puccinia bupleuri.** (D. C.)

*Æcidiospores*—Pseudoperidia uniformly scattered over the whole leaf-surface, flat, with torn white edges. Spores polygonal, nearly smooth, 14–21$\mu$ in diameter.

*Uredospores*—Sori scattered or circinating, small, roundish or irregular. Spores very few, subglobose, verrucose, yellowish brown, 17–23$\mu$ in diameter.

*Teleutospores*—Sori numerous, scattered, dark brown, oblong, surrounded by the ruptured epidermis, rather small. Spores broadly elliptical or oblong, apical thickening slight or absent, rounded at both ends, broadly constricted, smooth, dark brown, 26–42 × 17–30$\mu$. Pedicels long, deciduous.

*Synonyms.*

*Æcidium falcariæ*, var. *Bupleuri falcati.* D. C., "Flore franç.," vol. vi. p. 91.

*Puccinia Bupleuri falcati* (D. C.). Winter in Rabh., "Krypt. Flor.," vol. i. p. 212. Cooke, "Grevillea," vol. xvi. p. 47.

On *Bupleurum tenuissimum.*

Walton-on-the-Naze. August, 1887.

**Puccinia pimpinellæ.** (Strauss.)

*Æcidiospores*—Pseudoperidia mostly on distortions of the stems, at first hemispherical, then rounded or elongate, edges slightly torn. Spores roundish, yellow, finely verrucose, 18–35 × 16–21μ.

*Uredospores*—Sori chestnut or cinnamon brown, rounded or elongated, pulverulent. Spores rounded, shortly elliptical or pyriform, pale brown, echinulate, 23–32 × 20–24μ.

*Teleutospores*—Sori blackish, rounded or elongated, pulverulent, often on the stems, which are then much distorted. Spores elliptical or ovate, scarcely constricted, rounded at both extremities, membrane reticulated with broad net-like ridges, chestnut-brown, 25–35 × 17–26μ. Pedicels long, deciduous, colourless.

*Synonyms.*

*Puccinia pimpinellæ*, Strauss. Winter in Rabh., "Krypt. Flor.," vol. i. p. 212.

*Uredo pimpinellæ*, Strauss. "Wett. Ann.," vol. ii. p. 102.

*Æcidium bunii*, D. C. Grev., "Flor. Edin.," p. 445. Berk., "Eng. Flor.," p. 370, in part.

*Æcidium pimpinellæ*, Kirch. Cooke, "Micro. Fungi," 4th edit., p. 196.

*Uredo heraclei*, Grev. Berk., "Eng. Flor.," vol. v. p. 380.

*Trichobasis heraclei.* Cooke, "Hdbk.," p. 502; "Micro. Fungi," 4th edit., p. 225.

*Trichobasis angelicæ*, Schum. Cooke, "Micro. Fungi," 4th edit., p. 224.

*Trichobasis pimpinellæ*, Strauss. Cooke, "Micro. Fungi," 4th edit., p. 224.

*Puccinia umbelliferarum*, D. C. Grev., "Flor. Edin.," p. 431. Berk., "Eng. Flor.," vol. v. p. 316. Johnst., "Flor. Berw.," vol. ii. p. 196. Cooke, "Hdbk.," p. 501, in part.

*Puccinia chærophylli.* Purton, "Midl. Flor.," vol. iii., No. 1553.

*Puccinia heraclei.* Grev., "Scot. Crypt. Flor.," t. 42. Cooke, "Hdbk.," p. 502; "Micro. Fungi," 4th edit., p. 208.

*Puccinia angelicæ*, Fckl. Cooke, "Micro. Fungi," 4th edit., p. 208.

*Puccinia pimpinellæ*, Link. Cooke, "Micro. Fungi," 4th edit., p. 209.

*Exsiccati.*

Vize, "Micro. Fungi Brit.," 124, 425.

On *Pimpinella saxifraga*, *Heracleum sphondylium*, *Anthriscus sylvestris*, *Myrrhis odorata.*

May to October.

**Puccinia apii.** (Wallr.)

*Æcidiospores*—Spots rounded or irregular, yellow above, paler below, causing elongated yellow swellings on the stems. Pseudoperidia subcylindrical, with torn, white, everted edges. Spores subglobose, orange-yellow, finely echinulate, about 25$\mu$ in diameter.

*Uredospores*—Sori rather large, cinnamon-brown, pulverulent, round or elongate, scattered, sometimes circinating. Spores globose, oval, or subpyriform, pale brown, finely verrucose, 30–40 × 20–30$\mu$.

*Teleutospores*—Sori rounded, elongated, or irregular, long covered by the epidermis, blackish brown. Spores oval or clavate, smooth, dark brown, scarcely constricted, apex not thickened, 30–50 × 15–20$\mu$. Pedicels hyaline, rather persistent, not very short.

*Synonyms.*

*Uredo apii.* Wallr., "Flor. Crypt. Germ.," vol. ii. p. 203.

*Puccinia apii.* Corda, "Icones," vol. vi. p. 30, fig. 11. Cooke, "Hdbk.," p. 502; "Micro. Fungi," 4th edit., p. 208.

*Exsiccati.*

Cooke, i. 40A. Vize, "Fungi Brit.," 54, 127; "Micro. Fungi Brit.," 554.

On *Apium graveolens.*

May to September.

BIOLOGY.—The æcidiospores occur at the latter part of May, and are by no means profuse, although the yellow spots on which they occur are very conspicuous. On May 27, 1888, I applied the æcidiospores and uredospores freely to a plant of *Pimpinella saxifraga* in my garden, but without effect. This is the only Uredine affecting the Umbelliferæ with which I have made any experimental cultures, and I have therefore followed Winter and Schröter in grouping the various spore-forms together under *Puccinia pimpinellæ* and *bullata;* but it is probable that biological research will show several of them to be good species.

**Puccinia menthæ.** Pers.

*Æcidiospores*—Pseudoperidia immersed, flat, opening irregularly, edges torn; principally on the stems, which are much swollen, more rarely on concave spots on the leaves. Spores subglobose or polygonal, coarsely granular, pale yellowish, $17-26 \times 26-35\mu$.

*Uredospores*—Sori small, roundish, soon pulverulent and confluent, cinnamon-brown. Spores irregularly rounded or ovate, echinulate, pale brown, $17-28 \times 14-19\mu$.

*Teleutospores*—Sori black-brown, roundish, pulverulent. Spores elliptical, oval, or subglobose, central constriction slight or absent, apex with a hyaline or pale brown papilla, verrucose, deep brown, $26-35 \times 19-23\mu$. Pedicels long, delicate, colourless.

*Synonyms.*

*Puccinia menthæ.* Pers., "Syn.," p. 227. Winter in Rabh., "Krypt. Flor.," vol. i. p. 204.

*Æcidium menthæ*, D. C. Berk., "Eng. Flor.," vol. v. p. 369. Cooke, "Hdbk.," p. 544; "Micro. Fungi," 4th edit., p. 199.

*Trichobasis labiatarum*, Lév. Cooke, "Hdbk.," p. 496; "Micro. Fungi," 4th edit., p. 224.

*Trichobasis clinopodii*, D. C. Cooke, "Micro. Fungi," 4th edit., p. 224.

*Puccinia clinopodii*, D. C. Cooke, "Micro. Fungi," 4th edit., p. 205.

*Puccinia menthæ*, Pers. Berk., "Eng. Flor.," vol. v. p 364. Grev., "Flor. Edin.," p. 430. Johnst., "Flor. Berw.," vol. ii. p. 195. Cooke, "Hdbk.," p. 496; "Micro. Fungi," 4th edit., p. 204, t. iv. figs. 69, 70.

*Uredo menthæ.* Purton, "Midl. Flor.," vol. iii., No. 1550. Sow., t. 398, fig. 3, *sub* Æcidium.

*Uredo labiatarum*, D. C. Grev., "Flor. Edin.," p. 436. Berk., "Eng. Flor.," vol. v. p. 378. Johnst., "Flor. Berw.," vol. ii. p. 203.

*Exsiccati.*

Berk., 217, 232, 233. Cooke, i. 30; ii. 33, 60, 444. Vize, "Fungi Brit.," 14, 15, 75; "Micro. Fungi Brit."

On *Mentha rotundifolia, sylvestris, aquatica, arvensis, viridis, Origanum vulgare, Calamintha clinopodium.*

May to October.

BIOLOGY.—The æcidiospore mycelium is diffused and probably perennial—at any rate, lasting for more than one year; at least, this appears to be the case with *M. viridis*, which I have cultivated for a period of three years. The mycelium of the uredospores and teleutospores is purely local.

Johnoston records this species on *Ajuga reptans.*

**Puccinia ægra.** Grove

*Æcidiospores*—Pseudoperidia on all green parts of the plants, scattered, not collected on swollen roundish or elliptical spots, with torn, white, somewhat recurved margins. Spores roundish or oblong, angular, smooth, orange-yellow, 17–21 × 14–16μ.

*Uredospores*—Sori numerous, amphigenous, on yellow spots, not very small, scattered or collected in groups, roundish, flatly onvex, covered with the silvery, shining, persistent epidermis. Spores elliptical or obovate, finely echinulate, brown, 28–30μ.

*Teleutospores*—Sori resembling those of the uredospores. Spores elliptical, oblong, or roundish, very irregular, rounded or tapering at the base or apex, sometimes truncated, smooth, not constricted, dark brown, 22–30 × 18–24μ. Pedicels short, hyaline.

*Synonyms.*

*Puccinia ægra.* Grove, *Jour. of Bot.*, vol. xxi. (1883), p. 274.

*Æcidium depauperans,* Vize. Cooke, "Micro. Fungi," 4th edit., p. 195.

*Exsiccati.*

Vize, "Micro. Fungi Brit.," 56.

On *Viola cornuta, lutea* var. *amæna.*

May to September.

BIOLOGY.—The æcidiospore mycelium is perennial, and the affected plants are altered in habit, being paler in colour and more elongated in their stems and petioles. The mycelium of the uredospores and teleutospores is purely local.

**Puccinia primulæ.** (D. C.)

*Æcidiospores*—Pseudoperidia hypophyllous, in roundish clusters, shortly cylindrical, with whitish torn edges. Spores subglobose, finely echinulate, orange-yellow, 17–23 × 12–18μ.

*Uredospores*—Sori small, hypogenous, brown or yellowish, rounded. pulverulent. Spores ovoid, brown, echinulate, 20–23 × 15–18μ.

*Teleutospores*—Sori hypogenous, greyish brown. Spores ovate or oblong, slightly constricted, apex thickened, smooth, brown, 22–30 × 15–17μ. Pedicels short, deciduous.

*Synonyms.*

*Puccinia primulæ* (D. C.). Winter in Rabh., "Krypt. Flor.," vol. i. p. 203.

*Uredo primulæ.* D. C., "Flore franç.," vol. vi. p. 68. Berk., "Eng. Flor.," p. 379. Grev., "Flor. Edin.," p. 435.

*Æcidium primulæ*, D. C. Berk., "Eng. Flor.," p. 369. Cooke, "Hdbk.," p. 544; "Micro. Fungi," 4th edit., p. 199.

*Trichobasis primulæ.* Cooke, "Micro. Fungi," 4th edit., p. 227.

*Puccinia primulæ.* Grev., "Flor. Edin.," p. 432. Berk., "Eng. Flor.," vol. v. p. 364. Cooke, "Hdbk.," p. 496; "Micro. Fungi," 4th edit., p. 204.

*Exsiccati.*

Berk., 341. Cooke, i. 28, 296; ii. 85, 138, 141; "L. F.," 6. Vize, "Fungi Brit.," 80, 115; "Micro. Fungi Brit.," 121, 233, 429.

On *Primula officinalis, facchini.*

May to July.

**Puccinia soldanellæ.** (D. C.)

*Æcidiospores*—Pseudoperidia hypophyllous, scattered, shortly cylindrical, edges toothed. Spores subglobose, polygonal, finely granular, yellow, 20–26 × 17–20μ. Mycelium diffused.

*Uredospores*—Sori brown, scattered irregularly or circinate, often confluent, brown. Spores round or ovate, echinulate, brown, 20–30 × 19–30 .

*Teleutospores*—Mixed with the uredospores, elliptical or broadly constricted, base rounded, apex thickened, smooth, chestnut-brown, 35–50$\mu$ × 20–35$\mu$. Pedicels rather long, deciduous.

*Synonyms.*

*Puccinia soldanellæ* (D. C.). Winter in Rabh., "Krypt. Flor.," vol. i. p. 202.

*Uredo soldanellæ.* D. C., "Flore franç.," vol. vi. p. 85.

*Æcidium soldanellæ*, Hornsch. Berk., "Eng. Flor.," p. 369. Cooke, "Hdbk.," p. 537; "Micro. Fungi," 4th edit., p. 195.

On *Soldanella alpina.*

**Puccinia saniculæ.** Grev.

*Æcidiospores*—Pseudoperidia on purple, thickened spots, hypogenous, cup-shaped, with torn white edges. Spores subglobose, yellow, verrucose, 20–27$\mu$ in diameter.

*Uredospores*—Sori scattered, small, round. Spores roundish or clavate, echinulate, brown, 25–40 × 15–26.

*Teleutospores*—Sori small, blackish, scattered, pulverulent. Spores obtuse, very slightly constricted, apex slightly thickened, smooth, brown, 26–45 × 17–26$\mu$. Pedicels very long, colourless, deciduous.

*Synonyms.*

*Puccinia saniculæ.* Grev., "Flor. Edin.," p. 431. Berk., "Eng. Flor.," vol. v. p. 366. Cooke, "Hdbk.," p. 503; "Micro. Fungi," 4th edit., p. 208.

*Æcidium saniculæ*, Carm. Cooke, *Seem. Jour. Bot.*, vol. ii. p. 39, t. 14, fig. 1; "Hdbk.," p. 543; "Micro. Fungi," 4th edit., p. 198.

*Exsiccati.*

Berk., 315. Cooke, i. 14, 41; ii. 136; "L. F." Vize, "Micro. Fungi," 160; "Micro. Fungi Brit.," 32.

On *Sanicula europæa.*

May to October.

### Puccinia vincæ. (D. C.)

*Spermogonia*—Scattered over the whole leaf-surface and stems, yellow, sweet-scented.

*Æcidiospores*—Sori pulvinate, not cup-shaped, but discoidal, solid, 2–3mm. in diameter, dark brown, with a greyish lustre. Spores globose, finely echinulate, colourless, 10–12$\mu$ in diameter.

*Uredospores—Primary:* Sori irregular, elongate, crowded, soon naked and often confluent, brown, occurring with the spermogonia and æcidiospores. *Secondary:* Sori on roundish brown spots, brown, long covered by the epidermis. Spores subglobose, or obovate, brown, echinulate, 20–38 × 18–25$\mu$.

*Teleutospores*—Sori small, on discoloured spots, roundish, scattered, surrounded by the ruptured epidermis, blackish. Spores elliptical or oval, slightly constricted, lower cell sometimes attenuated, sometimes rounded, upper cell thickened above, often surmounted with a pale papilla, chestnut-brown, verrucose, 38–56 × 17–28$\mu$. Pedicels rather long, deciduous.

*Synonyms.*

*Uredo vincæ.* D. C., "Flore franç.," vol. vi. p. 70. Berk., "Eng. Flor.," vol. v. p. 378.

*Trichobasis vincæ,* Berk. Cooke, "Micro. Fungi," 4th edit., p. 226, t. vi. figs. 130, 131.

*Puccinia vincæ,* Berk. Cooke, "Hdbk.," p. 497; "Micro. Fungi," 4th edit., p. 205, t. vi. fig. 132. Winter in Rabh., "Krypt. Flor.," vol. i. p. 187.

*Exsiccati.*

Berk., 234. Cooke, i. 32; ii. 3, 331; "L. F.," 10. Vize, "Micro. Fungi," 118; "Micro. Fungi Brit.," 29, 227.

Æcidiospores, March to May; uredospores, May to July; teleutospores, July to October, and lasting through the winter.

On *Vinca major.*

BIOLOGY.—The life-history of this species is peculiar. The mycelium of the æcidiospores is perennial, and causes the affected plant to produce shorter and thicker leaves. The æcidia are not cups, but convex or flat sori, more resembling a Coryneum than an æcidium. The spores are colourless, comparatively small, and produced from the hyphæ in basipetal chains (Plate II. Figs. 11–14). They emit a germ-tube, which usually exhibits circumnutatory motions. The primary uredospores occur with the æcidiospores, and resemble the secondary in form, but are much more profuse. The secondary uredospores and teleutospores have localized mycelia (Plow., *Gard. Chron.*, July 25, 1885, p. 108, figs. 22, 23).

B. HETEROPUCCINIA. Schröt.

Spermogonia and æcidiospores on one host-plant; the uredospores and teleutospores on another, belonging to a different genus.

### Puccinia graminis. Pers.

*Æcidiospores*—Spots generally circular, thick, swollen, reddish above, yellow below. Pseudoperidia cylindrical, with whitish torn edges. Spores subglobose, smooth, orange-yellow, 15–25μ in diameter.

*Uredospores*—Sori orange-red, linear, but often confluent, forming very long lines on the stems and sheaths, pulverulent. Spores elliptical, ovate, or pyriform, with two very marked, nearly opposite germ-pores, echinulate, orange-yellow, 25–38 × 15–20μ.

*Teleutospores*—Sori persistent, naked, linear, generally forming lines on the sheaths and stems, often confluent. Spores fusiform or clavate, constricted in the middle, generally attenuated below, apex much thickened (8–10μ), rounded or pointed, smooth, chestnut-brown, 35–65 × 15–20μ. Pedicels long, persistent, yellowish brown.

*Synonyms.*

*Puccinia graminis.* Pers., "Disp. Meth.," p. 39, t. 3, fig. 3. Winter in Rabh., "Krypt. Flor.," vol. i. p. 217. Cooke, "Hdbk.," p. 493; "Micro. Fungi," 4th edit., p. 202, t. iv. figs. 57–59. Grev., Berk. "Eng. Flor.," vol. v. p. 363; "Flor. Edin.," p. 433. Johnst., "Flor. Berw.," vol. ii. p. 195.

*Uredo frumenti.* Sow, t. 140.

*Uredo linearis*, Pers. Berk., "Eng. Flor.," vol. v. p. 375. Grev., "Flor. Edin.," p. 440. Johnst., "Flor. Berw.," vol. ii. p. 198.

*Trichobasis linearis*, Lév. Cooke, "Micro. Fungi," 4th edit., p. 223, t. vii. figs. 143, 144.

*Æcidium berberidis*, Pers. Berk., "Eng. Flor.," vol. v. p. 372. Grev., t. 97; "Flor. Edin.," p. 446. Johnst., "Flor. Berw.," vol. ii. p. 207. Sow., t. 397, fig. 5. Cooke, "Hdbk.," p. 538; "Micro. Fungi," 4th edit., p. 195, t. i. figs. 7–9.

*Exsiccati.*

Cooke, i. 24, 441; ii. 93, 121, 122, 124. Vize, "Micro. Fungi Brit.," 453, 456; "Fungi Brit.," 76, 78.

Æcidiospores on *Berberis vulgaris* and *Mahonia ilicifolia*, chiefly on the berries, May to July.

Teleutospores on *Triticum vulgare, repens, Secale cereale, Dactylis glomerata, Festuca gigantea, Alopecurus pratensis, Agrostis alba, Avena sativa, elatior*, July and August, and throughout the winter.

BIOLOGY.—The mycelium of all spore-forms is strictly localized. The æcidiospores on the berries of *Mahonia ilicifolia* I found, in May, 1883, readily produced the uredospores on wheat in about eight or ten days. This is a fungus which varies very much in frequency; some years almost every straw in a wheat-field is affected, in others scarcely one can be found attacked. Certain conditions render the wheat plant more susceptible to the parasite, foremost amongst which is a too large supply of nitrogen, as evinced by the uniformity with which wheat plants growing on manure-heaps are attacked with the parasite; so are the plants grown on the ground where a manure-heap has stood, and also plants growing where an old ditch has been filled up, although perhaps to a less degree. A very thin, or what is termed in some parts of England "a gathering crop," one in which the plants are far apart, and which consequently throw out a large number of lateral shoots, is also liable to become affected by the parasite. It is interesting to remark that a Puccinia occurs on *Berberis glauca* in Chili, which is accompanied by an Æcidium. In some parts of Europe an Æcidium, having a perennial mycelium (*Æ. magelhænicum*, Berk. in Hooker's "Flor. Antarctica," pp. 1, 2. London: 1844–1847), occurs on *Berberis vulgaris*, which has been shown by Magnus to have a distinct life-history.

**Puccinia coronata.** Corda.

*Æcidiospores*—Pseudoperidia often on very large orange swellings, causing great distortions on the leaves and peduncles, cylindrical, with whitish torn edges. Spores subglobose, very finely verrucose, orange-yellow, 15–25 × 12–18$\mu$.

*Uredospores*—Sori orange, pulverulent, elongated or linear, often confluent. Spores globose or ovate, with three or four germ-pores, echinulate, orange-yellow, 20–28 × 15–20$\mu$.

*Teleutospores*—Sori persistent, black, linear, often confluent, long covered by the epidermis. Spores subcylindrical or cuneiform, attenuated below, constriction slight or absent, apex truncate, somewhat thickened, with six or seven curved blunt processes, brown, 40–60 × 12–20$\mu$. Pedicels short, thick.

*Synonyms.*

*Puccinia coronata.* Corda, "Icones," vol. i. p. 6, t. ii. fig. 96. Winter in Rabh., "Krypt. Flor.," vol. i. p. 218. Schröt., "Krypt. Flor. Schl.," p. 323. Cooke, "Hdbk.," p. 494; "Micro. Fungi," 4th edit., p. 203, t. iv. figs. 60–62.

*Æcidium crassum*, Pers. Berk., "Eng. Flor.," vol. v. p. 373. Pers., "Ic. et Descrip.," t. 10, figs. 1, 2. Cooke, "Hdbk.," 538; "Micro. Fungi," 4th edit., p. 196.

*Æcidium rhamni.* Pers., "Obs.," t. 2, fig. 4. Purton, "Midl. Flor.," vol. iii. No. 1538.

*Exsiccati.*

Berk., 110. Cooke, i. 7, 26; ii. 94, 95; "L. F.," 53. Vize, "Micro. Fungi," 116, 155; "Micro. Fungi Brit.," 210, 234.

Æcidiospores on *Rhamnus frangula* and *catharticus*, May and June.

Teleutospores on *Holcus mollis*, *Avena elatior*, *sativa*, *Festuca sylvatica*, *Lolium perenne*, *Dactylis glomerata*, July to September, and throughout the winter.

Biology.—Nielsen found that when the æcidiospores from *Rhamnus frangula* and *cathartica* were placed on *Lolium perenne*, the latter only gave rise to uredospores, and that these uredospores, put on *Avena sativa*, reproduced the uredospores and subsequently the teleutospores of *P. coronata.* I have found, by numerous cultures, that the teleutospores from *Dactylis glomerata* and *Festuca sylvatica* readily produced the æcidium on *R. frangula*, but I have failed to produce on *R. frangula* the æcidium from the teleutospores on *Lolium perenne.* I think two species are confounded under the name *P. coronata.* As this fungus occurs on *Lolium perenne*, which it does abundantly in this neighbourhood, it is accompanied by a profuse development of uredospores, and only in the autumn, from September to November; whereas the *P. coronata* on *Dactylis* is an early summer species, with a much less free development of uredospores. *P. coronata* is distinguished from all other British species by its coronate teleutospores, which are surmounted by a crown of digitate processes; these, being outgrowths of the spore-membrane, and not of the epispore, are dark brown in colour. What their precise function may be is unknown, but they appear, for one thing, to hold the spores together, so that they may be regarded in part as organs of attachment. Sometimes these processes are not confined to the summit of the spore, but extend downwards, gradually becoming smaller, as far as the septum. Such spores show the transition towards the American species, *P. aculeata*, Schw. Winter has described a coronate Uromyces, *U. digitata*, from South Australia; and Ellis an American Puccinia, *P. digitata*, on *Rhamnus crocea* from California.

## **Puccinia sessilis.** Schneider.

*Ecidiospores*—Pseudoperidia on pale roundish or elongated spots, circinating, cup-shaped, with broad, torn white edges. Spores subglobose, smooth, orange-yellow, 18–25 × 13–20$\mu$.

*Uredospores*—Sori small, elliptical or linear, scattered, yellowish-brown. Spores roundish or elliptical, echinulate, pale brown, 20–28 × 20–23$\mu$, without paraphyses.

*Teleutospores*—Sori very small, linear, numerous, covered by the epidermis, black, scattered or in linear series. Spores oblong or wedge-shaped, apex slightly thickened, generally truncate, base attenuated, lower cell often paler, central constriction slight, smooth, brown, 25–40 × 15–20$\mu$. Pedicels very short or absent.

*Synonyms.*

*Puccinia sessilis*, Schneider. Schröt., "Brandp.," p. 119. Winter in Rabh., "Krypt. Flor.," vol. i. p. 222.

*Puccinia linearis*, Rob. Cooke, "Micro. Fungi," 4th edit., p. 203.

*Æcidium allii.* Grev., "Flor. Edin.," p. 447. Berk., "Eng. Flor.," vol. v. p. 369. Cooke, "Hdbk.," p. 545; "Micro. Fungi," 4th edit., p. 200. Johnst., "Flor. Berw.," vol. ii. p. 207.

*Exsiccati.*

Cooke, i. 16. Vize, "Micro. Fungi Brit.," 59, 322.

Æcidiospores on *Allium ursinum*, June and July.

Teleutospores on *Phalaris arundinacea*, from August, throughout the winter.

BIOLOGY.—The life-history of this fungus was worked out by Winter. I have been able to repeat Winter's culture of *P. sessilis* by placing the germinating teleutospores which Mr. Soppitt sent me in 1888 on *Allium ursinum* and *Arum maculatum*. On the first-named plant they produced the æcidium, but they had no effect upon the Arum.

### **Puccinia phalaridis.** Plow.

*Æcidiospores*—Pseudoperidia mostly hypophyllous, in circular clusters on yellowish spots, not very prominent, with whitish torn edges. Spermogonia appearing first on the upper surface of the leaf, centrally then surrounded by the cups. Spores roundish, echinulate, orange-yellow, 20–25$\mu$ in diameter.

*Uredospores*—Sori orange, small, elongated, oval, oblong or linear, sometimes confluent, at length naked. Spores subglobose or oval, yellow, echinulate, 20–25$\mu$ in diameter.

*Teleutospores*—Sori minute, punctate or linear, black, covered by the epidermis, persistent. Spores smooth, brown, sessile, quadrate or attenuated below, truncate or rounded above, apex thickened, sometimes obliquely hooded, slightly constricted, 40–50 × 15–18$\mu$. Pedicels very short.

*Synonyms.*

*Æcidium ari.* Desm., "Catal. des Plant. omiss" (1823), p. 26. Cooke, "Hdbk.," p. 545; "Micro. Fungi," 4th edit., p. 199. Berk., "Eng. Flor.," vol. v. p. 369.

*Puccinia phalaridis.* Plow., *Jour. Linn. Soc.*, vol. xxxiv. p. 88.

*Exsiccati.*

Cooke, i. 534; ii. 84. Vize, "Micro. Fungi Brit.," 165, 559.

Æcidiospores on *Arum maculatum*, L., May and June.

Uredospores, teleutospores, on *Phalaris arundinacea*, L., June and July, and from July to May.

BIOLOGY.—The teleutospores are surrounded by a bed of dark-brown tissue, the individual fibres of which are not easily demonstrable. Numerous cultures in 1885-6 proved the connection between the *Æcidium ari* and this Puccinia. The teleutospores of this species placed on *Allium ursinum* produce no effect.

### Puccinia rubigo-vera. (D. C.)

*Æcidiospores*—Spots large, generally circular, discoloured, generally crowded. Pseudoperidia flat, broad, with torn white edges. Spores subglobose, verrucose, orange-yellow, 20–25$\mu$ in diameter.

*Uredospores*—Sori oblong or linear, scattered, yellow, pulverulent. Spores mostly round or ovate, echinulate, with three or four germ-pores, yellow, 20–30 × 17–24$\mu$.

*Teleutospores*—Sori small, oval, or linear, black, covered by epidermis, surrounded by a thick bed of brown paraphyses. Spores oblong or elongate, cuneiform, slightly constricted, the lower cell generally attenuated, apex thickened, truncate or often obliquely conical. Spores smooth, brown, variable in size, 40–60 × 15–20$\mu$. Pedicels short.

*Synonyms.*

*Puccinia rubigo-vera* (D. C.). Winter in Rabh., "Krypt. Flor.," vol. i. p. 217.

*Uredo rubigo-vera.* D. C., "Flore franç.," vol. vi. p. 83. Berk., "Eng. Flor.," vol. v. p. 375.

*Trichobasis rubigo-vera*, Lév. Cooke, "Micro. Fungi," 4th edit., p. 222, t. viii. figs. 140–142.

*Trichobasis glumarum*, Lév. Berk., "Outl.," p. 332. Cooke, "Hdbk.," p. 529; "Micro. Fungi," 4th edit., p. 223.

*Puccinia straminis*, De Bary. Cooke, "Micro. Fungi," 4th edit., p. 202.

*Æcidium asperifolii*, Pers. Cooke, "Hdbk.," p. 541; "Micro. Fungi," 4th edit., p. 197.

*Exsiccati.*

Cooke, i. 325; ii. 48. Vize, "Fungi Brit.," 7; "Micro. Fungi Brit.," 431, 440, 441.

Æcidiospores on *Anchusa arvensis, Echium vulgare*, September and October.

Teleutospores on *Bromus mollis, Triticum vulgare, repens, sativum, Secale cereale, Hordeum vulgare, Holcus lanatus, mollis, Avena sativa, elatior*, from November to August.

BIOLOGY.—Mesospores are not uncommonly mixed with the normal teleutospores, especially when the fungus occurs on any of the genus Hordeum = var. *simplex*, Körnicke, and *P. anomala*, Rost. The mesospores are about 45μ long. The teleutospores germinate in the autumn of the year in which they are produced; hence the æcidium is to be found on *Anchusa arvensis* in September and October. The knowledge of this fact only came recently under my notice. In 1885, I laid a bundle of wheat straw affected with *P. rubigo-vera* in my garden, near some plants of Anchusa, in August, with the intention of doing some cultures in the following spring. In September, I found to my surprise the æcidium appearing on the Anchusa. This was the more remarkable, because I had never previously found the æcidium, and the plants of Anchusa had been growing all the year in the garden perfectly healthy. The uredospores are exceedingly abundant in spring, on wheat plants, in many parts of England, but the Puccinia, unlike *P. graminis*, has very little prejudicial effect upon the crop. Nielsen also finds *P. rubigo-vera* germinates in autumn. It is very possible that more than one species is included under *P. rubigo-vera.*

## **Puccinia poarum.** Nielsen.

*Æcidiospores*—Spots yellow above, often surrounded by a violet margin, slightly thickened, concave. Pseudoperidia flat, with whitish torn edges. Spores subglobose, finely echinulate, orange-yellow, 18–24 × 15–18μ in diameter.

*Uredospores*—Sori small, round or elliptical, orange. Spores spherical or elliptical, echinulate, yellow, 20–30μ in diameter. Without paraphyses.

*Teleutospores*—Sori small, black, persistent, usually circinating, covered by the epidermis. Spores elliptical or clavate, very variable, apex flattened or conically thickened, dark brown, 35–45 × 18–25μ. Pedicels very short, persistent, brownish.

*Synonyms.*

*Puccinia poarum.* Nielsen, "Bot. Tids.," vol. ii. p. 26. Winter in Rabh., "Krypt. Flor.," vol. i. p. 220.

*Æcidium compositarum*, Mart., var. *tussilaginis*, Pers. Cooke, "Hdbk.," p. 542; "Micro. Fungi," 4th edit., p. 198. Berk., "Eng. Flor.," vol. v. p. 370, in part. Sow., t. 397, fig. 1. Grev., "Flor. Edin.," p. 447. Johnst., "Flor. Berw.," vol. ii. p. 207, in part.

*Exsiccati.*

Cooke, i. 12; ii. 89. Vize, "Fungi Brit.," 70; "L. F.," 58.

Æcidiospores on *Tussilago farfara*, June and July; October and November.

Teleutospores on *Poa trivialis, pratensis, nemoralis*, and *annua*, but rarely on the latter, July and August, October to December.

BIOLOGY.—Schröter states that between the uredospores are numerous stiff, capitate paraphyses. The teleutospores germinate after a very short period of rest, and produce generally two crops of æcidia in spring and autumn. In 1882, I found the æcidiospores produced the uredospores on *Poa annua* in ten or twelve days.

**Puccinia caricis.** (Schum.).

*Æcidiospores*—On roundish yellow spots on the leaves, and elongated swellings on the stems, causing much thickening and distortion. Pseudoperidia flat, broad, with torn white edges. Spores subglobose, very finely verrucose, orange-yellow, 15–25$\mu$ in diameter.

*Uredospores*—Sori small, brown, linear, often confluent in long lines. Spores ovate or elliptical, yellowish brown, echinulate, 20–30 × 15–25$\mu$.

*Teleutospores*—Sori dark brown, linear, often confluent in long lines. Spores elongate, oblong, or clavate, persistent, apex truncate or rounded, much thickened, base attenuated, smooth, brown, 40–70 × 14–20$\mu$. Pedicels short, brownish, persistent.

*Synonyms.*

*Puccinia caricis*, Schum. Winter in Rabh., "Krypt. Flor.," vol. i. p. 222.

*Uredo caricis.* Schum., "Enum. Plant. Sæll.," vol. ii. p. 231. Berk., "Eng. Flor.," vol. v. p. 376. Grev., t. 12; "Flor. Edin.," p. 437, *sub U. oblongata*.

*Puccinia caricina.* Grev., "Flor. Edin.," p. 433.

*Trichobasis caricina*, Berk. Cooke, "Hdbk.," p. 493; "Micro. Fungi," 4th edit., p. 223, t. viii. figs. 170, 171.

*Puccinia striola*, Link. Cooke, "Hdbk.," p. 493; "Micro. Fungi," 4th edit., p. 203. Berk., "Eng. Flor.," vol. v. p. 363.

*Æcidium urticæ*, D. C. Berk., "Eng. Flor.," vol. v. p. 374. Grev., "Flor. Edin.," p. 445. Johnst., "Flor. Berw.," vol. ii. p. 206. Purton, "Midl. Flor.," vol. iii. p. 294. Cooke, "Hdbk.," p. 541; "Micro. Fungi," 4th edit., p. 197, t. i. figs. 10, 11.

*Exsiccati.*

Berk., 112. Cooke, i. 67, 634; ii. 86, 317, 634; "L. F.," 33, 35. Vize, "Fungi Brit.," 121, 159; "Micro. Fungi Brit.," 220, 235.

Æcidiospores on *Urtica dioica*, May and June.

Teleutospores on *Carex hirta*, *riparia*, *paludosa*, *binervis*, July to April.

BIOLOGY.—In many cultures I have found the teleutospores from *Carex hirta* produce the æcidiospores on *Urtica dioica*, and the æcidiospores produce the uredospores in from eight to fourteen days. It is interesting to note that Surgeon-Major Barclay has described a Leptopuccinia on *Urtica parvifolia* in India ("Scientific Memoirs by Medical Officers of the Army in India" (1887), p. 10, t. v. figs. 18, 19).

## **Puccinia arenariicola.** Plow.

*Æcidiospores*—Spots circular on the leaves, elongated on the stems, yellow, surrounded by a purple margin. Pseudoperidia mostly hypophyllous, rarely epiphyllous, edges whitish, prominent, torn, everted. Spores subglobose or polygonal, nearly smooth, yellow, 15–20$\mu$ in diameter.

*Uredospores*—Sori brown, erumpent, linear or oblong, about 5 mm. long, surrounded by the torn epidermis, seated on yellowish spots. Spores globose or ovate, brown, finely echinulate, 18–20$\mu$ in diameter.

*Teleutospores*—Sori small, black, naked, persistent, linear or elongate, mostly hypophyllous. Spores smooth, dark brown, oblong or cuneiform, the upper cell usually darker, rounded, and thickened above, lower cell somewhat attenuated, constriction slight, 40–50 × 20$\mu$. Pedicels long, persistent.

*Synonym.*

*Puccinia arenariicola.* Plow., *Jour. Linn. Soc. Bot*, vol. xxiv. p. 90. Vize, "Micro. Fungi Brit." (1888), 549.

Æcidiospores on *Centaurea nigra*, L., May and June.

Uredospores, teleutospores, on *Carex arenaria*, L., July to May.

On the seashore at Hemsby, Norfolk.

BIOLOGY.—The above life-history was worked out by a series of cultures (1885), by which it was proved that this species is distinct from the other Pucciniæ on Carices, *P. schœleriana* and *caricis*, for details of which see *Linn. Soc. Jour. Bot.*, vol. xxiv. pp. 90–93.

This may be the *Puccinia tenuistipes*, Rost. (Schröt., "Krypt. Flor. von Schlesien," vol. iii. p. 329) on *C. muricata*, which has its æcidiospores on *Centaurea jacea*.

**Puccinia schœleriana.** Plow. and Mag.

*Æcidiospores*—Spots yellow, circular, mostly on the radical leaves. Pseudoperidia hypophyllous, with reflected torn edges, spermogonia on the upper surface of the leaves. Spores rounded, yellow, finely echinulate, 15–20$\mu$ in diameter.

*Uredospores*—Spots yellow, pallid. Sori elongated or subrotund, surrounded by the torn epidermis, brown, generally hypophyllous. Spores subglobose or ovate, yellowish brown, 25–30 × 14–20$\mu$.

*Teleutospores*—Sori erumpent, oblong or elongate, large, prominent, persistent, black, hypophyllous, naked. Spores slightly constricted, upper cell subglobose, ovate or attenuated upwards, apex much thickened, rounded or pointed, lower cell cuneiform, often paler than the upper, rich brown, smooth, 60–80 × 15–20$\mu$. Pedicels long, firm.

*Synonyms.*

*Puccinia schœleriana.* Plow and Mag., *Quart. Jour. Micro. Science*, vol. xxv., new series, p. 170.

*Æcidium jacobææ.* Grev., "Flor. Edin.," p. 445.

*Æcidium compositarum*, Mart. Cooke, "Hdbk.," p. 542; "Micro. Fungi," 4th edit., p. 198.

*Exsiccati.*

Vize, "Micro. Fungi," 156; "Micro. Fungi Brit.," 434.

Æcidiospores on *Senecio jacobæa*, May and June.

Uredospores and teleutospores on *Carex arenaria*, July to May.

BIOLOGY.—The life-history of this species was worked out in a series of numerous cultures, by which it proved to be distinct not only from *P. caricis*, but also from *P. arenariicola;* for the details of which see *Quart. Jour. Micro. Science*, vol. xxv., new series, pp. 167 and 170; and also *Linn. Jour. Soc. Bot.*, vol. xxiv. pp. 90-93. It has been found by me at North Wootton, Norfolk; at Skegness, Lincolnshire; and by Professor J. W. H. Trail near Aberdeen.

## **Puccinia sylvatica.** Schröt.

*Æcidiospores*—Pseudoperidia on round or elongate patches, which are often surrounded by a violet margin, crowded, short, with white torn edges. Spores polygonal, smooth, orange-yellow, 14–21$\mu$ in diameter.

*Uredospores*—Sori small, scattered, dark brown. Spores globose, elliptical, or ovate, brown, echinulate, 22–26 × 15–22$\mu$.

*Teleutospores*—Sori small, round, persistent, pulvinate. Spores clavate, apex much thickened, 6–8$\mu$, rounded, base attenuated, somewhat constricted, 35–44 × 12–20$\mu$. Pedicels persistent, about the length of the spores.

*Synonyms.*

*Puccinia sylvatica.* Schröt., "Cohn's Beiträge," vol. iii. p. 68. Winter in Rabh., "Krypt. Flor.," vol. i. p. 223.

*Æcidium compositarum*, Mart., var. *Taraxaci.* Grev., "Flor. Edin.," p. 444. Cooke, "Hdbk.," p. 542; "Micro. Fungi," 4th edit., p. 198, in part.

*Exsiccati.*

Vize, "Micro. Fungi," 166.

Æcidiospores on *Taraxacum officinale*, June and July.

Uredospores and teleutospores on *Carex divulsa, brizoides, præcox, leporina, remota, rigida, pallescens, panicea, ericetorum, pilulifera, flava.*

BIOLOGY.—The admission of this species into the British flora requires confirmation, as the teleutospores have not yet been found in this country. Mr. Grove also thinks that there are two æcidia on

Taraxacum, one of which belongs to *P. variabilis*, and the other to *P. sylvatica*

The two æcidia on Taraxacum have both been found by Mr. Soppitt, who kindly sent me specimens; one has numerous small clusters of pseudoperidia scattered over the leaves = *Æc. grevillei*, Grove. This Mr. Soppitt laid upon a plant of Taraxacum, and in a fortnight he found the Taraxacum affected with Puccinia. I have repeated this experiment with æcidiospores found near King's Lynn with success.

In the other æcidium (*P. sylvaticæ*) the pseudoperidium occurs in large clusters, and is the one described above.

The teleutospore host-plants are those given by Schröter in Cohn's "Krypt. Flor. von Schlesien," vol. iii. p. 328.

## **Puccinia dioicæ.** Magnus.

*Æcidiospores*—Spermogonia in small crowded groups, yellow. Pseudoperidia in one or many circles, crowded on roundish spots, flat, with torn white edges. Spores orange-yellow, as much as 25μ in diameter.

*Uredospores*—Sori small, chestnut-brown, roundish. Spores globose, elliptical, or ovate, as much as 25μ long, pale brown, echinulate.

*Teleutospores*—Sori thick, roundish, persistent, black, pulvinate. Spores clavate, apex much thickened, rounded or pointed, constriction slight, base attenuated, smooth, brown, 35–55 × 17–20μ. Pedicels persistent, 40μ long.

### *Synonyms.*

*Puccinia dioica.* Magnus, *Tagebl. der Naturf. vers. zu München* (1877), p. 200. Schröt., "Krypt. Flor. Schl.," p. 329.

*Æcidium cirsii.* D. C., "Flore franç," vol. v. p. 94.

Æcidiospores on *Carduus palustris*, July—Professor J. W. H. Trail.

Teleutospores on *Carex dioica, davalliana.*

BIOLOGY.—Schröter worked out the biology of this species in 1880 ("Krypt. Flor. Schl.," vol. iii. p. 330), by placing the teleutospores of *P. dioicæ* on *Cirsium oleraceum.* Johanson and Rostrup, in June, 1883, during an excursion in Jutland, found the æcidium on *C. palustris, lanceolatum,* and *arvense*, growing in company with *P. dioicæ.* The teleutospores have not at present been found in Britain, but Professor Trail found the æcidiospores in July.

### Puccinia paludosa. Plow.

*Æcidiospores*—Pseudoperidia irregularly clustered on the leaves, petioles, peduncles, stems, and sometimes even on the calices, sub-immersed, with torn white edges. Spores subglobose, echinulate, hyaline, with orange contents, 15–28 × 10–16μ.

*Uredospores*—Sori small, oval or elongated, on yellowish spots, light brown, soon naked, and pulverulent. Spores round, rarely oval, brown, rough with small, sharply pointed, not very numerous warts, 20–25μ.

*Teleutospores*—Sori small, oval or elongate, rarely roundish, often in linear series, black, naked. Spores, upper cell rounded and thickened above, often obliquely, lower cell subcylindrical or cuneiform, constriction rather marked, smooth, dark brown, 50–60 × 18–20μ. Pedicels rather long, brownish, persistent.

*Synonyms.*

*Æcidium pedicularis.* Libosch., "Mém. de Moscou," vol. v. p. 76, t. 5, fig. 1. Berk., *Ann. Nat. Hist.*, No. 254. Cooke, "Hdbk.," p. 544; "Micro. Fungi," 4th edit., p. 199.

*Puccinia caricis.* Rabh., *Fung. Europ.*, No. 1589.

*Exsiccati.*

Cooke, i. 105. Vize, "Fungi Brit.," 168; "Micro. Fungi Brit.," 555, 557. Rabh., 1589.

Æcidiospores on *Pedicularis palustris*, June and July.

Uredospores on *Carex vulgaris*, *stricta*, *fulva*, and *panicea* (?), July and August; teleutospores, August to June.

Biology.—In June, 1888, I found the æcidiospores at Irstead, Norfolk, in company with the last year's teleutospores. The latter germinated freely, and, when applied to *Pedicularis palustris*, give rise to the æcidiospores. Conversely, the æcidiospores, applied to *Carex vulgaris*, give rise to the uredospores and teleutospores. Professor Trail has found this species in Orkney.

### Puccinia obscura. Schröt.

*Æcidiospores*—Pseudoperidia on large roundish or elongate discoloured spots, mostly epiphyllous, shortly cylindrical, with torn white edges. Spores irregularly globose, finely echinulate, 18–20μ in diameter.

*Uredospores*—Sori oval or linear, generally surrounded by a red margin, long covered by the epidermis, yellow. Spores orange, spherical, elliptical, or ovate, echinulate, 19–26 × 17–20$\mu$.

*Teleutospores*—Sori elliptical or elongate, pulvinate. Spores oblong, clavate, apex rounded, truncate or conical, thickened (5–9$\mu$), base attenuated, somewhat constricted in the middle, brown, smooth, 30–45 × 14–20$\mu$. Pedicels rather long, persistent. Mesospores clavate, pretty abundant.

*Synonyms.*

*Puccinia obscura.* Schröt. in *Nuov. Giorn. Bot. Ital.*, vol. ix. p. 256. "Krypt. Flor.," vol. iii. p. 330. Plow., *Jour. Linn. Soc. Bot.*, vol. xx. pp. 511, 512.

*Æcidium compositarum*, Mart., var. *bellidis*, D. C. Cooke, "Hdbk.," p. 543.

*Exsiccati.*

Berk., 225. Cooke, i. 327; ii. 90; "L. F.," 57. Vize, "Micro. Fungi," 156.

Æcidiospores on *Bellis perennis*, October to December.

Uredospores and teleutospores, on *Luzula campestris*, L., June to November.

BIOLOGY.—The teleutospores are not formed until August or September, and germinate after a very short period of rest. The life-history was first worked out in November, 1883, and June, 1884. Further cultures were made in 1887.

### **Puccinia phragmitis.** (Schum.)

*Æcidiospores*—Pseudoperidia on circular red spots (·5 to 1·5 cm. in diameter), shallow, edges white, torn. Spores white, subglobose, echinulate, 15–26$\mu$ in diameter.

*Uredospores*—Sori rather large, dark brown, elliptical, pulverulent, without paraphyses. Spores ovate or elliptical, echinulate, brown, 25–35 × 15–23$\mu$.

*Teleutospores*—Sori large, long, sooty black, thick, often confluent. Spores elliptical, rounded at both ends, markedly constricted in the middle, dark blackish brown, smooth, 45–65 × 16–25$\mu$. Pedicels very long, 150–200 × 5–8$\mu$, yellowish, firmly attached.

*Synonyms.*

*Puccinia phragmitis* (Schum.). Schröt., "Krypt. Flor.," vol. iii. p. 331.

*Uredo phragmitis.* Schum., "Enum. Plant. Sæll.," vol. ii. p. 231.

*Puccinia arundinacea*, Hedw. Cooke, "Hdbk.," p. 491; "Micro. Fungi," 4th edit., p. 202.

*Æcidium rumicis*, Pers. Purton, vol. iii., No. 1540, t. 26.

*Æcidium rubellum*, Pers. Sow., t. 405. Grev., "Flor. Edin.," p. 447. Berk., "Eng. Flor.," vol. v. p. 369, t. 26. Cooke, "Hdbk.," p. 544; "Micro. Fungi," 4th edit., p. 199.

*Exsiccati.*

Cooke, i. 15, 25; ii. 81, 123; "L. F.," 59. Vize, "Micro. Fungi Brit.," 161, 438, 439; "Fungi Brit.," 124.

Æcidiospores on *Rumex conglomeratus, obtusifolius, crispus, hydrolapathum, Rheum officinale*, May and June.

Uredospores and teleutospores on *Phragmites communis*, July to May.

BIOLOGY.—The life-history of this fungus has been the subject of considerable confusion, owing to its having been confounded with *P. magnusiana*. In a series of more than thirty cultures, made during the years 1882 and 1883, it was proved that *P. phragmitis* has its æcidiospores on the above-named Rumices and Rheum, and it was, moreover, shown by several serial cultures that it does not affect *Rumex acetosa*. For details, see *Proc. Royal Soc.*, No. 228, 1883, pp. 47–50; and *Quar. Jour. Micro. Science*, vol. xxv., new series, pp. 156–161.

**Puccinia trailii.** Plow.

*Æcidiospores*—Spots rounded, often irregular, purple-red, surrounded by a yellow margin. Pseudoperidia cup-shaped, wide, with torn, white, everted edges. Spores round, oval, or irregular, white, finely echinulate, 20–40$\mu$ in diameter.

*Uredospores*—Sori rather large, elliptical or elongate, reddish brown, pulverulent, amphigenous, without paraphyses. Spores oval, subpyriform, sometimes nearly globose, brown, echinulate, 25–35 × 20–25$\mu$.

*Teleutospores*—Sori blackish, large (2–4 mm. long by ·5 mm. high), compact, pulvinate, elongate, with numerous smaller ones scattered between. Spores oval, fusiform, or subcylindrical, brown, granular, markedly constricted. Upper cell slightly thickened above, rounded above and below, lower cell similar in form, but more attenuated below, 50–60 × 20–23$\mu$. Pedicels brown, stout, persistent, 75–100 × 6–8$\mu$.

*Synonyms.*

*Æcidium rumicis*, Grev., in part.
*Æcidium rubellum*, Pers., in part.
*Puccinia phragmitis* (Schum.), in part.

Æcidiospores on *Rumex acetosa.*

Uredospores and teleutospores on *Phragmites communis.*

Æcidiospores, May and June; uredospores, July and August; teleutospores, August to June.

BIOLOGY.—While studying the life-history of *Puccinia phragmitis* and *magnusiana*, I became convinced that a third species occurred upon the common reed, which had its æcidiospores on *Rumex acetosa.* Professor Trail was fortunate enough to find near Aberdeen a Puccinia on reed, and in close proximity to it the Æcidium on *R. acetosa*, and, although other Rumices were growing equally near, none of them had the æcidiospores on them. From specimens he was kind enough to send me in July, 1888, the above descriptions have been drawn up. The teleutospores, while closely resembling those of *P. phragmitis*, differ in having a granular spore-membrane, and in being borne on stouter pedicels. The last year's teleutospores collected by him in June and July were found to germinate freely, and on July 29 were applied to *R. acetosa* and *crispus* (Exp. 950, 951), and on August 1 to *R. acetosa*, *crispus*, *obtusifolia*, and *Rheum officinale.* The æcidiospores were produced in due course upon *R. acetosa*, but not upon the other plants.

### **Puccinia magnusiana.** Körn.

*Æcidiospores*—Pseudoperidia in small circular groups on the leaves, mostly hypophyllous, with everted, toothed, whitish edges. Spores subglobose, 15–25$\mu$ in diameter.

*Uredospores*—Sori orange-brown, small, elliptical or linear. Spores globose, ovate or elliptical, finely echinulate, orange-yellow,

21–35 × 12–20μ, intermixed with great numbers of hyaline, clavate paraphyses.

*Teleutospores*—Sori black, persistent, very numerous, small, scattered, elliptical or linear, on the stems and sheaths confluent into very long, black lines. Spores oblong or clavate, base attenuated, apex much thickened, rounded and darker, truncate or sometimes conical, constriction slight, chestnut-brown, 30–55 × 16–26μ. Pedicels firm, about the length of the spores.

*Synonyms.*

*Puccinia magnusiana.* Körnicke, "Hedwigia" (1876), p. 179. Winter in Rabh., "Krypt. Flor.," vol. i. p. 221.

*Puccinia graminis*, Pers., var. *arundinis.* Cooke, "Hdbk.," p. 493; "Micro. Fungi," 4th edit., p. 202. Berk., "Eng. Flor.," vol. v. p. 363. Grev., "Flor. Edin," p. 433. Johnst., "Flor. Berw.," vol. ii. p. 195.

*Æcidium ranunculacearum*, D. C. Cooke, "Hdbk.," p. 539; "Micro. Fungi," 4th edit., p. 196, in part. Berk., "Eng. Flor.," vol. v. p. 370. Grev., "Flor. Edin.," p. 446. Johnst., "Flor. Berw.," vol. ii. p. 206, in part.

*Exsiccati.*

Vize, "Micro. Fungi Brit.," 560.

Æcidiospores on *Ranunculus repens* and *bulbosa*, April to June.

Uredospores and teleutospores on *Phragmites communis*, June to April.

BIOLOGY.—The life-history of this fungus was worked out by a long series of cultures, more than fifty in number; for the details of which, see *Quar. Jour. Micro. Science*, vol. xxv., new series, pp. 156–161. These included the sowing of the promycelial spores on barberry, which was without result; on various Ranunculi, which was also without result, excepting upon *R. repens* and *bulbosa*, and the converse cultures of the æcidiospores on Phragmites. No result was obtained on *Rumex conglomeratus, obtusifolium, crispus, hydrolapathum*, nor on *Rheum.*

### Puccinia moliniæ. Tul.

*Æcidiospores*—Pseudoperidia on round discoloured spots, flat, with torn, white edges. Spores subglobose, finely verrucose, orange-yellow, 17–25 × 15–21μ.

*Uredospores*—Sori chestnut-brown, elliptical or linear, often confluent. Spores globose or shortly elliptical, very thickly but finely echinulate, yellowish brown, 24–28μ in diameter.

*Teleutospores*—Sori black-brown, elliptical or linear, pulvinate, naked, often confluent. Spores elliptical, central constriction, slight, base rounded, apex rounded, much thickened, smooth, brown, 30–56 × 20–26μ. Pedicels brownish, permanent, very long (100μ).

*Synonyms.*

*Puccinia moliniæ.* Tul., *Mem. Ured. Ann. Sc. Nat.*, series iv., vol. ii. p. 141. Winter in Rabh., "Krypt. Flor.," vol. i. p. 219. Cooke, "Micro. Fungi," 4th edit., p. 203.

*Æcidium orchidearum*, Field. Cooke, "Hdbk.," p. 545; "Micro. Fungi," 4th edit., p. 200.

*Exsiccati.*

Cooke, i. 106. Vize, "Micro. Fungi Brit.," 558 (?).

Æcidiospores on *Orchis latifolia*, May and June.

Uredospores and teleutospores on *Molinia cærulea*, July to October.

BIOLOGY.—The life-history was worked out by Rostrup, and while there is no reason to doubt the accuracy of his conclusions, yet I was unable to produce the uredospores on *M. cærulea* from the æcidiospores on *O. latifolia* from Irstead. Neither have I been able to find *P. moliniæ* at Irstead, although the Æcidium is abundant there during May and June on *O. latifolia*. It is therefore probable that two species have their æcidiospores on *O. latifolia*.

### Puccinia perplexans. Plow.

*Æcidiospores*—Pseudoperidia in roundish or elongate clusters on thickened yellow, often concave spots, shortly cylindrical, with torn white everted edges. Spores subglobose or polygonal, finely echinulate, orange-yellow, 24–28 × 20–25μ.

*Uredospores*—Sori linear, oblong or rounded, sometimes confluent, amphigenous, golden-yellow. Spores globose, oval or subovate, orange, finely echinulate, with or without capitate paraphyses, 20–25 × 30–35$\mu$.

*Teleutospores*—Sori small, sometimes round, but generally elongate, oblong, or linear, covered by the cuticle, black, often in clusters, sometimes confluent. Spores very irregular in form and size, clavate, subfusiform, oblong, upper cell rounded, truncate or attenuated, often obliquely, central constriction little or none, lower cell generally cuneiform, clear brown, epispore often appearing granular, 40–60 × 10–12$\mu$. Pedicels very short.

*Synonym.*

*Æcidium ranunculacearum*, D. C. Cooke, "Micro. Fungi," 4th edit., p. 196, in part; "Hdbk.," p. 539. Johnst., "Flor. Berw.," vol. ii. p. 206, in part.

*Exsiccati.*

Vize, "Micro. Fungi Brit.," 443, 548.

Æcidiospores on *Ranunculus acris*.
Uredospores and teleutospores, on *Alopecurus pratensis*.

BIOLOGY.—The life-history of this fungus was worked out in a series of more than thirty cultures. The uredo, when produced directly from the æcidiospores, is without paraphyses; but these appear later on in the year. For details, see *Quar. Jour. Micro. Science*, vol. xxv., new series, pp 164–166. I have found this species near King's Lynn; and Mr. George Brebner has met with it at Aboyne, Scotland.

### **Puccinia persistens.** Plow.

*Æcidiospores*—On thickened spots, which are purple-brown above, yellow below, and surrounded by brownish margins. Pseudoperidia cylindrical, short, yellow, with torn white edges. Spores subglobose or oval, finely echinulate, orange-yellow, 20–30 × 17–20$\mu$.

*Uredospores*—Sori small, orange, round or elongate, on yellowish spots. Spores subglobose, hyaline, finely echinulate, with orange-yellow contents, 25–30$\mu$ in diameter.

*Teleutospores*—Sori small, black, oval, oblong, or linear, long covered by the epidermis. Spores cylindrical or clavate, apex slightly thickened, rounded, truncate, or obliquely attenuated, constriction sometimes marked, sometimes almost absent, lower cell attenuated below, smooth, brown, 50–60 × 15–20$\mu$. Pedicels short, very persistent.

*Synonym.*

*Æcidium ranunculacearum*, var. *thalictri-flavi*. D. C., "Flore franç.," vol. vi. p. 97. Winter, "Krypt. Flor.," p. 269.

Æcidiospores on *Thalictrum flavum* and *minus* (?), May and June.

Uredospores on *Triticum repens* (*Avena elatior*?), July and August; teleutospores, August to July.

BIOLOGY.—The æcidiospores were found by Mr. T. Birks near Goole, in June, 1884, growing near a plant of *Avena elatior*, affected with an orange Uredo with paraphyses. I was, however, unable to produce the Uredo on *A. elatior*, either from Mr. Birks' æcidiospores, or from specimens collected near King's Lynn in 1888. The teleutospores were found in June, 1888, on a piece of dead grass of the previous season, germinated, and when applied to a plant of *Thalictrum flavum*, which had been growing in my garden since 1884, the Æcidium appeared in due course. In July, 1888, the æcidiospores, applied to a plant of *Triticum repens*, gave rise to the Uredo in twelve days.

### **Puccinia extensicola.** Plow

*Æcidiospores*—Pseudoperidia on paler spots (which are more or less circular on the leaves, and elongate on the stems), erect, circinating or scattered, often amphigenous, margins torn, white, everted. Spores globose, very finely echinulate, orange-yellow, 20–25$\mu$ in diameter.

*Uredospores*—Sori small, reddish brown, elongate, elliptical or linear, on rather extensive though pale discoloured spots. Spores subglobose or ovoid, rather irregular, very finely echinulate, brownish with a tinge of yellow, 25–30 × 15–20$\mu$.

*Teleutospores*—Sori small, ·5 to 1 mm. long, elongated, erumpent, black, surrounded by the torn epidermis. Spores ovate or subclavate, brown, smooth, apex thickened, often hooded, rounded or conical, lower cell attenuated below, slightly constricted, 40–60 × 18–20$\mu$. Pedicels short, persistent.

*Exsiccati.*

Vize, "Micro. Fungi Brit.," 544, 552.

Æcidiospores on *Aster tripolium*, June and July.
Teleutospores on *Carex extensa*, August to June.
Wells-next-the-Sea; Norfolk.

BIOLOGY.—The life-history of this species was worked out by a series of experimental cultures during the year 1888.

## II. BRACHYPUCCINIA. Schröt.

Having spermogonia, uredospores, and teleutospores.

### **Puccinia suaveolens.** (Pers.)

*Spermogonia*—Pale yellow, scattered over the whole leaf-surface, sweet-scented.

*Uredospores—Primary:* Sori hypogenous, large, crowded, soon confluent, very pulverulent. Spores echinulate, globose, brown, 22–20$\mu$ in diameter. *Secondary:* Sori small, scattered, discrete, rarely confluent. Spores globose or broadly elliptical, echinulate, epispore brown, 20–30$\mu$ in diameter.

*Teleutospores*—Sori small, roundish or elongate, blackish brown. Spores oval or oblong, rounded above, echinulate, pale brown, 25–40 × 17–25$\mu$.

*Synonyms.*

*Puccinia suaveolens* (Pers.). Winter in Rabh., "Krypt. Flor.," vol. i. p. 189.

*Uredo suaveolens.* Pers., "Syn.," p. 221. Berk., "Eng. Flor.," vol. v. p. 379. Purton, "Midl. Flor.," vol. iii., No. 1548. Grev., "Flor. Edin.," p. 434. Johnst., "Flor. Berw.," vol. ii. p. 202. Sow., t. 398, fig. 5.

*Trichobasis suaveolens*, Lév. Berk., "Outl.," p. 208. Cooke, "Hdbk.," p. 530; "Micro. Fungi," 4th edit., p. 226, t. vii. figs. 151–153.

*Puccinia compositarum*, Mart. Cooke, "Hdbk.," p. 498, in part.

*Puccinia obtegens*, Fckl. Vize, Exs.

*Exsiccati.*

Cooke, i. 73; ii. 54. Vize, "Fungi Brit.," 64, 147: "Micro. Fungi Brit.," 323.

On *Carduus arvensis*, April to November. A variety occurs on *Centaurea cyanus*.

BIOLOGY.—The life-history of this fungus has been worked out by Rostrup. The first generation consists of roundish uredospores, spermogonia, and sometimes a few stray teleutospores. The mycelium invades the whole of the infected plant, and hybernates in the upper part of the root-stock, so that every shoot sent up from this root contains the mycelium, and bears spermogonia and the primary uredo (= *Uredo suaveolens*, Pers.). The affected plants appear sooner than the healthy ones, have a sickly pale-green colour, and do not bear flowers. The second generation consists of uredospores and teleutospores, and has only a localized mycelium (= *P. obtegens*, Fckl.). The variety on *Centaurea cyanus* was first described by Magnus, who does not consider it a distinct species, but that it is modified by the host-plant. It is in this country not very rare on cultivated Centaureæ in gardens. I have gathered it in the public park at Aberdeen, and the Rev. Canon Du Port has also sent me specimens.

## Puccinia bullata. (Pers.)

*Spermogonia*—Honey-coloured, in small roundish clusters.

*Uredospores*—Sori on small, roundish, pale-brown spots, roundish on the leaves, elongated on the stems, sometimes circinating. Spores ovate, generally attenuated below, with two lateral, markedly thickened germ-pores. Membrane ochraceous, echinulate, thickened and hooded above, contents pale reddish, 23–35 × 20–25$\mu$.

*Teleutospores*—Sori roundish, or more often elongate, dark brown. Spores elliptical or clavate, rounded above, often attenuated below, apex slightly thickened, smooth, brown, 30–40 × 20–25$\mu$. Pedicels deciduous.

*Synonyms.*

*Puccinia bullata* (Pers.). Schröt., "Krypt. Flor.," vol. iii. p. 335. Winter in Rabh., "Krypt. Flor.," vol. iii. p. 191.

*Uredo bullata.* Pers., "Obs. Myc.," vol. i. p. 68.

*Uredo æcidiiformis.* Grev., "Flor. Edin.," p. 441.

*Uredo umbellatarum.* Johnst., "Flor. Berw.," vol. ii. p. 202. Berk., "Eng. Flor.," vol. v. p. 380.

*Trichobasis umbelliferarum*, Lév. Cooke, "Micro. Fungi," 4th edit., p. 225.

*Trichobasis conii*, Strauss. Cooke, "Micro. Fungi," 4th edit., p. 225.

*Trichobasis cynapii*, D. C. Cooke, "Micro. Fungi," 4th edit., p. 224.

*Trichobasis petroselini.* Cooke, "Micro. Fungi," 4th edit., p. 223.

*Puccinia umbelliferarum*, D. C. Berk., "Eng. Flor.," vol. ii. p. 366. Cooke, "Hdbk.," p. 501; "Micro. Fungi," 4th edit., p. 208, t. iv. figs. 70, 71. Grev., "Flor. Edin.," p. 431. Johnst., "Flor. Berw.," vol. ii. p. 196.

*Puccinia conii*, Fckl. Cooke, "Micro. Fungi," 4th edit., p. 209.

*Puccinia æthusæ*, Link. Cooke, "Micro. Fungi," 4th edit., p. 209.

*Puccinia tumida.* Grev., "Flor. Edin.," p. 430. Berk., "Eng. Flor.," vol. v. p. 366.

*Puccinia silai*, Fckl. Cooke, "Grevillea," vol. xiv. p. 39.

*Puccinia bullaria*, Link. Berk., "Eng. Flor.," vol. v. p. 366. Cooke, "Hdbk.," p. 503. Pers., "Obs.," vol. i. t. ii. fig. 5. Johnst., "Flor. Berw.," vol. ii. p. 197.

*Exsiccati.*

Berk., 57, 221. Cooke, i. 40, 40*a*, 42, 42*b*; ii. 319, 328; "L. F.," 15, 16, 38. Vize, "Micro. Fungi Brit.," 123; "Fungi Brit.," 9, 54, 127.

On *Petroselinum sativum*, *Æthusa cynapium*, *Conium maculatum*, *Silaus pratensis.*

BIOLOGY.—I have followed Winter and Schröter in uniting the above species, but I am convinced biological research will show many of them to be distinct.

## **Puccinia hieracii.** (Schum.)

*Spermogonia*—In small clusters, which are roundish on the leaves, but more or less elongated on the stems, yellowish.

*Uredospores*—Sori dark brown, small, roundish, scattered or circinating, soon pulverulent. Spores round or shortly elliptical, brown, with two or three germ-pores, 20–30 × 15–25μ.

*Teleutospores*—Sori blackish, small, round, soon pulverulent, elliptical or ovate, constriction slight, apex generally rounded. Epispore dark brown, when seen in water usually punctate, 25–40 × 20–25μ. Pedicels colourless, deciduous, delicate.

*Synonyms.*

*Puccinia flosculosorum* (Alb. and Schw.). Winter in Rabh., "Krypt. Flor.," vol. i. p. 206.

*Uredo hieracii.* Schum., "Enum. Plant. Sæll.," vol. ii. p. 223.

*Puccinia compositarum.* Schlecht. "Flor. Berol.," vol. ii. p. 133. Cooke, "Hdbk.," p. 449; "Micro. Fungi," 4th edit., p. 206, t. 4. figs. 67, 68. Johnst., "Flor. Berw.," vol. ii. p. 196. Berk., "Eng. Flor.," vol. v. p. 365.

*Trichobasis hieracii*, Schum. Cooke, "Micro. Fungi," 4th edit., p. 224.

*Puccinia hieracii*, Mart. Cooke, "Micro. Fungi," 4th edit., p. 207.

*Puccinia clandestina*, Carm. Cooke, "Hdbk.," p. 498: "Micro. Fungi," 4th edit., p. 205. Berk., "Eng. Flor.," vol. v. p. 365.

*Uredo cichoracearum.* Grev., "Flor. Edin.," p. 435. Johnst., "Flor. Berw.," vol. ii. p. 201.

*Exsiccati.*

Berk., 219. Cooke, i. 33, 34, 68; ii. 53, 56, 638, 639; "L. F.," 11, 12, 36. Vize, "Micro. Fungi Brit.," 120, 428; "Fungi Brit.," 23, 62.

On *Carlina vulgaris, Arctium lappa, Carduus crispus, lanceolatus, palustris, Cichorium intybus, Leontodon autumnalis, hispidus, Hieracium vulgatum, murorum, corymbosum, boreale, pilosella, Scabiosa succisa* (?) *Crepis virens.*

Biology.—I have followed Winter and Schröter in uniting the above species in the absence of biological information, although I feel sure many of them are distinct species.

### **Puccinia centaureæ.** Mart.

*Spermogonia*—On yellow spots, which are round on the leaves, elliptical or lanceolate on the petioles, where they are surrounded by a purple margin.

*Uredospores—Primary:* Sori few, surrounded by the spermogonia, large, round or elliptical. Spores produced singly on basidia, globose or subovate, echinulate, brown, about 25$\mu$ across. *Secondary:* Sori small, very numerous, scattered, mostly hypophyllous, dark brown, soon pulverulent. Spores subglobose or elliptical, brown, echinulate, 20–25 × 16–23$\mu$.

*Teleutospores*—Sori very small, amphigenous, roundish or oval, black, pulverulent, often confluent. Spores elliptical, pyriform, or subglobose, not constricted, both cells equal or nearly so, dark brown, smooth, 30–50 × 20–28$\mu$. Pedicels hyaline, short.

*Synonyms.*

*Puccinia centaureæ.* Mart., "Flor. Mosq.," p. 226. Cooke, "Micro. Fungi," 4th edit., p. 207. Grev., "Flor. Edin.," p. 430.

*Puccinia compositarum*, Mart., in part. Berk., "Eng. Flor.," vol. v. p. 365. Johnst., "Flor. Berw.," vol. ii. p. 196. Cooke, "Hdbk.," p. 499.

*Exsiccati.*

Vize, "Micro. Fungi Brit.," 120.

On *Centaurea nigra.*

BIOLOGY.—The primary uredospores, accompanied by the spermogonia, occur early in May, and are in very much larger sori than the secondary uredospores. I found by experimental culture that the primary uredospores produced the secondary in about fourteen days (Exp. 851, 852), but that no result followed when they were applied to *Taraxacum officinale* (853), *Leontodon autumnalis* (923), and *Hieracium pilosella* (854, 855).

### **Puccinia taraxaci.** Plow.

*Spermogonia*—On yellow oval or rounded spots. Paraphyses not conspicuous. Spermatia globose or oval, 1–2$\mu$ in diameter.

*Uredospores—Primary:* Sori scanty, large, dark brown, elongated or circinating. Spores ovate, round, or subpyriform, echinulate, brown, 25–30 × 25μ. *Secondary:* Sori small, very profuse, round, cinnamon-brown, soon pulverulent, often confluent. Spores subglobose, brown, echinulate, 20–25μ in diameter.

*Teleutospores*—Sori amphigenous, minute, blackish, round, pulverulent, surrounded by the ruptured epidermis. Spores obtuse, shortly oval, ovoid, or even subglobose, constriction almost none, brown, echinulate, especially above, 30–40 × 20–25μ. Pedicels short, hyaline, deciduous.

*Synonyms.*

*Puccinia variabilis.* Grev., "Flor. Edin.," p. 431; "Flor. Scot.," t. 75. Cooke, "Hdbk.," p. 500; "Micro. Fungi," 4th edit., p. 207, t. 4, figs. 82, 83. Johnst., "Flor. Berw.," vol. vi. p. 196, all in part.

*Puccinia flosculosorum* (Alb. and Schw.). Winter in Rabh., "Krypt. Flor.," p. 206, in part.

*Exsiccati.*

Cooke, i. 539; ii. 128. Vize, "Micro. Fungi Brit.," 53.

On *Taraxacum officinale.*
April to November.

BIOLOGY.—The spermogonia and primary uredospores occur early in the year, about the end of April. Mr. Grove considers this species has a true Æcidium (*Æc. grevillei*, Grove), which is scattered over the leaves in small clusters. If this had been the case, I think I must have met with it, as I have for many years searched the Taraxacum in the neighbourhood of King's Lynn for the Æcidium. It has been asserted that the Pucciniæ on the Compositæ belong to one species, but this is clearly incorrect. I found, for instance, that the æcidiospores of *P. lapsanæ*, placed on *Taraxacum officinale* and *Lapsana communis* in a duplicated culture, produced the uredospores on the latter in twenty days, but had no effect upon Taraxacum (Exp. 497, 498), and conversely the germinating teleutospores of *P. lapsanæ* produced the æcidiospores on *L. communis* in twenty days, but had no effect upon the Taraxacum (Exp. 499, 500). I also found that the teleutospores on *Leontodon autumnalis* (Exp. 620), and the uredospores on *Centaurea nigra* (923), when applied to *Taraxacum officinale*, produced no effect, although they readily infected their respective host-plants.

*P. taraxaci* is a much more common species than *P. variabilis.*

III. HEMIPUCCINIA. Schröt.

Having uredospores and teleutospores, the latter germinating only after a period of rest.

## **Puccinia polygoni.** Pers.

*Uredospores*—Sori at first yellowish brown, then dark brown, roundish, scattered irregularly or arranged in circular groups, soon pulverulent. Spores spherical, elliptical, or ovate, echinulate, brown, 17–30 × 16–20μ.

*Teleutospores*—Sori compact, blackish, generally crowded, rounded or elliptical, on the stems, elongated, often confluent, brown. Spores cuneiform or clavate, constriction slight or absent, base attenuated towards the stem, apex much thickened (4–7μ), rounded, truncate, capitate or conical, smooth, brown, 30–50 × 14–20μ. Pedicels about as long as the spores, persistent.

*Synonyms.*

*Puccinia polygoni.* Pers., "Syn.," p. 227.

*Puccinia polygonorum*, Link. Cooke, "Micro. Fungi," 4th edit., p. 203; "Hdbk.," p. 495. Berk., "Eng. Flor.," vol. v. p. 363. Johnst., "Flor. Berw.," vol. ii. p. 195; Grev., "Flor. Edin.," p. 430.

*Puccinia amphibii*, Fckl. Cooke, "Micro. Fungi," 4th edit., p. 204.

*Uredo polygonorum*, D. C. Grev., t. 80. Berk., "Eng. Flor.," vol. v. p. 377. Johnst., "Flor. Berw.," vol. ii. p. 201.

*Trichobasis polygonorum*, Berk. Cooke, "Micro. Fungi," 4th edit., p. 226.

*Exsiccati.*

Berk., 231, 216. Cooke, i. 27; "L. F.," 5, 42. Vize, "Micro. Fungi Brit.," 28; "Fungi Brit.," 111.

On *Polygonum convolvulus*, *amphibium*, *lapathifolium*.
June to November.

BIOLOGY.—Winter (Rabh., "Krypt. Flor.," vol. i. pp. 185, 186) separates the form on *P. amphibium*, following Fuckel, the principal morphological difference being that in this form the teleutospores remain covered by the epidermis. This is also the case when the fungus occurs on *P. lapathifolium*. On *P. convolvulus* the sori are pulverulent. In the absence of biological information, I have followed Schröter in uniting these species.

### Puccinia tanaceti. (D. C.)

*Uredospores*—Sori pale brown, small, round or oval, pulverulent, elongated. Spores round or ovate, pale brown, with three germ-pores, echinulate, 20–35 × 15–20$\mu$.

*Teleutospores*—Sori compact, pulvinate, black. Spores elliptical or clavate, base attenuated, summit thickened (5–7$\mu$), hooded, smooth, or slightly verrucose towards the summit, deep brown, 40–50 × 20–28$\mu$. Pedicels long, permanent.

*Synonyms.*

*Puccinia tanaceti*, D. C. Winter in Rabh., "Krypt. Flor.," vol. i. p. 209, in part.

*Uredo tanaceti.* D. C., "Encyc.," vol. viii. p. 224.

*Trichobasis artemisiæ*, Berk. Cooke, "Micro. Fungi," 4th edit., p. 223.

*Uredo artemisiæ.* Berk., "Outl.," p. 332.

*Puccinia tanaceti*, D. C. Cooke, "Micro. Fungi," 4th edit., p. 207.

*Puccinia discoidearum*, Link. Cooke, "Micro. Fungi," 4th edit., p. 206; "Hdbk.," p. 499.

*Exsiccati.*

Berk. Cooke, i. 115, 35, 437; ii. 126.

On *Artemisia absinthium, maritima, Tanacetum vulgare.*

### Puccinia iridis. (D. C.)

*Uredospores*—Sori hypophyllous or amphigenous, reddish brown, scattered irregularly, elongated or roundish, long covered by the epidermis, then pulverulent. Spores elliptical or ovate, with three germ-pores, brown, echinulate, 20–35 × 16–26$\mu$.

*Teleutospores*—Sori black, linear, soon naked, compact, persistent. Spores oblong, apex very much thickened (10–12$\mu$), rounded or obliquely conical, seldom truncate, base wedge-shaped, constriction slight, smooth, brown, 35–45 × 15–22$\mu$. Pedicels rather long, often brown, persistent.

*Synonyms.*

*Puccinia iridis*, D. C. Winter in Rabh., "Krypt. Flor.," vol. i. p. 184.

*Uredo iridis.* D. C., "Encyc.," vol. viii. p. 224. Berk., "Eng. Flor.," vol. v. p. 376.

*Trichobasis iridis.* Cooke, "Micro. Fungi," 4th edit., p. 227.

*Puccinia truncata.* B. and Br., No. 754. Cooke, "Hdbk.," p. 494; "Micro. Fungi," 4th edit., p. 203.

*Exsiccati.*

Berk., 59. Cooke, i. 77; "L. F.," 28. Vize, "Fungi Brit.," 122.

On *Iris fœtidissima, pseudo-acorus.*
July to December.

BIOLOGY.—The uredospores which occur on certain cultivated species of iris, as *Iris iberica, tolmicana*, etc., produced no effect upon *I. fœtidissima* and *pseudo-acorus*. (See *Uredo iridis*, p. 258.)

## Puccinia oblongata. (Link.)

*Uredospores*—Sori reddish brown, bullate, scattered, roundish, oblong, long covered by the epidermis. Spores generally ovate or pyriform, epispore very thick, quite smooth, very pale yellow, generally 30–42 × 12–15$\mu$.

*Teleutospores*—Sori black, elliptical or linear, compact, sooner naked and surrounded by the ruptured epidermis. Spores clavate, slightly constricted, base attenuated, apex thickened (10–20$\mu$), hooded, round or conical, smooth, brown, 40–60 or even 80 × 17–23$\mu$. Pedicels short, firmly attached.

*Synonyms.*

*Puccinia oblongata* (Link.). Winter in Rabh., "Krypt. Flor.," vol. i. p. 183.

*Cæoma oblongatum.* Link., "Obs.," vol. ii. p. 27.

*Uredo oblongata.* Grev., t. 12. Berk., "Eng. Flor.," vol. v. p. 376; "Outl.," p. 208. Cooke, "Hdbk.," p. 529. Grev., "Flor. Edin.," p. 437. Johnst., "Flor. Berw.," vol. ii. p. 202.

*Tricholbasis oblongata*, Berk. Cooke, "Micro. Fungi," 4th edit., p. 223, t. 7, figs 158, 159.

*Puccinia luzulæ*, Lib. Cooke, "Micro. Fungi," 4th edit., p. 203.

*Exsiccati.*

Cooke, i. 535. Vize, "Fungi Brit.," 61.

On *Luzula pilosa, campestris, maxima.* May to November.

BIOLOGY.—This is probably a heterœcious species.

### Puccinia scirpi. D. C.

*Uredospores*—Sori reddish brown, elliptical, often numerous and confluent. Spores ovate or subcuneiform or oblong, echinulate, yellowish brown, 20–32 × 12–24μ.

*Teleutospores*—Sori small, blackish, sometimes circinating, at length naked. Spores clavate, attenuated below, constriction slight, apex thickened, generally roundish, conical, smooth, brown, 30–60 × 10–20μ. Pedicels long, brownish, persistent.

*Synonym.*

*Puccinia scirpi*, D. C. Winter in Rabh., "Krypt. Flor.," vol. i. p. 182. "Flore franç.," vol. ii. p. 223.

*Exsiccati.*

Cooke, ii. 636. Vize, "Micro. Fungi Brit.," 122.

On *Scirpus lacustris.* King's Lynn.
July to November.

BIOLOGY.—This is most likely a heterœcious species, the life-history of which is unknown. Mesospores are often abundantly mixed with the normal teleutospores. They measure from 25 to 40μ in length.

### Puccinia baryi. (B. and Br.)

*Uredospores*—Sori small, scattered or crowded, elongate or linear, reddish yellow, usually occurring in long linear series on the leaves. Spores globose or ovate, finely echinulate, orange-yellow, 20–25μ in diameter. Paraphyses numerous, capitate, hyaline, about 20μ long.

*Teleutospores*—Sori small, blackish, crowded in linear series, long covered by the epidermis. Spores irregularly elliptical or cuneiform, constriction slight or absent, summits generally truncate, often obliquely, rarely rounded, usually thickened, attenuated below, smooth, brown, 25–35 × 15–25μ. Pedicels very short, often obsolete.

*Synonyms.*

*Puccinia baryi* (B. and Br.). Winter in Rabh., "Krypt. Flor.," vol. i. p. 178.

*Epitea baryi.* B. and Br., *Ann. Nat. Hist.*, No. 755.

*Lecythea baryi*, Berk. Cooke, "Micro. Fungi," 4th edit., p. 222; "Hdbk.," p. 532. Berk., "Outl.," p. 334.

On *Brachypodium sylvaticum* and *pinnatum*.
June to November.

BIOLOGY.—This is probably a heterœcious species. Mr. Soppitt considers that it also occurs on *Aira cæspitosa*.

## **Puccinia bistortæ.** D. C.

*Uredospores*—Sori yellow, small, irregularly rounded, soon naked, scattered. Spores rounded or elliptical, finely echinulate, yellow, 20–28 × 18–20μ.

*Teleutospores*—Sori black-brown, very small, often confluent. Spores elliptical or ovate, very slightly constricted, apex not thickened, rounded, smooth, brown, 24–38 × 15–24μ. Pedicels rather long, deciduous.

*Synonym.*

*Puccinia bistortæ*, D. C. Winter in Rabh., "Krypt. Flor.," vol. i. p. 186. Cooke, "Micro. Fungi," 4th edit., p. 204.

On *Polygonum viviparum.*
July to September.

## **Puccinia pruni.** Pers.

*Uredospores*—Sori light brown, small, round, crowded, pulverulent, often confluent. Spores ovate or subpyriform, apex darker, thickened, bluntly conical, closely echinulate, brown, 20–35 × 12–16$\mu$, mixed with numerous capitate brownish paraphyses.

*Teleutospores*—Sori pulverulent, dark brown, almost black. Spores consisting of two spherical cells, flattened at their point of union, the lower cell often being smaller and paler. Epispore uniformly thick, chestnut-brown, thickly studded with short stout spines. Spores 30–45 × 17–25$\mu$. Pedicels short, colourless.

*Synonyms.*

*Puccinia pruni-spinosæ*, Pers. Winter in Rabh., "Krypt. Flor.," vol. i. p. 193.

*Puccinia pruni.* Pers., "Syn.," p. 226.

*Puccinia prunorum*, Link. Berk., "Eng. Flor.," vol. v. p. 368. Cooke, "Hdbk.," p. 507; "Micro. Fungi," 4th edit., p. 211. Purton, "Midl. Flor.," vol. iii. No. 1552.

*Trichobasis rhamni.* Cooke, "Hdbk.," p. 507.

*Exsiccati.*

Baxt., 83. Cooke, i. 51; ii. 139. Vize, "Micro. Fungi," 128.

On *Prunus spinosa, domestica, Rhamnus catharticus.*

BIOLOGY.—The teleutospores are very apt to fall asunder at the septum. The frequently imperfect development of the lower cell shows the relationship which exists between the present genus and Uromyces. The form on Rhamnus requires to be experimentally examined.

**Puccinia argentata.** (Schultz.)

*Uredospores*—Sori yellowish brown, small, rounded, circinate or scattered, often confluent. Spores spherical or elliptical, finely echinulate, yellow, 15–20 × 14–18$\mu$.

*Teleutospores*—Sori rounded, chestnut-brown, soon pulverulent. Spores ovate or clavate, rounded or attenuated above, constriction slight or absent, 25–35 × 12–17$\mu$. Pedicels hyaline, very deciduous.

*Synonyms.*

*Puccinia argentata* (Schultz). Winter in Rabh., "Krypt. Flor.," vol. i. p. 194.

*Æcidium argentatum*, Schultz. "Prod. Flor. Starg.," p. 454.

*Trichobasis impatiens*, Rabh. Cooke, "Micro. Fungi," 4th edit., p. 225.

*Puccinia noli-tangeris*, Corda. B. and Br., No. 1044. Cooke, "Hdbk.," p. 504; "Micro. Fungi," 4th edit., p. 210.

*Exsiccati.*

Cooke, i. 44; "L. F.," 19.

On *Impatiens noli-me-tangere.*
April to September.

### **Puccinia anthoxanthi.** Fckl.

*Uredospores*—Sori elliptical or linear, dusky orange, soon naked. Spores elliptical or ovate, finely echinulate, pale yellowish brown, 20–30 × 15–20$\mu$. Paraphyses hyaline, capitate.

*Teleutospores*—Sori scattered, very small, mostly linear, dark brown. Spores elliptical or subpyriform, slightly constricted, sometimes rounded below, rarely cuneiform, summits thickened or slightly apiculated, smooth, chestnut-brown, 25–40 × 15–20$\mu$. Pedicels long, firmly attached.

*Synonym.*

*Puccinia anthoxanthi.* Fckl., "Symb. Myc.," vol. ii. p. 15. Winter in Rabh., "Krypt. Flor.," vol. i. p. 180.

*Exsiccati.*

Vize, "Micro. Fungi Brit.," p. 436.

On *Anthoxanthum odoratum.*

BIOLOGY.—This is probably a heterœcious species, the life-history of which has not yet been worked out. The uredospores are usually mixed with a large number of capitate, hyaline paraphyses.

### **Puccinia oxyriæ.** Fckl.

*Uredospores*—Sori cinnamon-brown, roundish or irregular, crowded, often confluent. Spores rounded or pyriform, finely echinulate, brown, with orange-yellow contents, 25–30 × 20–25$\mu$.

*Teleutospores*—Sori amphigenous, black-brown, rounded on the leaves, more often on the stems, petioles, and peduncles, where they are elongated. Spores elliptical or ovate, slightly constricted, base rounded, apex a little thickened, rounded or hooded, smooth, brown, 30–45 × 15–25$\mu$. Pedicels rather long, colourless.

*Synonym.*

*Puccinia oxyriæ.* Fckl., "Symb. Myc.," vol. iii. p. 14. Cooke, "Grevillea," vol. xi. p. 15. Winter in Rabh., "Krypt. Flor.," vol. i. p. 186.

On *Oxyria reniformis.*

August.

The apex of the spores has often an irregular outline.

### Puccinia hydrocotyles. (Link.)

*Uredospores*—Sori small, scattered, epiphyllous, surrounded by the ruptured epidermis, rarely confluent, sometimes circinating round a central larger sorus. Spores irregularly oval, subpyriform or subglobose, pale brown, finely echinulate, 20–30$\mu$ in diameter.

*Teleutospores*—Few, mixed with uredospores. Spores oblong or oval, slightly constricted, upper cell rounded at the apex, generally larger, smooth, brown, 40–45 × 20–25$\mu$. Pedicels rather long, hyaline.

*Synonyms.*

*Cæoma hydrocotyles.* Link., "Sp. Plant.," vol. ii. p. 22.

*Trichobasis hydrocotyles*, Cooke. Cooke, "Micro. Fungi," 4th edit., p. 225, t. 8, figs. 168, 169; "Hdbk.," p. 530.

*Exsiccati.*

Cooke, i. 69; ii. 59; "L. F.," 44. Vize, "Micro. Fungi Brit.," 228.

On *Hydrocotyle vulgaris.* Epping Forest.

July to October.

**Puccinia sonchi.** Rob.

*Uredospores*—Sori small, brown, mostly hypophyllous, at first bullate, then naked, roundish, pulverulent. Spores roundish or ovate, verrucose, 23–35 × 16–21μ.

*Teleutospores*—Sori black, compact, roundish, surrounded by dark brown, clavate paraphyses, 80–100 × 12–15μ. Spores oblong or ovate, constricted, rounded at both ends, smooth, brown, 35–60 × 20–30μ. Mesospores numerous, of similar form, but unicellular, apex more distinctly thickened (45–55μ). Pedicels long, persistent.

*Synonym.*

*Puccinia sonchi*, Rob. Winter in Rabh., "Krypt. Flor.," vol. i. p. 189. Desmaz., *Ann. Sc. Nat.*, 3rd series, vol. ii. p. 274. Grove, *Science Gossip*, 1885, vol. xxi. p. 9, figs. 6–9.

On *Sonchus arvensis;* Mr. H. Hawkes.
October.

**Puccinia lychnidearum.** Link.

*Uredospores*—Sori on pallid yellowish spots, subrotund, flat, scattered, often arranged in a circular manner, hypophyllous, cinnamon-brown. Spores subglobose or ovate, pale brown, echinulate, 18–20 × 23–30μ.

*Teleutospores*—Sori dark brown, on pale spots, oblong or roundish, generally circinating, soon naked, but persistent. Spores pale brown, ovate, clavate, or subfusiform, apex not thickened, smooth, 35–40 × 15–18μ. Pedicels long, persistent, hyaline.

*Synonyms.*

*Puccinia lychnidearum*, Link. Linn., "Sp. Plant.," vol. ii. p. 80. Berk., "Eng. Flor.," vol. v. p. 367. Cooke, "Hdbk.," p. 505; "Micro. Fungi," 4th edit., p. 210.

*Trichobasis lychnidearum*, Lév. Cooke, "Hdbk.," p. 505; "Micro. Fungi," 4th edit., p. 224.

*Exsiccati.*

Berk., 224. Cooke, i. 47; ii. 129. Vize, "Micro. Fungi Brit.," 31, 130; "Micro. Fungi," 125.

On *Lychnis diurna.*
Uredospores, April to June ; teleutospores, July to September.

## IV. PUCCINIOPSIS. Schröt.

Æcidiospores and teleutospores on the same host-plant. Not infrequently a few isolated uredospores are found concealed between the teleutospores, but they do not form proper uredospore sori.

### Puccinia liliacearum. Duby.

*Æcidiospores*—Pseudoperidia scattered on the leaves, deeply imbedded, opening above by a small orifice. Spores rounded, polygonal, isodiametric, rarely elliptical, orange-yellow, 16–23 × 14–17μ.

*Teleutospores*—Sori scattered in groups, imbedded in the leaf-tissue, covered by the epidermis, which becomes at length ruptured by a small rift above. Spores oblong, slightly or not at all constricted, apex not thickened, rounded, but more often attenuated and surmounted by a conical point, base attenuated, smooth, pale brown, 40–70 × 22–35μ. Pedicels rather long, thick, colourless.

*Synonym.*

*Puccinia liliacearum.* Duby, "Bot. Gall.," vol. ii. p. 891. Winter in Rabh., "Krypt. Flor.," vol. i. p. 194. Schröt., "Krypt. Flor. Schl.," vol. iii. p. 342.

On *Ornithogalum umbellatum.*
Near Carlisle ; Rev. Hilderic Friend.
April and May.

BIOLOGY.—Schröter remarks that the æcidia are always few, and often altogether absent, but that the spermogonia are abundant, especially on the ends of the affected leaves.

### Puccinia tragopogi. (Pers.)

*Æcidiospores*—Pseudoperidia on the whole plant—leaves, stems, bracts, receptacles—shortly cylindrical, at first mammæform, with whitish torn edges. Spores rounded, verrucose, orange-yellow, 18–27$\mu$, sometimes as much as 35$\mu$ long. Mycelium diffused throughout the host-plant.

*Teleutospores*—Sori brown, few, small, scattered, elliptical or elongate, long covered by the epidermis. Spores broadly oval, often almost globose, slightly constricted, apex not thickened, thickly verrucose, brown, 26–48 × 20–35$\mu$. Pedicels short, colourless, deciduous. Mycelium localized.

*Synonyms.*

*Puccinia tragopogi* (Pers.). Winter in Rabh., "Krypt. Flor.," vol. i. p. 209.

*Æcidium tragopogi.* Pers., "Syn.," p. 211. Berk., "Eng. Flor.," vol. v. p. 370. Sow., t. 397, fig. 2. Cooke, "Hdbk.," p. 537; "Micro. Fungi," 4th edit., p. 195, t. 1, figs. 1–3.

*Æcidium cichoracearum*, D. C. Johnst., "Flor. Berw.," vol. ii. p. 205.

*Puccinia sparsa.* Cooke, "Hdbk.," p. 498; "Micro. Fungi," 4th edit., p. 205.

*Puccinia syngenesiarum*, Link. Johnst., "Flor. Berw.," vol. ii. p. 197, in part.

*Exsiccati.*

Cooke, i. 5, 330; ii. 79; "L. F.," 51. Vize, "Micro. Fungi," 158; "Micro. Fungi Brit.," 133.

On *Tragopogon pratensis.*
April to September.

BIOLOGY.—The life-history of this species was first worked out by De Bary ("Champ. parasit.," *Ann. Sc. Nat.*, 1861, 4th series, t. xx., pp. 76 and 87–88 of the reprint). He found the æcidiospores to have a colourless epispore, with three germ-pores; that when they were sown upon the young leaves of *Tragopogon pratensis* and *porrifolius* they gave rise to the teleutospores, with a localized mycelium at the place where the æcidiospores were placed. Mixed with the teleutospores are a very few uredospores. The teleutospores give rise, when placed upon young plants, to a mycelium, which pervades the whole of the plant. The infected plants produce the æcidiospores on the leaves, involucre, and receptacle. The mycelium is to be found in the upper part of the root-stock.

I have cultivated this fungus for several years. Sometimes the young seedlings will produce a crop of æcidia in the autumn, but as a rule they do not do so until the following spring, after the mycelium has hybernated in them during the winter. I have sown the seeds from æcidiiferous plants, in which the æcidiospores were produced upon the receptacle; most of these seeds failed to germinate, but those which did produced healthy plants. I do not find the spermogonia at all frequently; as a rule they are absent.

### **Puccinia smyrnii.** Corda.

*Æcidiospores*—Pseudoperidia hypophyllous or cauline, yellow, at first closed hemispherical, then bursting at the apex and becoming cupulate, edges not torn. Spermogonia epiphyllous. Spores oval or oblong, golden yellow, finely echinulate, 25–30 × 10–15$\mu$.

*Teleutospores*—Sori small, pulverulent, scattered, black, hypophyllous. Spores ovoid obtuse or subpyriform, constriction slight or absent, upper cell generally the larger, outline irregular from the epispore being covered by very large discrete tubercles, rich brown, 30–50 × 20–25$\mu$. Pedicels hyaline, short.

*Synonym.*

*Puccinia smyrnii.* Corda, vol. i. fig. 67. B. and Br., *Ann. Nat. Hist.*, No. 469. Cooke, "Hdbk.," p. 503; "Micro. Fungi," 4th edit., p. 209, t. 3, figs. 55, 56.

*Exsiccati.*

Cooke, i. 320; ii. 440. Vize, "Fungi Brit.," 22; "Micro. Fungi Brit.," 562, 563.

On *Smyrnium olusatrum*.
Æcidiospores, May and June.
Teleutospores, June and July.

## V. MICROPUCCINIA. Schröt.

Having teleutospores only, which do not germinate until after a period of rest.

### **Puccinia betonicæ.** (Alb. and Schw.)

*Teleutospores*—Sori dark brown, numerous, small, roundish, confluent, covering the whole surface of the leaf. Spores broadly elliptical or ovate, generally rounded at both ends, apices surmounted by a brown wart-like papilla, constriction slight, smooth, yellow-brown, 28–45 × 15–25μ. Pedicels hyaline, deciduous.

*Synonyms.*

*Puccinia betonicæ* (Alb. and Schw.). Winter in Rabh., "Krypt. Flor.," vol. i. p. 172.

*Puccinia anemones, β betonicæ.* Alb. and Schw., "Consp.," p. 131.

*Puccinia betonicæ*, D. C. Berk., "Eng. Flor.," vol. v. p. 364. Grev., "Flor. Edin.," p. 431. Cooke, "Hdbk.," p. 477; "Micro. Fungi," 4th edit., p. 205.

*Exsiccati.*

Berk., 218. Cooke, i. 108; ii. 532. Vize, "Fungi Brit.," 16.

On *Betonica officinalis.*

BIOLOGY.—The mycelium appears to be perennial. Mr. W. B. Grove figures in the *Gardener's Chronicle* (August 8, 1885) an abnormal condition of the teleutospores in which they are variously septate and distorted.

**Puccinia campanulæ.** Carm.

*Teleutospores*—Sori brown, small, rounded, long covered by the epidermis, often circinate. Spores elliptical or ovate, summits thickened, conical, with a flat or wart-like (usually brown) incrassation, constricted in the middle, smooth, pale brown, 26–45 × 12–21μ. Pedicels very deciduous.

*Synonym.*

*Puccinia campanulæ*, Carm. Berk., "Eng. Flor.," vol. v. p. 365. Cooke, "Hdbk.," p. 498; "Micro. Fungi," 4th edit., p. 205. Winter in Rabh., "Krypt. Flor.," vol. i. p. 173.

*Exsiccati.*

Cooke, i. 109.

On *Campanula rapunculus* and *rotundifolia.*
June to August.

## Puccinia schneideri. Schröt.

*Teleutospores*—Mycelium pervading the stems and leaves of the affected plants, causing an increase in length of the former, and more or less crumpling of the latter. Sori small, black-brown, for some time covered by the epidermis, usually elongated, pulverulent. Spores elliptical, rounded at both ends, apex thickened, medial constriction pronounced, smooth, chestnut-brown, 25–28 × 15–18$\mu$. Pedicels deciduous, colourless.

*Synonyms.*

*Puccinia schneideri.* Schröt., "Krypt. Flor. Schl.," vol. iii. p. 344.

*Puccinia caulincola.* Schneider, 48, *Jahresber d. Schles. Ges.*, 1870, p. 120.

On *Thymus serpyllum.*
June to October.
Links., Aberdeen ; Prof. J. W. H. Trail.

BIOLOGY.—The presence of the mycelium in the affected stems causes them to assume a more erect habit of growth, so that they can be distinguished from the healthy plants by the naked eye.

## Puccinia ægopodii. (Schum.)

*Teleutospores*—Sori on brown spots, on the leaves, small, roundish, crowded, on the petioles generally larger, at first yellow, then brown. Spores elliptical, oval, or subpyriform, constriction little or none, apex generally surmounted by a paler papilla, dark chestnut-brown, 30–40 × 15–23$\mu$. Pedicels colourless, deciduous.

*Synonyms.*

*Puccinia ægopodii* (Schum.). Winter in Rabh., "Krypt. Flor.," vol. i. p. 174.

*Uredo ægopodii.* Schum., "Enum. Plant Sæll.," vol. ii. p. 233.

*Puccinia ægopodii,* Link. Berk., "Eng. Flor.," vol. v. p. 366. Cooke, "Hdbk.," p. 502 ; "Micro. Fungi," 4th edit., p. 208. Grev., "Flor. Edin.," p. 429.

*Exsiccati.*

Cooke, i. 540; ii. 439. Vize, "Micro. Fungi Brit.," 34, 426.

On *Ægopodium podagraria.*

April to August.

BIOLOGY.—The presence of the mycelium in the stems and midribs causes them to be much swollen and distorted.

**Puccinia epilobii.** D. C.

*Teleutospores*—Sori small, roundish, rather crowded, but seldom confluent, soon naked, surrounded by the torn epidermis, dark brown. Spores oblong or elliptical, often irregular, rounded at both ends, not thickened above, much constricted, brown, very minutely verrucose, 27–40 × 17–25$\mu$. Pedicels hyaline, deciduous, rather short.

*Synonym.*

*Puccinia epilobii.* D. C., "Flore franç.," vol. vi. p. 61. Grev., "Flor. Edin.," p. 431. Berk., "Eng. Flor.," vol. v. p. 368. Cooke, "Hdbk.," p. 506; "Micro. Fungi," 4th edit., p. 211. Johanson, *Botan. Centralblatt*, bd. 28 (1886), p. 395.

*Exsiccati.*

Berk., 348.

On *Epilobium palustre.*

**Puccinia asarina.** Kze.

*Teleutospores*—Sori small, brown, roundish, long covered by the epidermis, often circinate, at length naked. Spores ovate, elliptical, or fusiform, constriction slight, surmounted by a conical paler point, apex scarcely, base slightly (if at all) attenuated, pale brown, 30–40 × 14–24$\mu$. Pedicels very long, deciduous.

*Synonym.*

*Puccinia asarina.* Kze. and Schm., "Mykol. Hefte," i. p. 70. Cooke, "Hdbk.," p. 504. Winter in Rabh, "Krypt. Flor.," vol. i. p. 172.

*Exsiccati.*

Cooke, i. 10.

On *Asarum europæum.*
June to October.
The sori are generally confluent in large circular patches.

**Puccinia paliformis.** Fckl.

*Teleutospores*—Sori small, thick, pulvinate, round or oblong, black. Spores clavate or fusiform, central constriction slight, base wedge-shaped, apex much thickened, generally truncate, rarely rounded, occasionally conical, brown, 23–52μ × 10–16μ. Pedicels long, firm.

*Synonym.*

*Puccinia paliformis.* Fckl., "Symb. Myc.," p. 59. Winter in Rabh., "Krypt. Flor.," vol. i. p. 224.

On *Kœleria cristata.*
September and October.
Prof. J. W. H. Trail, near Aberdeen.

**Puccinia virgaureæ.** (D. C.).

*Teleutospores*—Sori black, shining, punctiform, generally in large numbers on orbicular, pallid, yellowish spots, crowded in the centre, scattered towards the periphery. Spores fusiform, pyriform, or clavate, constriction slight or absent, base attenuated, apex thickened, rounded or truncate, often with a central oblique point, hooded or conically attenuated, smooth, pale brown, 30–56 × 12–20μ. Sori surrounded by a thick bed of dark-brown paraphyses. Pedicels hyaline.

*Synonyms.*

*Puccinia virgaureæ* (D. C.). Winter in Rabh, "Krypt. Flor.," vol. i. p. 173.

*Xyloma virgaureæ.* D. C., "Syn.," No. 821

*Puccinia virgaureæ*, Lib. Cooke, "Hdbk.," p. 500; "Micro. Fungi," 4th edit., p. 206.

*Exsiccati.*

Cooke, i. 45.

On *Solidago virgaurea.*
August and September.

**Puccinia andersoni.** B. and Br.

*Teleutospores*—Spots epiphyllous, orbicular, surrounded by a brown border. Sori hypophyllous, compact, minute, but crowded into large circular patches, often almost concealed by the pubescence of the leaf, dark violet. Spores oblong, fusiform, or clavate, constricted, smooth, brown, upper cell often the darker, apex rounded or conically thickened (8 to 10$\mu$), 45–55 × 20$\mu$. Pedicels long, stout, pale brown, 60–70 × 8–9$\mu$.

*Synonym.*

*Puccinia andersoni.* B. and Br., *Ann. Nat. Hist.*, No. 1464. Cooke, "Micro. Fungi," 4th edit., p. 204.

On *Carduus heterophyllus.*
September and October.
Den of Airlie, Scotland. Yorkshire; Mr. Soppitt.

Biology.—A very striking species. The sori are crowded together in large round patches, often from 1 to 1.5 cm. in diameter; dark violet in colour, and are without paraphyses.

**Puccinia umbilici.** Guép.

*Teleutospores*—Sori seated upon pallid spots, round, often circinating, convex, compact, then pulverulent, at length confluent in large orbicular patches. Spores dark brown, not constricted, smooth, globose or ovate, 28–30 × 24–26$\mu$. Pedicels short, hyaline.

*Synonym.*

*Puccinia umbilici*, Guép. Duby, "Bot. Gall.," p. 890. B. and Br., *Ann. Nat. Hist.*, No. 470. Cooke, "Hdbk.," p. 505; "Micro. Fungi," 4th edit., p. 211, t. iv. figs. 80, 81.

*Exsiccati.*

Berk., 329. Cooke, i. 48; ii. 132. Vize, "Fungi Brit.," 29.

On *Cotyledon umbilicus.*

May and June.

BIOLOGY.—The small sori are surrounded by a pseudoperidium, from which the spores eventually fall out, leaving a honeycomb-like matrix.

**Puccinia fusca.** (Relhan.)

*Teleutospores*—Sori black-brown, scattered equally over the whole surface of the leaves, small, round, pulverulent, often confluent. Spores formed of two almost spherical cells, which are flattened at their point of contact, hence the spores are strongly constricted and easily fall apart, thickly covered with large warts, brown, 30–50 × 15–24$\mu$. Pedicels stout, hyaline. Mycelium perennial.

*Synonyms.*

*Puccinia fusca* (Relh.). Winter in Rabh., "Krypt. Flor.," vol. i. p. 199.

*Æcidium fuscum.* Relh., "Flor. Cantab. Suppl.," ii.

*Puccinia anemones.* Pers., "Syn.," p. 226. Berk., "Eng. Flor.," vol. v. p. 367. Cooke, "Hdbk.," p. 503; "Micro. Fungi," 4th edit., p. 209, t. iv. figs. 64, 65.

*Exsiccati.*

Baxter, 82. Berk., 222. Cooke, i. 43; ii. 530; "L. F.," 18. Vize, "Fungi Brit.," 26.

On *Anemone nemorosa.*

April to July.

BIOLOGY.—The mycelium is perennial, and was found by De Bary in the upper part of the rhizome. The teleutospores are accompanied by spermogonia. The æcidiospores (*Æ. leucospermum*), which continental botanists regard as belonging to this Puccinia, are much less common than the teleutospores, and further experimental culture is desirable as to the life-history of this well-known and widely distributed species.

### Puccinia bunii. (D. C.)

*Teleutospores*—Sori brownish black, roundish, often cauline, elongated, usually confluent, long covered by the epidermis. Spores elliptical or subpyriform, central constriction slight or absent, punctate, brown, 25–45 × 15–23μ. Pedicels rather long, deciduous.

*Synonyms.*

*Puccinia bunii* (D. C.). Winter in Rabh., "Krypt. Flor.," vol. i. p. 197.

*Puccinia bulbocastani*, Fckl. Cooke, "Micro. Fungi," 4th edit., p. 209.

*Puccinia tumida.* Grev., "Flor. Edin.," p. 430.

*Exsiccati.*

Cooke, i. 39; ii. 327. Vize, "Micro. Fungi Brit.," 216.

On *Conopodium denudatum* (*Bunium flexuosum*).

Biology.—The mycelium is perennial. I have cultivated in my garden for four years plants of Bunium in a flower-pot, which are annually affected with the fungus, but they have never produced the æcidiospores. Although this fungus is very common around King's Lynn, yet I have never found the *Æcidium bunii*, which continental botanists consider to be connected with it.

### Puccinia thalictri. Chevall.

*Teleutospores*—Sori dark brown, very crowded, covering the whole surface of the leaf, soon pulverulent, small, rounded. Spores deeply constricted, both cells rounded, flattened at their point of contact. Epispore covered with pointed warts, dark brown, 25–50 × 15–25μ. Pedicels long, deciduous.

*Synonym.*

*Puccinia thalictri.* Chevall., "Flor. Paris," vol. i. p. 417. Winter in Rabh., "Krypt. Flor.," vol. i. p. 177.

On *Thalictrum flavum*, *minus* var. *montanum.*
September and October.
Rannoch, near Perth.

BIOLOGY.—This plant appears to me to have a perennial mycelium, but I have not had the opportunity of cultivating it.

### Puccinia fergussoni. B. and Br.

*Teleutospores*—Sori on rounded yellowish or paler spots, crowded in orbicular clusters, long covered by the epidermis. Spores oblong, slightly constricted, apex often obtusely and conically apiculate, smooth, yellow-brown, 23–38 × 13–20$\mu$. Pedicels short, deciduous.

*Synonym.*

*Puccinia fergussoni*, B. and Br. Cooke, "Micro. Fungi," 4th edit., p. 210. Winter in Rabh., "Krypt. Flor.," vol. i. p. 176.

On *Viola palustris*.
Scotland and Wales.

### Puccinia rhodiolæ. B. and Br.

*Teleutospores*—Spots orbicular, brown. Sori minute, crowded. Spores shortly pedicellate, articulations depressed, sometimes spuriously subdivided.

*Synonym.*

*Puccinia rhodiolæ*, Berk. Gardiner's "Flora of Forfar.," p. 296. B. and Br., *Ann. Nat. Hist.*, No. 468. Cooke, "Hdbk.," p. 506; "Micro. Fungi," 4th edit., p. 211.

On *Sedum rhodiola*.
Summer.

### Puccinia adoxæ. D. C.

*Teleutospores*—Sori black, small, rounded, circinate on the leaves, scattered on the swollen stems, confluent, pulverulent. Spores oblong, elliptical, or ovate, extremities attenuated, central constriction very slight or absent, apiculate above, smooth, brown, 25–35 × 15–20$\mu$. Pedicels hyaline, as long as, or longer than, the spores, deciduous.

*Synonyms.*

*Puccinia adoxæ*. D. C., "Flore franç.," vol. ii. p. 220. Winter in Rabh., "Krypt. Flor.," vol. i. p. 211, in part. Cooke, "Micro. Fungi," 4th edit., p. 209. Grev., "Flor. Edin.," p. 432.

*Puccinia saxifragarum*, Schlecht. Berk., "Eng. Flor.," vol. v. p. 367. Cooke, "Hdbk.," p. 506. W. G. Smith, *Gard. Chron.*, July 4, 1885, p. 21, fig. 7.

*Exsiccati.*

Cooke, ii. 531. Vize, "Fungi Brit.," 117; "Micro. Fungi Brit.," 217.

On *Adoxa moschatellina.*
March to May.

BIOLOGY.—The mycelium of this fungus is perennial; affected plants, cultivated by Professor Trail at Aberdeen, and by myself at King's Lynn, year after year, produced only teleutospores. Mr. Soppitt placed some over-wintered teleutospores in active germination on healthy plants, in March, 1888; in ten days the teleutospores had reproduced themselves without the intervention of either uredospores or æcidiospores. He had previously satisfied himself of the distinctness of this species from *P. albescens* by his observations of the fungi as they occurred in a state of nature in Yorkshire.

## **Puccinia saxifragæ.** Schlecht.

*Teleutospores*—Sori amphigenous, dark chestnut or blackish brown, on numerous discoloured spots, irregular, soon pulverulent, confluent. Spores elliptical or ovate, slightly constricted, apex surmounted with a conical or wart-like pale papilla, faintly reticulate, yellow-brown, 26–45 × 14–20$\mu$. Pedicels short, deciduous.

*Synonym.*

*Puccinia saxifragæ*. Schlecht., "Flor. Berol.," vol. ii. p. 134. Cooke, "Micro. Fungi," 4th edit., p. 209. Winter in Rabh., "Krypt. Flor.," vol. i. p. 174.

On *Saxifraga granulata, stellaris.*
April to August.

**Puccinia senecionis.** Lib.

*Teleutospores*—Sori small, crowded, chestnut-brown, at first hemispherical, then depressed in the centre, becoming perforate, at length pulverulent. Spores irregular in form, subovoid or elliptical, lower cell sometimes attenuated, upper generally rounded, but often pointed, brown, smooth or granular, 40–50 × 18–28μ. Pedicels short, hyaline, deciduous.

*Synonym.*

*Puccinia senecionis*, Lib. Cooke, "Micro. Fungi," 4th edit., p. 207.

*Exsiccati.*

Cooke, i. 37; ii. 236. Vize, "Fungi Brit.," 21.

On *Senecio aquaticus*.
September.

BIOLOGY.—The mycelium has a great tendency to follow the venation. The sori are enclosed in an indistinct pseudoperidium, which remains as a honeycomb-like matrix in the centre of the sori-clusters after the spores have fallen off. The sori are brown, not black as in *P. glomerata*.

**Puccinia glomerata.** Grev.

*Teleutospores*—Sori small, black, circinating, crowded, soon pulverulent, in roundish or elongated clusters. Spores elliptical, elongated, or irregular, lower cell often smaller, constriction slight, upper cell rounded or attenuated, usually surmounted by a colourless papilla, smooth or granular when old, dark brown, 30–40 × 20–25μ.

*Synonym.*

*Puccinia glomerata*. Grev., "Flor. Edin." Berk., "Eng. Flor.," vol. v. p. 365. Cooke, "Hdbk.," p. 500; "Micro. Fungi," 4th edit., p. 206.

*Exsiccati.*

Berk., 220.

On *Senecio jacobæa*.
August to November.

BIOLOGY.—The presence of the mycelium causes elongated fusiform swellings on the petioles and midribs. It differs from *P. senecionis* in the darker colour of the sori, which are almost black.

### VI. LEPTOPUCCINIA. Schröt.

Having teleutospores only, which germinate upon the living plant as soon as they have arrived at maturity. The germ-tubes of the promycelial spores enter the host-plant either through the stomata or between the epidermal cells.

### Puccinia arenariæ. (Schum.)

*Teleutospores*—Sori compact, pulvinate, roundish, scattered, often circinate. Spores broadly fusiform or pyriform, summits pointed or rounded, often thickened, base rounded or attenuated, slightly constricted, smooth, pale yellowish brown, 30–50 × 10–20$\mu$. Pedicels hyaline, colourless, as long as the spores.

*Synonyms.*

*Puccinia arenariæ* (Schum.). Winter in Rabh., "Krypt. Flor.," vol. i. p. 169.

*Uredo arenariæ.* Schum., "Enum. Plant. Sæll.," vol. ii. p. 232,

*Puccinia lychnidearum*, Link. Berk., "Eng. Flor.," vol. v. p. 367, in part. Cooke, "Hdbk.," p. 505, in part; "Micro. Fungi," 4th edit., p. 210, in part.

*Puccinia mœhringiæ*, Fckl. Cooke, "Micro. Fungi," 4th edit., p. 210.

*Puccinia stellariæ*, Duby. Vize, Exs.

*Puccinia saginæ*, Fckl. Vize, Exs.

*Puccinia dianthi*, D. C. Vize, Exs.

*Puccinia spergulæ*, D. C. Cooke, "Micro. Fungi," 4th edit., p. 210.

*Exsiccati.*

Cooke, i. 8, 297; ii. 129, 130, 321, 432. Vize, "Micro. Fungi," 35, 125; "Micro. Fungi Brit.," 32, 35.

On *Dianthus barbatus, Mœhringia trinervia, Stellaria media, uliginosa, holostea, Sagina nodosa, procumbens, Spergula arvensis.*

May to November.

BIOLOGY.—Cornu found that the teleutospores from *M. trinervia* reproduced themselves in twenty-nine days on *Alsine media* and

*Stellaria holostea.* De Bary found *P. dianthi* from *D. barbatus* would not reproduce itself on *Silene inflata* and *Lychnis diurna*, and that the germ-tube of the promycelial spores entered the host-plant through its stomata.

### **Puccinia chrysosplenii.** Grev.

*Teleutospores*—Sori compact, very small, scattered or collected into concentric groups, roundish, often confluent. Spores fusiform or clavate, summits much thickened, conical or rounded, base attenuated, very slightly or not at all constricted in the middle, smooth, pale yellow-brown, 28–45 × 10–16$\mu$. Pedicels long, firm.

*Synonym.*

*Puccinia chrysosplenii.* Grev., "Flor. Edin.," p. 429. Berk., "Eng., Flor.," vol. v. p. 367. Cooke, "Hdbk.," p. 506; "Micro. Fungi," 4th edit., p. 210. Winter in Rabh., "Krypt. Flor.," vol. i. p. 165

*Exsiccati.*

Cooke, ii. 322; Vize, "Micro. Fungi.," 119; "Micro. Fungi Brit.," 218.

On *Chrysosplenium alternifolium* and *oppositifolium*.
April and May.

### **Puccinia veronicæ.** (Schum.)

*Teleutospores*—Sori at first yellowish, then brown, generally circinating, pulverulent. Spores fusiform, apex rounded, slightly constricted in the middle, pale yellowish, smooth, 28–36 × 10–12$\mu$. Pedicels as a rule not quite so long as the spores.

*Synonyms.*

*Puccinia veronicæ* (Schum.). Schröt., "Krypt. Flor. Schl.," vol. iii. p. 347. Var. a *fragilipes*—Winter in Rabh., "Krypt. Flor.," vol. i. p. 166.

*Uredo veronicæ.* Schum., "Enum. Plant. Sæll.," vol. ii. p. 288.

On *Veronica montana* and *alpina*.
June to October.

**Puccinia valantiæ.** Pers.

*Teleutospores*—Sori compact, round, pulvinate, on the stems elongated and often confluent, at first yellow, then brown, then almost black. Spores fusiform, slightly constricted, summits much thickened, smooth, pale yellow-brown, 35–65 × 12–17μ. Pedicels firm, long.

*Synonyms.*

*Puccinia valantiæ*, Pers. "Obs. Myc.," vol. ii. p. 25. Winter in Rabh., "Krypt. Flor.," vol. i. p. 167. Berk., "Eng. Flor.," vol. v. p. 366. Cooke, "Hdbk.," p. 500; "Micro. Fungi," 4th edit., p. 207. Grev., "Flor. Edin.," p. 432, in part.

*Puccinia acuminata*, Fckl. Cooke, "Micro. Fungi," 4th edit., p. 207.

*Puccinia galii-cruciati.* Johnst., "Flor. Berw.," vol. ii. p. 196.

*Exsiccati.*

Cooke, i. 38; ii. 437; "L. F.," 14. Vize, "Fungi Brit.," 24; "Micro. Fungi Brit.," 27, 213.

On *Galium cruciata*, *saxatile*.
June to September.

**Puccinia malvacearum.** Mont.

*Teleutospores*—Sori greyish brown, compact, round, pulvinate, elongate on the stems, scattered, seldom confluent, pale reddish brown. Spores fusiform, attenuated at both extremities, apex sometimes rounded, constriction slight or absent, apical thickening slight, smooth, yellow-brown, 35–75 × 15–25μ. Pedicels firm, long, sometimes measuring 120μ.

*Synonym.*

*Puccinia malvacearum* (Mont.). Gay, "Hist. Chili," vol. viii. p. 43. Cooke, "Micro. Fungi," 4th edit., p. 205. Winter in Rabh., "Krypt. Flor.," vol. i. p. 168.

*Exsiccati.*

Cooke, i. 630; ii. 137. Vize, "Fungi Brit.," 27.

On *Malva moschata*, *sylvestris*, *rotundifolia*, *Althæa rosea*.

BIOLOGY.—The sori occur on yellow spots on the leaves, which often, as the leaves expand, fall out and leave circular perforations. On the stems the sori are elongated, often with pointed extremities; they fall off as the stems grow, and leave elliptical wounds, at the bottom of which the woody parts of the stem are exposed. The sori often occur on the calyces and on the young fruit. The teleutospores readily germinate in water. I have infected cotton plants with the promycelial spores, but obtained no result.

This fungus was first described by Montagne in 1852, from a specimen from Chili. I have in my herbarium a specimen from Melbourne, Australia, gathered in 1865. It is said, however, to have been found in Algeria at an early date on *Lavatera cretica.* In 1869, it appeared in Spain.* In April, 1873, Durieu found it near Bordeaux, and in the same month Decaisne at Montpellier. In June and July, it appeared in England, and did great damage to the hollyhocks. It was found by Messrs. Roper, Hussey, Paxton, and Parfitt at Exeter, Salisbury, Chichester, and soon after by myself at King's Lynn. In October, Schröter found it in Bavaria; in January, 1874, Beltrani-Pisani met with it at Rome; and in April it was seen at Panisperma, in June at Erlangen, in July at Dusseldorf, and in the course of a year or two spread all over Europe from Athens to Denmark and Finland. So virulently did it attack the hollyhocks that for several years they almost disappeared from our gardens. It seems to have spent its energy, as these plants are again beginning to be cultivated. Kellerman, in 1874, pointed out that germ-tubes of the promycelial spores insinuated themselves between the epidermal cells, and he described the haustoria on the mycelium. He found that on plants cultivated indoors spore-formation continued throughout the winter.

## **Puccinia circææ.** Pers.

*Teleutospores*—Sori compact, pulvinate, round, at first yellowish, then brown, often circinate. Spores two kinds, similar in form, but differing in colour, those formed earlier in the year being paler, those formed later being a darker brown, fusiform, with thick (6 or $7\mu$) conical apices, attenuated towards the stem, very slightly constricted in the middle, $23-40 \times 10-14\mu$. Pedicels hyaline.

*Synonym.*

*Puccinia circææ*, Pers. "Disp. Meth.," p. 39. Winter in

* For an account of the spread of this fungus in Europe, see Egon Ihne, "Studien zur Pflanzengeographie." 1880.

Rabh., "Krypt. Flor.," vol. i. p. 168. Berk., "Eng. Flor.," vol. v. p. 368. Cooke, "Hdbk.," p. 507; "Micro. Fungi," 4th edit., p. 211.

*Exsiccati.*

Berk., 319. Cooke, i. 50; ii. 131. Vize, "Fungi Brit.," 16, 120; "Micro. Fungi Brit.," 125.

On *Circæa lutetiana.*

BIOLOGY.—Schröter pointed out the two spore-forms, and that, while the paler teleutospores germinated at once on the living host-plant, the darker ones only did so after a period of rest. The resting spores are often produced on the stems, petioles, and midribs.

### Puccinia veronicarum. D. C.

*Teleutospores*—Sori pulvinate, compact, firm, dark chestnut-brown, often greyish. Spores fusiform, surmounted with a thickened conical point, constricted in the middle, 35–40 × 15–20$\mu$. Epispore smooth, thick, pale or dark brown, 6–9$\mu$ thick at the apex of the spore, colourless above. Pedicels firm, yellowish, 24–46$\mu$ long (Schröt.).

*Synonyms.*

*Puccinia veronicarum.* D. C., "Flore franç.," vol. ii. p. 597. Pers., "Syn.," p. 45. Winter in Rabh., "Krypt. Flor.," vol. i. p. 166, in part. Schröt., "Krypt. Flor. Schl.," vol. iii. p. 348. Johnst., "Flor. Berw.," vol. ii. p. 194. Berk., "Eng. Flor.," vol. v. p. 364. Cooke, "Hdbk.," p. 496; "Micro. Fungi," 4th edit., p. 204.

*Puccinia veronicæ, forma β persistens.* Körn.

*Exsiccati.*

Cooke, ii. 112; "L. F.," 7. Vize, "Fungi Brit.," 28; "Micro. Fungi Brit.," 124.

On various species of Veronicæ.
July to September.

### Puccinia glechomatis. D. C.

*Teleutospores*—Sori yellowish, then brown, becoming darker, compact, round, pulvinate, single or circinating, causing elongated swellings on the stems, petioles, and nerves. Spores elliptical or fusiform, central constriction slight, summits surmounted by a pointed erect or oblique papilla (8–10$\mu$ high), smooth, chestnut-brown, 30–50 × 16–24$\mu$. Pedicels very long, rather firm.

*Synonyms.*

*Puccinia glechomatis.* D. C., "Encycl.," vol. viii. p. 245. Berk., "Eng. Flor.," vol. v. p. 364. Cooke, "Hdbk.," p. 496; "Micro. Fungi," 4th edit., p. 204, t. iv. figs. 73, 74.

*Puccinia verrucosa* (Schultz). Winter in Rabh., "Krypt. Flor.," vol. i. p. 166.

*Exsiccati.*

Cooke, i. 29; ii. 438, 635; "L. F.," 8. Vize, "Fungi Brit.," 114; "Micro. Fungi Brit.," 126.

On *Glechoma hederacea.* Prof. J. W. H. Trail has met with this species (?) on *Prunella vulgaris,* on Ben Lawers, in September.

June to October.

**Puccinia asteris.** Duby.

*Teleutospores*—Sori rather large, confluent into pulvinate masses, amphigenous, brown, with a greyish tinge. Spores cylindrical, fusiform, or clavate, smooth, brown, apex rounded or truncate, rarely attenuated into a conical point, lower cell cuneiform, attenuated below, 45–60 × 20–25$\mu$. Pedicels long, stout.

*Synonyms.*

*Puccinia asteris,* Duby. "Bot. Gall.," vol. ii. p. 888.

*Puccinia tripolii,* Wallr. Cooke, "Micro. Fungi," 4th edit., p. 207.

*Exsiccati.*

Cooke, i. 631; ii. 127. Vize, "Fungi Brit.," 25.

On *Aster tripolium.*

June to September.

**Puccinia millefolii.** Fckl.

*Teleutospores*—Sori small, rounded, but generally irregular and following the structure of the foliage, soon naked, persistent.

Spores ovate or fusiform, apex generally rounded and thickened, often subtruncate, pale brown, smooth, 45–55 × 18–20$\mu$. Pedicels not very long.

*Synonym.*

*Puccinia millefolii.* Fckl., "Symb. Myc.," p. 55. Cooke, "Micro. Fungi," 4th edit., p. 207.

*Exsiccati.*

Cooke, ii. 633. Vize, "Fungi Brit.," 33.

On *Achillea millefolium.*

August to October.

BIOLOGY.—This species is, by both Winter and Schröter, united with *P. asteris;* but I found that, by placing the promycelial spores of *P. millefolii* on *Aster tripolium*, no effect was produced. Neither did a plant of Achillea, which was richly covered by the teleutospores and planted close to two plants of Aster, so that the diseased foliage of the former touched the latter, cause them to become diseased, although they grew together for a period of two months.

## **Puccinia cardui** (nov. sp.).

*Teleutospores*—Sori small, circinating, crowded and confluent in large clusters, 3 or 4 mm. across, hypophyllous, long covered by the epidermis; spots pale on the opposite surface of the leaf. Spores fusiform, subcylindrical, or clavate, markedly constricted, smooth, pale brown, base attenuated, apex generally thickened and rounded, 45–50 × 16–18$\mu$. Pedicels pale brown, persistent, as much as 50$\mu$ long.

*Synonyms.*

*Puccinia syngenesiarum*, Link. Johnst., "Flor. Berw.," vol. ii. p. 97. Berk., "Eng. Flor.," vol. v. p. 365. Cooke, "Hdbk.," p. 499; "Micro. Fungi," 4th edit., p. 206, t. iv. figs. 63, 64.

*Puccinia cirsii*, Fckl. Exs. No. 340 (?).

On *Carduus lanceolatus, crispus.*

August to October.

BIOLOGY.—This species has the appearance of a Leptopuccinia, but it may belong to the previous group. It is clearly not the plant described by Link ("Sp. Plant.," vol. vi. pt. ii. p. 74), which has very short pedicels, and occurred on *Tussilago alpina* and *Centaurea alpina*. It may be Fuckel's *P. cirsii*.

## Puccinia buxi. D. C.

*Teleutospores*—Sori chestnut-brown, compact, hemispherical, cushion-shaped, soon naked, amphigenous. Spores oblong or elliptical, deeply constricted, upper cell obovate, apex rather thickened, lower cell attenuated below, cuneiform, generally longer than the upper, brown, smooth, 55–90 × 25–35$\mu$. Pedicels very long.

*Synonym.*

*Puccinia buxi.* D. C., "Flore franç.," vol. vi. p. 60. Winter in Rabh., "Krypt. Flor.," vol. i. p. 164. Sow., t. 439; Berk., "Eng. Flor.," vol. v. p. 369. Cooke, "Hdbk.," 508; "Micro. Fungi," 4th edit., p. 212.

*Exsiccati.*

Berk., 109. Cooke, i. 52; ii. 140. "L. F.," 23. Vize, "Fungi Brit., 11.

On *Buxus sempervirens.*
April to May.

Biology.—The spores have a tendency to fall in halves at the septum. The sori occur on both surfaces of the leaves, and are accompanied by slight yellowish or brown discolorations. Schröter was unable to get this fungus to reproduce itself by applying the promycelial spores to the foliage of box plants. It seems to me probable that the germ-tubes enter the leaves and give rise to a mycelium which remains in a quiescent state until the following spring. This, however, is only an opinion, and has not been proved by experimental culture.

## Puccinia annularis. (Strauss).

*Teleutospores*—Sori small, compact, round, confluent in subrotund patches, hypogenous, at first greyish brown, then cinnamon, at length brown from the rupture of the epidermis. Spores oblong, slightly constricted in the middle, summits rather strongly thickened, rounded, rarely truncate, sometimes attenuated, smooth, very pale yellowish brown, 30–50 × 15–20$\mu$. Pedicels very long, persistent, hyaline.

*Synonyms.*

*Puccinia annularis* (Strauss). Winter in Rabh., "Krypt. Flor.," vol. i. p. 165.

*Uredo annularis.* Strauss (in "Wetter. Annal.," vol. ii. p. 106).

*Puccinia scorodiniæ*, Link. Cooke, "Hdbk.," p. 497; "Micro. Fungi," 4th edit., p. 205. Berk., "Eng. Flor.," vol. v. p. 364 Johnst., "Flor. Berw.," p. 194.

*Exsiccati.*

Cooke, i. 31; ii. 329; "L. F.," 9. Vize, "Fungi Brit.," 17; "Micro. Fungi Brit.," 123.

On *Teucrium scorodoniæ*.
September and October.

Biology.—The sori occur on the under side of the leaves, on brown concave spots. As I understand Schröter, he considers the production of those spores which germinate at once ceases with the cold weather, and that the spores produced under the influence of a low temperature, as well as those which are found late in the year, surrounded by a circumferential zone of dead leaf-tissue, retain their power of germination until the following year.

## TRIPHRAGMIUM. Link.

Teleutospores separate, pedicellate, composed of three cells placed laterally, which are triangular in form and firmly held together, each cell having a single germ-pore.

### BRACHYTRIPHRAGMIUM.

Having spermogonia, uredospores, and teleutospores.

### Triphragmium ulmariæ. (Schum.).

*Primary uredospores*—Vernal, spermogonia flattish, spermatia about 6$\mu$ long. Sori hypophyllous, very large, pulverulent, mostly on the petioles and venation, causing elongated swellings, which greatly distort the affected leaves by preventing their expansion, more rarely in compact, extended sori on the under surface of the leaves. Spores brilliant orange, finely verrucose, globose, oval, or ovate, pedicellate, 18–24 × 17–22μ.

*Secondary uredospores*—In summer and autumn. Sori hypophyllous, small, round, orange. Spores globose or elliptical, finely echinulate, orange-yellow, 18–30 × 18–28μ.

*Teleutospores*—Sori small, round, black, persistent, but pulverulent, on the leaves and petioles. Spores globose, rough, with obtuse warts, chestnut-brown, 35–50μ in diameter. Pedicels colourless, persistent.

*Synonyms.*

*Triphragmium ulmariæ* (Schum.). Winter in Rabh., "Krypt. Flor.," vol. i. p. 225.

*Uredo ulmariæ.* Schum., "Enum. Plant. Sæll.," vol. ii. p. 227.

*Uredo effusa.* Berk., "Eng. Flor.," vol. v. p. 381. Grev., t. 19.

*Puccinia ulmariæ*, D. C. "Eng. Flor.," vol. v. p. 368. Grev., "Flor. Edin.," p. 433. Johnst., "Flor. Berw.," vol. ii. p. 194.

*Puccinia spirææ.* Purton, "Midl. Flor.," vol. iii. p. 304.

*Uromyces ulmariæ*, Lév. Cooke, "Micro. Fungi," 4th edit., p. 212, t. vii. figs. 147, 148.

*Triphragmium ulmariæ*, Link. Cooke, "Hdbk.," p. 492; "Micro. Fungi," 4th edit., p. 202, t. iii. fig. 48.

*Exsiccati.*

Berk., 343. Cooke, i. 23, 75; ii. 146, 212; "L. F.," 4. 25. Vize, "Fungi Brit.," 136.

On *Spiræa ulmaria.*
May to October.

BIOLOGY.—It is not difficult to find the teleutospores still attached to the foliage in spring. These will germinate very readily if placed in water.

## Triphragmium filipendulæ. (Lasch.)

*Primary uredospores*—Spores oblong or pyriform, often as much as 35μ long.

*Secondary uredospores*—Sori scattered, roundish, at first covered by the epidermis, then surrounded by it, orange. Spores globose or ovate, pedicellate, orange.

*Teleutospores*—Sori scattered, roundish, soon pulverulent, brownish, then black. Spores globose, at first yellowish, then brown, smooth. Pedicels hyaline, rather long.

*Synonyms.*

*Uredo filipendulæ*, Lasch in Klotzsch. Rabh., "Herb. Myc.," i. No. 580.

*Triphragmium filipendulæ*, Pass. "Nuovo Giorn. Bot. Ital.," vol. vii. p. 255. Cooke, "Grevillea," vol. xi. p. 15.

On *Spiræa filipendula.*

Mount Caburn, Lewes, September 14, 1862. Herb. Currey.

## PHRAGMIDIUM. Link.

Teleutospores separate, pedicellate, consisting of from three to ten superimposed cells, the uppermost of which has a single apical germ-pore, the others about four each, placed laterally. Uredospores single; æcidiospores in basipetal chains. The last two spore-forms in pulverulent sori, surrounded by clavate or capitate, hyaline paraphyses.

### Phragmidium fragariastri. (D. C.)

*Æcidiospores*—Spots roundish, especially on the stems and veins, irregular, scattered, often confluent, orange-yellow. Spores globose or elliptical, produced in chains, verrucose, surrounded by clavate, curved paraphyses, orange-yellow, $17–26 \times 14–20\mu$.

*Uredospores*—Sori orange-yellow, small, with clavate paraphyses. Spores globose, ovate, or elliptical, echinulate, orange-yellow, $17–24 \times 14–20\mu$.

*Teleutospores*—Sori brownish black, scattered, small, round. Spores cylindrical, rounded at both ends, three or four celled, brown, coarsely verrucose, especially towards the apex, $46–60 \times 22–26\mu$. Pedicels short, colourless, deciduous, $22\mu$ long.

*Synonyms.*

Schrot., "Krypt. Flor. Schl.," vol. iii. p. 351.

*Puccinia fragariastri*. D. C., "Flore franç.," vol. vi. p. 55.

*Uredo poterii*, Spreng. Berk., "Eng. Flor.," vol. v. p. 385.

*Uredo potentillæ*, D. C. Grev., "Flor. Edin.," p. 438. Johnst., "Flor. Berw.," vol. ii. p. 199.

*Uredo fragariæ.* Purton, "Midl. Flor.," vol. iii. p. 299.

*Puccinia fragariæ.* Purton, "Midl. Flor.," vol. iii. p. 304. Johnst., "Flor. Berw.," vol. ii. p. 193.

*Uredo potentillæ.* Grev., "Flor. Edin.," p. 428; "Scot. Crypt. Flor.," t. 57, in part.

*Aregma acuminatum*, Fries. Berk., "Eng. Flor.," vol. v. p. 358.

*Phragmidium acuminatum*, Fries. Cooke, "Hdbk.," p. 490; "Micro. Fungi," 4th edit., p. 201, t. iii. fig. 32.

*Exsiccati.*

Cooke, i. 19; ii. 211. Vize, "Micro. Fungi Brit.," 109.

On *Potentilla fragariastrum.*
May to October.

**Phragmidium sanguisorbæ.** (D. C.).

*Æcidiospores*, *Uredospores* } Similar to those of *Phragmidium fragariastri.*

*Teleutospores*—Sori punctiform, black. Cells in the spores four or five, very seldom three, the four cells measuring 44–55 × 20–22μ, cylindrical, apex rounded; the fifth cell 66–70μ long, usually with a pointed apex. Epispore dark brown, smooth, or with a few scattered obtuse warts. Pedicels delicate, 22μ long.

*Synonyms.*

Schröt., *loc. cit.*, p. 352.

*Puccinia sanguisorbæ.* D. C., "Flor. franç.," vol. vi. p. 54.

*Cæoma poterii.*, Schl.

*Phragmidium poterii*, Fckl.

*Phragmidium sanguisorbæ* (D. C.). Schröt., *loc. cit.*, p. 352.

On *Poterium sanguisorba.*
June to October.

**Phragmidium potentillæ.** (Pers.)

*Æcidiospores*—Like those of *Ph. fragariastri.*

*Uredospores*—Sori orange-red, roundish, often confluent Spores spherical, elliptical, ovate, generally 20–22 × 16–20μ, finely echinulate. Paraphyses abundant, swollen above, curved.

*Teleutospores*—Sori black, pulvinate. Spores cylindrical, cells four to six, 50–70 × 20–22μ, constricted, apex rounded or slightly pointed, sometimes surmounted with a brown blunt point. Pedicels 100–150μ long, colourless, rather firm.

*Synonyms.*

Schröt., *loc. cit.*, p. 352.

*Puccinia potentillæ.* Pers., "Syn.," p. 229.

*Uredo obtusa*, Strauss.

*Uredo potentillarum*, D. C.

*Phragmidium potentillæ*, Winter.

*Phragmidium obtusatum*, Fries. Cooke, "Micro. Fungi," 4th edit., p. 201, in part.

*Phragmidium potentillæ.* Schröt., *loc. cit.*, p. 352.

On *Potentilla argentea* and various cultivated species.

May to November.

**Phragmidium tormentillæ.** Fckl.

*Æcidiospores*—Resembling those of *Ph. fragariastri*.

*Uredospores*—Sori round, small, orange-red, punctiform. Spores spherical or ovate, finely echinulate, reddish orange, 20–23 × 17–20μ.

*Teleutospores*—Sori small, round, punctiform, pale brown. Spores cylindrical, cells three to ten, generally five to eight, often curved, clear brown, smooth, apex rounded or pointed, 100–160 × 16–22μ. Pedicels long (100μ), equally thick, colourless.

*Synonyms.*

Schröt., *loc. cit.*, p. 352.

*Phragmidium tormentillæ.* Fckl., "Symb. Myc.," p. 46.

*Uredo potentillarum*, D. C. Berk., "Eng. Flor.," vol. v. p. 382. Sow., t. 398, fig. 2.

*Puccinia potentillæ.* Grev., t. 37, in part.

*Aregma obtusatum*, Fries. Berk., "Eng. Flor.," vol. v. p. 359.

*Phragmidium obtusatum*, Fries. Cooke, "Hdbk.," p. 491; "Micro. Fungi," 4th edit., p. 201, t. iii. fig. 35.

*Exsiccati.*

Berk., 105. Cooke, i. 22, 67; ii. 100. Vize, "Fungi Brit.," 5.

On *Potentilla tormentilla.*

BIOLOGY.—I have followed Schröter in the arrangement of these species, but am by no means sure this is distinct from the preceding. Dietel states ("Beiträge zur Morph. Biol. der Uredineen," 1887, p. 9) that the teleutospores of *P. obtusatum* germinate in the autumn, and they resemble Puccinia teleutospores in having an apical germ-pore in the upper cell, and only one lateral germ-pore in each of the other cells.

**Phragmidium violaceum.** (Schultz.)

*Æcidiospores*—Sori hypophyllous, roundish or elongate, often in circular clusters on irregularly rounded spots, above which the corresponding upper surface of the leaf is reddish, and surrounded by a broad, irregular, violet-red margin. Paraphyses very few. Spores in short chains, globose or elliptical, echinulate, orange-yellow, 11–30 × 17–24$\mu$.

*Uredospores*—Sori greenish yellow, roundish, pulverulent. Spores globose, ovate or elliptical, echinulate, verrucose, yellow, 19–25 × 15–20$\mu$.

*Teleutospores*—Sori thick, black, rather persistent. Spores cylindrical, blunt or rounded at both ends, opaque, cells three to five, mostly four, verrucose with colourless hemispherical flat warts, apex generally provided with a wart-like brown point, 70–80 × 25–30$\mu$. Pedicels long, slightly clavate below.

*Synonyms.*

Schröt., *loc. cit.*, p. 353. Winter, *loc. cit.*, p. 231.

*Puccinia violacea.* Schultz, "Prodr. Flor. Starg.," p. 459.

*Lecythea ruborum*, Lév. Cooke, "Micro. Fungi," 4th edit., p. 221, in part.

*Exsiccati.*

Vize, "Micro. Fungi Brit.," 442.

On *Rubus fructicosus.*

## **Phragmidium rubi.** (Pers.)

*Æcidiospores* } As in *Ph. violaceum.*
*Uredospores* }

*Teleutospores*—Sori black, small, loose, round, often confluent. Spores cylindrical, rounded, verrucose, cells six to eight (but mostly five to six), apex an awl-shaped, hyaline point (5–10$\mu$), 77–100 × 25–28$\mu$. Pedicels long, bulbous, 70–80$\mu$ long, 15$\mu$ wide.

*Synonyms.*

Schröt., *loc. cit.*, p. 353. Winter, *loc. cit.*, p. 230.

*Puccinia mucronata, β rubi.* Pers., "Disp. Meth.," p. 38.

*Puccinia rubi*, Sow., t. 400, fig. 9. Purton, "Midl. Flor.," vol. ii. p. 726; iii. p. 507. Grev., "Flor. Edin.," p. 428.

*Puccinia ruborum*, D. C. Grev., "Flor. Edin.," p. 438. Johnst., "Flor. Berw.," vol. ii. p. 199. Berk., "Eng. Flor.," vol. v. p. 382, *pro parte.*

*Æcidium rubi.* Sow., t. 398, fig. 1.

*Lecythea ruborum*, Lév. Cooke, "Micro. Fungi," 4th edit., p. 221, in part.

*Aregma bulbosum*, Fries. Berk., "Eng. Flor.," vol. v. p. 358.

*Phragmidium bulbosum*, Fries. Cooke, "Hdbk.," p. 491; "Micro. Fungi," 4th edit., p. 201.

*Exsiccati.*

Baxter, 33. Cooke, i. 18, 20; ii. 65, 99, 209. Vize, "Micro. Fungi Brit.," 443, 449.

On *Rubus fructicosus, cæsius, saxatilis.*

BIOLOGY.—The teleutospores often remain attached to those leaves, which survive the winter and may be gathered in spring, when the teleutospores will germinate very readily if placed in water. The æcidiospores have six germ-pores.

## **Phragmidium subcorticatum.** (Schrank.)

*Æcidiospores*—Similar to those of the preceding species, but when they occur on the peduncles and stems they form large pulverulent orange sori. Spores rounded or ovate, finely echinulate, orange-yellow, 17–25 × 12–20$\mu$.

*Uredospores*—Sori rounded, soon naked, yellowish orange, often confluent. Spores globose, elliptical, or ovate, echinulate, yellow, 18–30 × 15–25$\mu$.

*Teleutospores*—Sori black, small, loose, round, scattered over the whole leaf. Spores verrucose, apex attenuated into a hyaline point, 10–12$\mu$ long, base rounded, dark brown, cells seven to nine, 80–100 × 25–30$\mu$. Pedicels bulbous, very long, 100–110$\mu$.

*Synonyms.*

Schröt., *loc. cit.*, p. 353. Winter, *loc. cit.*, p. 228.

*Lycoperdon subcorticatum.* Schrank. in Hoppe's *Bot. Taschb.*, 1793, p. 68.

*Uredo effusa*, Strauss. Grev., "Flor. Edin.," p. 439; "Scot. Crypt. Flor.," t. 19. Johnst., "Flor. Berw.," vol. ii. p. 199. Berk., "Eng. Flor.," vol. v. p. 381, in part.

*Uredo aurea.* Purton, "Midl. Flor.," vol. ii. p. 725.

*Uredo rosæ*, D. C. Grev., "Flor. Edin.," p. 348. Berk., "Eng. Flor.," vol. v. p. 381.

*Puccinia rosæ.* Grev., "Flor. Edin.," p. 428; "Scot. Crypt. Flor.," t. 15. Purton, "Midl. Flor.," vol. iii. t. 28, No. 1551, p. 301. Johnst., "Flor. Berw.," vol. ii. p. 193.

*Coleosporium pingue*, Lév. Cooke, "Hdbk.," p. 520; "Micro. Fungi," 4th edit., p. 217.

*Coleosporium miniatum*, Pers. Cooke, "Micro. Fungi," 4th edit., p. 217.

*Lecythea rosæ*, Lév. Cooke, "Micro. Fungi," 4th edit., p. 221, t. iii. fig. 37.

*Aregma mucronatum*, Fries. Berk., "Eng. Flor.," vol. v. p. 385; "Outl.," p. 329. Grev., t. 15.

*Phragmidium bullatum*, West. Cooke, "Micro. Fungi," 4th edit., p. 202.

*Phragmidium mucronatum*, Fries. Cooke, "Hdbk.," p. 490; "Micro. Fungi," 4th edit., p. 201, t. iii. fig. 38.

*Exsiccati.*

Cooke, i. 17; ii. 66, 98. Vize, "Fungi Brit.," 2, 40, 138; "Micro. Fungi Brit.," 450. Baxter, 37.

On *Rosa canina, centifolia.*
June to October.

**Phragmidium rosæ-alpinæ.** (D. C.)

*Æcidiospores*—Sori of two kinds, on the leaves, small and punctiform; on the stems, large and erumpent. Spores polygonal or ovate, echinulate, orange-yellow, 17–28 × 15–20μ. Paraphyses globose, colourless.

*Uredospores*—Sori small, round, pale yellow. Spores rounded or oval, echinulate, yellow, 15–20μ in diameter.

*Teleutospores*—Sori very small, black. Spores rounded or oval, cell ten to thirteen, thickly verrucose, cylindrical, fusiform, attenuated at both ends, apex with a conical colourless papilla, dark brown, 115–120 × 18–25μ. Pedicels as long as the spores.

*Synonyms.*

Winter, *loc. cit.*, p. 227.

*Uredo pinguis, β Rosæ alpinæ.* D. C., "Flore franç.," vol. ii. p. 235.

*Phragmidium fusiforme.* Schröt., "Brand. und Rostp.," p. 24.

On *Rosa alpina.* Scotland.
July to October.

**Phragmidium rubi-idæi.** (Pers.)

*Æcidiospores*—Sori small, epiphyllous, greenish yellow, often circular, abundant, umbilicate. Spores in short chains, round or oval, echinulate, orange-yellow, 20–30μ in diameter. Paraphyses globose, orange-yellow.

*Uredospores*—Sori small, scattered, pale orange. Spores globose or ovate, echinulate, yellow, 16–22μ in diameter.

*Teleutospores*—Sori roundish, small, loose, scattered, black. Spores opaque, cylindrical, dark brown, apex with a blunt, conical hyaline papilla, cells eight to ten, 110–130 × 20–30μ. Pedicels long (120–135μ), bulbous.

*Synonyms.*

Winter, *loc. cit.*, p. 231.

*Uredo rubi-idæi.* Pers., "Obs. Myc.," vol. ii. p. 24.

*Uredo gyrosa*, Reb. Berk., "Eng. Flor.," vol. v. p. 384. Grev., "Flor. Edin.," p. 439.

*Lecythea gyrosa*, Berk. Cooke, "Micro. Fungi," 4th edit., p. 222.

*Puccinia gracilis.* Grev., "Flor. Edin.," p. 428. Johnst., "Flor. Berw.," vol. ii. p. 193.

*Aregma gracile*, Grev. Berk., "Eng. Flor.," vol. v. p. 358.

*Phragmidium gracile*, Berk. Cooke, "Hdbk.," p. 491; "Micro. Fungi," 4th edit., p. 201.

*Exsiccati.*

Cooke, i. 21, 64; ii. 68, 210. Vize, "Fungi Brit.," 4; "Micro. Fungi Brit.," 119. Baxter, 39.

On *Rubus idæus*.
July to October.

## XENODOCHUS.. Schlecht.

Teleutospores separate, pedicellate, cylindrical, multicellular (15–20). Æcidiospores in basipetal chains.

### Xenodochus carbonarius. Schlecht.

*Æcidiospores*—Sori orange-red, large, roundish, on the stems elongated. Spores in short chains, subglobose, orange-yellow, verrucose, 15–25 × 15–20$\mu$. Paraphyses clavate, with traces of orange-yellow endochrome.

*Teleutospores*—Sori jet-black, thick, pulvinate, roundish, often confluent. Spores long, cylindrical or vermiform, often curved, cells ten to twenty, much constricted, generally smooth, except the terminal, which are slightly verrucose, dark brown, cells 15–20$\mu$, the whole spore 250–300$\mu$ long. Pedicels very short, persistent.

*Synonyms.*

*Phragmidium carbonarium* (Schlecht.). Winter in Rabh., "Krypt. Flor.," vol. i. p. 227.

*Xenodochus carbonarius.* Schlecht., "Linnea," vol. i. p. 237, t. iii. fig. 3. B. and Br., No. 133. Currey, *Micro. Jour.*, vol. v. t. 8, fig. 34. Cooke, "Hdbk.," p. 489; "Micro. Fungi," 4th edit., p. 201, t. iii. fig. 29.

*Lecythea poterii*, Lév. Cooke, "Micro. Fungi," 4th edit., p. 221, t. iii. fig. 31.

*Uredo miniata.* Pers., "Syn.," p. 216.

*Exsiccati.*

Berk., 323. Cooke, i. 315 ; ii. 97. Vize, "Fungi Brit.," 1.

On *Sanguisorba officinalis.*
June to October.

**Xenodochus curtus.** Cooke.

*Teleutospores*—Sori scattered in very small tufts. Spores abbreviated, obtuse, broad, of from four to eight articulations.

*Synonym.*

*Xenodochus curtus.* Cooke, "Micro. Fungi," 4th edit., p. 201.

On leaves of *Valeriana officinalis* (?)
I have no knowledge of this species.

## ENDOPHYLLUM. Lév.

Teleutospores in basipetal chains, enclosed in a pseudoperidium of barren cells, and resembling the æcidiospores of Puccinia, but germinating by the protrusion of a promycelium and the abstriction of promycelial spores.

**Endophyllum euphorbiæ.** (D. C.)

Mycelium perennial. Spermogonia mostly on the upper surface of the leaves, yellow. Pseudoperidia hypophyllous, on all the foliage, round, immersed, wide, edges thick, erect. Spores subglobose, granular, orange, 16–26 × 12–18$\mu$.

*Synonyms.*

*Endophyllum euphorbiæ-sylvaticæ*, Lév. Winter in Rabh., "Krypt. Flor.," vol. i. p. 251.

*Æcidium euphorbiæ.* D. C., "Flore franç.," vol. ii. p. 241.

*Æcidium euphorbiæ*, Pers. Berk., "Eng. Flor.," vol. v. p. 374. Cooke, "Hdbk.," p. 537. "Micro. Fungi," 4th edit., p. 195. Purton, vols. ii. and iii., No. 1537.

*Exsiccati.*

Berk., 299. Cooke, i. 6; ii. 302. Vize, "Fungi Brit.," 154. "Micro. Fungi Brit.," 458.

On *Euphorbia amygdaloides.*

April to June.

BIOLOGY—The spores of this species germinate freely in water, and produce a promycelium with three or four promycelial spores. When placed on the cuticle of a leaf of the proper host-plant, these promycelial spores bore, by means of their germ-tubes, through the epidermal cells and enter the parenchyma of the leaf, between the cells of which they soon produce a richly branched and widely extending mycelium. If the entrance has been effected into an old leaf, the further development of the parasite ceases when the leaf falls off. The mycelium passes along the petiole and enters the stem, where it may be found, especially in the pith and inner bark. In the following spring, the foliage which is produced by an infected, is different from that which is produced by a healthy plant. The affected plant sends up longer shoots, with shorter and wider leaves, which have a paler green colour than the healthy foliage (De Bary, *Neue Untersuch.*, 1865, pp. 20, 21). I find that the promycelial spores often send out germ-tubes while still attached to the promycelium. I have always failed in permanently infecting old plants of Euphorbia; no matter what the age of the leaves may be, in the ensuing spring the foliage has always been healthy. But if a young seedling be infected shortly after it has come up—that is, while not more than a month or two old—the mycelium produced in its leaves readily gains an entrance into the stem. The foliage, and shoots sent up by it in the following year are pervaded by the perennial mycelium, and produce æcidia abundantly during the spring; but the late summer and autumn foliage differs little from healthy foliage, excepting that the leaves are somewhat shorter. The next vernal foliage is, however, æcidiiferous. The affected plants seldom flower.

**Endophyllum sempervivi.** (Alb. and Schw.)

Mycelium perennial. Spermogonia globose, then conical, yellow. Pseudoperidia scattered, immersed, at first papillæform, opening above by a small foramen, then broadly cup-shaped, with whitish edges. Spores subglobose or angular, verrucose, orange, 20–30$\mu$ in diameter, sometimes 30$\mu$ long.

*Synonyms.*

Winter in Rabh., "Krypt. Flor.," vol. i. p. 252.

*Uredo sempervivi.* Alb. and Schw., "Consp.," p. 126.

*Endophyllum sempervivi*, Lév. Berk., *Ann. Nat. Hist.*, No. 476. Cooke. "Hdbk.," p. 546; "Micro. Fungi," 4th edit., p. 200.

On *Sempervivum tectorum.*

April and May.

BIOLOGY.—The cups are produced early in the year; the spores germinate at once, and the promycelial spores enter all parts of the leaves, including the hairs, and produce mycelium in them as in the preceding species. During the summer the infected leaves and shoots maintain their normal appearance, but towards autumn the lower leaves fall off from the rosettes. The leaves produced during the summer after infection, and those produced from infected plants during the winter and spring, are more elongated in their contour, and, towards the base especially, have a paler, yellowish hue. The mycelium from the leaves reaches the stems, and all the leaves subsequently produced from the plant are pervaded by the mycelium (De Bary, *Neue Untersuch.*, 1865, p. 20).

It has been assumed that this Endophyllum on Sempervivum is identical with that on *Sedum acre*, but I have been unable to produce the Endophyllum on *Sedum acre* from the spores of *E. sempervivi*, although the same spores sown on *Sempervivum tectorum* always produced the Endophyllum in the following spring. I was further unable to infect *S. proliferum* and *californicum*. Mr. W. G. Smith (*Gard. Chron.*, 1880, pp. 660 and 725) reports this fungus on *S. montanum*, and Mr. Badger (p. 815) on *S. globiferum* and *calearum*.

## GYMNOSPORANGIUM. Hedw.

Teleutospores bicellular, united by a gelatinous matrix into variously shaped spore-masses. Each cell provided with from two to four germ-pores, placed laterally near the septum. Æcidiospores in basipetal chains with alternate barren cells, enclosed in a pseudoperidium.

### Gymnosporangium sabinæ. (Dicks.)

*Æcidiospores*—Spots at first orange, then reddish, thickened, generally circular. Pseudoperidia flask-shaped, pale brown, split to the base into laciniæ which remain for a long time united at the apex; the laciniæ are joined at intervals by short transverse trabeculæ. Spores subglobose, finely verrucose, brown, 25–40 × 18–25μ.

*Teleutospores*—Mycelium perennial, causing fusiform swellings on the branches of the host-plant. Spore-masses vernal, cylindrical or clavate, generally compressed laterally, 8–10 mm. high, at first blackish, firm, then red-brown and gelatinous. Spores broadly fusiform or bipyriform, central constriction almost none, yellowish brown or chestnut-brown, with four germ-pores, 38–50 × 23–26μ. (Plate IV. figs. 11 and 12.)

*Synonyms.*

Winter in Rabh., "Krypt. Flor.," vol. i. p. 232.

*Tremella sabinæ.* Dicks., "Plant. Crypt. Brit.," vol. i. p. 14; "Eng. Bot.," t. 710. With., vol. iv. p. 68. Purton, "Midl. Flor.," vols. ii., iii., No. 883.

*Podisoma sabinæ*, Fries. Berk., "Eng. Flor.," vol. v. p. 362. Pers., "Disp. Meth.," t. ii. fig. 1. "Eng. Bot.," t. 710. Berk., "Outl.," t. ii. fig 4. Cooke, "Hdbk.," p. 510; "Micro. Fungi," 4th edit., p. 214.

*Myxosporium colliculosum.* Berk., "Outl.," p. 325. Sow., t. 409.

*Ræstelia cancellata*, Reb. Berk., "Eng. Flor.," vol. v. p. 373. Sow., t. 409, 410. Cooke, "Hdbk.," p. 533; "Micro. Fungi," 4th edit., p. 193, t. ii. figs. 20, 21.

*Exsiccati.*

Berk., 58, 107. Cooke, i. 332. Vize, "Micro. Fungi Brit.," 37, 55.

Æcidiospores on *Pyrus communis*, July to September.
Teleutospores on *Juniperus sabina*, April to May.

BIOLOGY.—All the Gymnosporangia are easy of cultivation. If the gelatinous masses be placed in water they swell up, and in twelve or eighteen hours are covered by a golden-yellow powder, the promycelial spores. These, diffused in water and applied to the proper host-plant, are almost certain to produce æcidiospores. The teleutospores in the present species vary somewhat in shape, being often slightly contracted at both ends so as to resemble two pyriform bodies joined at their broad ends. The teleutospores are of two kinds, similar in form, but differing in colour. In the first kind the spore-walls are hyaline, and the contents orange ; in the second, the spore-walls are thicker, and brown or yellowish brown. The dark spores are the least abundant, and do not germinate so soon as the paler ones do, although they do so within a comparatively short period. So far as is at present known, *G. sabinæ* has its æcidiospores on the common pear only, which culture I have successfully made in seventeen experiments.

### **Gymnosporangium confusum.** Plow.

*Æcidiospores*—Pseudoperidia on thickened roundish spots, orange above, and often surrounded by a reddish or purple line, cylindrical or cylindrico-fusiform, opening by lateral longitudinal fissures, at length fimbriate. Spores subglobose, pale brown, verrucose, 15–20$\mu$ in diameter.

*Teleutospores*—Mycelium perennial. Spore-masses vernal, at first tuberculate, dark chocolate-brown, almost black, soon becoming cylindrical, often compressed, 5–8 mm. long, then rich chestnut-brown, swelling when moist, and speedily covered with golden-yellow promycelial spores. Spores smooth, oval or elliptical, generally acute at both ends, of two kinds, the more numerous with hyaline spore-walls and orange-yellow contents, the other with dark brown, thick walls, 40–50 × 20–25$\mu$, with from two to four germ-tubes. Pedicels long (80–100$\mu$), hyaline. (Plate IV. figs. 13 and 14.)

*Synonyms.*

*Æcidium penicillatum*, Mull. (?)
*Æcidium mespili*, D. C. Winter, *loc. cit.*, p. 266.

*Exsiccati.*

Vize, "Micro. Fungi Brit.," 454, 545, 551.

Æcidiospores on *Cratægus oxyacantha*, *Mespilus germanica*, *grandiflora* (?), *Pyrus vulgaris*, June to August.

Teleutospores on *Juniperus sabina*, April to May.

BIOLOGY.—This species has hitherto been confounded with *G. sabinæ*, which it resembles in many points. When the æcidiospores occur on *Cratægus oxyacantha*, the spermogonial spots are more brightly coloured than those of *G. clavariæforme*. The æcidiospores are slightly smaller than those of *G. clavariæforme*, and the cells of the pseudoperidia are delicately reticulated and longitudinally wrinkled. In thirty-six cultures I have produced the æcidiospores on *C. oxyacantha* from *G. confusum*; in six cultures on *Pyrus vulgaris* (quince), and in seven on *Mespilus germanica* (medlar). I have failed to produce any result on apple in three experiments, and on beam in one. Of these cultures several were serial. Thus, on May 16, 1887, three pears and one thorn were infected from the same material; on the three pears no result was obtained, but on the thorn æcidiospores were produced. In another series, on the same day, a pear, a quince, and a thorn were infected; the quince and thorn had spermogonia on the 30th, but no result was obtained on the pear. Again, on June 16, a quince, two pears, a thorn, a beam, and an apple were infected; the quince and thorn became infected with the æcidiospores, but the pears, apple, and beam remained free. On April 25, a pear and two thorns were infected; the two thorns produced the æcidiospores (spermogonia appearing on one on May 5, and on the other on the 10th), but the pear remained free. On May 7, 1885, the same infecting material was applied to a medlar and six thorns; both the medlar and all the thorns became affected by the 20th with the spermogonia. In due time on the thorns the æcidiospores were developed, and these were on June 25 applied to a healthy sabine bush. In September, it was noted that many of the leaves had turned yellow; these during the winter and following summer (1886), fell off, and in March, 1887, the *G. confusum* appeared on those parts of the branches from which the leaves had fallen off. It may be added, that in only one duplicated culture out of the hundred I have made in elucidating the life-history of the sabine *Gymnosporangia* have æcidiospores on both thorn and pear been produced from the same infecting material. This was doubtless caused by an accidental mixture of teleutospores.

### Gymnosporangium clavariæforme. (Jacq.)

*Æcidiospores*—Pseudoperidia on yellow, thickened spots, at first flask-shaped, then cylindrical, light brown, at length becoming fimbriate above. Spores subglobose, verrucose, pale brown, 20–40 × 20–28$\mu$.

*Teleutospores*—Mycelium perennial, causing fusiform swellings on the branches. Spore-masses ligulate, compressed, sometimes bifid, at first firm and cartilaginous, becoming gelatinous, pale orange, about 10 mm. long. Spores fusiform, constricted, dark yellow, 70–120 × 14–20$\mu$.

*Synonyms.*

Winter, *loc. cit.*, p. 233.

*Tremella clavariæformis.* Jacq., "Collect.," vol ii. p. 174.

*Podisoma juniperi*, Fries. Berk., "Eng. Flor.," vol. v. p. 362. Bull., t. 427, fig. 1. Cooke, "Hdbk.," p. 510; "Micro. Fungi," 4th edit., p. 214. Johnst., "Flor. Berw.," vol. ii. p. 146. Grev., "Flor. Edin.," p. 427.

*Æcidium laceratum*, Sow. Berk., "Eng. Flor.," vol. v. p. 373. Sow., t. 318. Grev., "Flor. Edin.," p. 447. Johnst., "Flor. Berw.," vol. ii. p. 107.

*Ræstelia lacerata*, Tul. Cooke, "Hdbk.," p. 534; "Micro. Fungi," 4th edit., p. 190, t. ii. figs. 22–26. Grev., t. 209.

*Exsiccati.*

Cooke, i. 2, 125; ii. 442, 640; "L. F.," 50. Berk., 106, 111. Vize, "Micro. Fungi," 129; "Micro. Fungi Brit.," 38, 69.

Æcidiospores on *Cratægus oxyacantha*, *Pyrus communis*, June to August.

Teleutospores on *Juniperus communis*, April to May.

BIOLOGY.—The teleutospores germinate within twenty-four hours after being placed in water, and the promycelial spores, when applied to Cratægus, give rise in ten or twelve days to the æcidiospores. This culture is very easy to make. I have done it sixteen times, and had no failure. A certain number of failures have followed my cultures of the promycelial spores on *Pyrus communis*, but still I have succeeded often enough to prove the metœcism. The pseudoperidia on pear are similar to those on thorn, and can be distinguished at a glance from *R. cancellata.* Ráthay states that æcidiospores occur also on *P. torminalis* and *Cratægus monogyna* in addition to the above-named plants ("Untersuch. über die Spermogonien der Rostpilze," pp. 20, 22. Wien: 1882).

The converse culture of the æcidiospores on juniper I made in 1884. On June 25, a small juniper was infected with the æcidiospores; on July 8, many of the leaves began to turn yellow, these during the summer and autumn fell off, leaving bare places on the branches, and giving the bush a very peculiar appearance; in December, 1885, these bare places began to swell; and on April 1, 1886, the teleutospore-masses were produced. Thus it will be seen that the æcidiospores require two years in which to perfect the development of perennial teleutospore mycelium, while the teleutospores only require two or three weeks in which to perfect the mycelium bearing the æcidiospores.

The germ-tubes of the æcidiospores will not enter old leaves, but only those produced the same year the experiment is made; or, if they do enter the former, the mycelium will not penetrate into the perennial tissues of the host-plant. Hence it is useless attempting to infect a bush which has recently been transplanted; but there is no difficulty with an established plant, which has thrown out a number of young leaves. Once established in the juniper, the mycelium lives in a vigorous condition, annually producing teleutospores for many years.

### Gymnosporangium juniperinum. (Linn.)

*Æcidiospores*—Pseudoperidia on orange or red, roundish spots, cylindrical, curved, yellowish brown, 8–10 mm. long, open above, fimbriate. Spores subglobose, brown, finely verrucose, $21–28 \times 19–24\mu$.

*Teleutospores*—Mycelium perennial. Spore-masses at first dark brown, then orange, soft, gelatinous, subglobose. Spores of two kinds: (1) fusiform, brown, with thick epispore, $75 \times 27\mu$; (2) yellow, with thinner epispore, about $66 \times 17\mu$, having six germ-pores.

*Synonyms.*

Winter, *loc. cit.*, p. 234.

*Tremella juniperina.* Linn., "Sp. Plant.," p. 1625.

*Gymnosporangium juniperi*, Link. Berk., "Eng. Flor.," vol. v. p. 361; "Outl.," t. ii. fig. 2. Cooke, "Hdbk.," p. 509; "Micro. Fungi," 4th edit., p. 214.

*Æcidium cornutum*, Pers. Berk., "Eng. Flor.," vol. v. p. 373. Johnst., "Flor. Berw.," vol. ii. p. 207. Sow., t. 319. Grev., t. 180; "Flor. Edin.," p. 447.

*Ræstelia cornuta*, Tul. Cooke, "Hdbk.," p. 534; "Micro. Fungi," 4th edit., p. 190, t. ii. figs. 18, 19.

*Exsiccati.*

Cooke i. 1; ii. 441. Vize, "Micro. Fungi Brit.," 54.

Æcidiospores on *Pyrus aucuparia*, July to October.
Teleutospores on *Juniperus communis*, May to June.

Biology.—The teleutospores do not generally appear until May. They readily produce promycelial spores in water, which, when applied to the foliage of the mountain ash, give rise to the spermogonia in from eight to fifteen days.

I have failed only twice in thirteen cultures, and I had to experiment with material sent from Scotland. Ráthay states ("Spermogonien der Rostpilze," 1882, pp. 20–22) that the æcidiospores of this species also occur on *Pyrus aria, malus*, and *Amelanchier vulgaris*. I have applied the promycelial spores to quince (*P. vulgaris*) in three experiments, to apple (*P. malus*) and to beam (*P. aria*) in one, but without success.

In the spring of 1884, a patch of spermogonia appeared on a mountain-ash, which had borne the æcidiospores in 1883; this is the only instance in which I have seen the mycelium of the æcidiospores survive the winter. Probably this arose from a bud being infected late the previous year. From the fact of my being unable to produce the æcidiospores on the host-plants with which Ráthay succeeded, viz. *Pyrus malus* and *aria*, it is clear that in Europe we have a second species of Gymnosporangium on *Juniperus communis*, namely *G. tremelloides*, Hartig. (Lehrb. d. Baumk, p. 55).

## MELAMPSORA. Castagne.

Teleutospores unicellular or rarely multicellular from transverse or longitudinal cleavage, in compact flat sori or crusts. Germination by a promycelium, as in Puccinia. Uredospores single, formed on sterigmata.

The Cæomata are considered by many botanists the æcidiospores of this genus. The promycelial spores, when applied to the teleutospore host-plant, do not give rise to the uredospores.

### I. MELAMPSORA.

Teleutospores formed outside the epidermal cells of the host-plant, and remaining single.

### Melampsora helioscopiæ. (Pers.)

*Uredospores*—Sori small, roundish, or irregular, soon pulverulent. Spores elliptical or ovate, finely echinulate, orange-yellow, $14–23 \times 10–17\mu$. Paraphyses abundant, between the spores, capitate, hyaline, $15–18\mu$ broad.

*Teleutospores*—At first deep orange, then blackish brown, rounded, on the stems elongate, scattered. Spores cuneiform, cylindrical, or prismatic, in section polygonal, dark brown, simple, intercellular, $30–35 \times 12–18\mu$.

*Synonyms.*

Winter in Rabh., "Krypt. Flor.," vol. i. p. 240.

*Uredo helioscopiæ.* Pers., "Disp. Meth.," p. 13. Grev., "Flor. Edin.," p. 440.

*Uredo euphorbiæ*, Reb. Berk., "Eng. Flor.," vol. v. p. 385. Purton, "Midl. Flor.," vol. iii. p. 397.

*Lecythea euphorbiæ*, Lév. Cooke, "Micro. Fungi," 4th edit., p. 221.

*Melampsora euphorbiæ*, Cast. Cooke, "Hdbk.," p. 523; "Micro. Fungi," 4th edit., p. 219, t. ix. figs. 193, 194.

*Exsiccati.*

Berk., 240. Cooke, i. 65, 439. Vize, "Fungi Brit.," 50, 141.

On *Euphorbia helioscopia*, *peplus*, *exigua*.
May to October.

**Melampsora lini.** (Pers.)

*Uredospores*—Sori scattered, small, roundish, soon pulverulent, orange. Spores round or ovate, echinulate, orange-yellow, 15–25 × 13–16$\mu$. Paraphyses markedly capitate, and especially abundant towards the margins of the sori.

*Teleutospores*—Sori rounded, reddish brown, then black, shining. Spores formed under the cuticle, cylindrico-prismatic, intercellular, in section polygonal, 45 × 20$\mu$.

*Synonyms.*

Winter, *loc. cit.*, p. 242.

*Uredo miniata*, var. *Lini.* Pers., "Syn.," p. 216.

*Uredo lini*, D. C. Berk., "Eng. Flor.," vol. v. p. 384; "Outl.," p. 334. Grev., t. 31.

*Lecythea lini*, Berk. Cooke, "Hdbk.," p. 532; "Micro. Fungi," 4th edit., p. 222, t. viii. figs. 165–167.

*Exsiccati.*

Berk., 118. Cooke, i. 446; ii. 70. Vize, "Fungi Brit.," 66.

On *Linum catharticum*.
June to September.

**Melampsora farinosa.** (Pers.)

*Uredospores*—Hypophyllous, roundish, scattered or clustered, often disposed in a circular manner, soon pulverulent, orange. Spores more or less globose, epispore echinulate, hyaline, contents golden-orange, germ-pores two, opposite each other, 17–22 × 13–15μ. Paraphyses very numerous, clavate, straight or slightly curved below, swollen and obtuse above, hyaline, 40–60 × 15–17μ, smooth.

*Teleutospores*—Epiphyllous, without spots, always covered by the epidermis, in clusters, often confluent and forming thick crusts, flat, firm, at first orange, then brown, at length blackish. Spores cylindrical, rather narrower below, smooth, pale brownish, 40–45 × 16–17μ.

*Synonyms.*

Schröt., "Krypt. Flor. Schl.," vol. iv. p. 360.

*Uredo farinosa*, Pers. "Syn.," p. 217. Purton, "Midl. Flor.," vol. iii. p. 298. Johnst., "Flor. Berw.," vol. ii. p. 200. Grev., "Flor. Edin.," p. 437.

*Uredo capreárum.* Berk., "Eng. Flor.," vol. v. p. 385.

*Lecythea caprearum*, Berk. Cooke, "Micro. Fungi," 4th edit., p. 222.

*Uredo saliceti*, Lév. Cooke, "Micro. Fungi," 4th edit., p. 221.

*Melampsora salicina*, Lév. Cooke, "Hdbk.," p. 522; "Micro. Fungi," 4th edit., p. 219, t. ix. figs. 191, 192.

*Melampsora salicis capreæ*, Pers. Winter, *loc. cit.*, p. 239.

*Exsiccati.*

Cooke, i. 55, 85; ii. 69, 155; "L. F.," 49. Vize, "Micro. Fungi," 51; "Micro. Fungi Brit.," 67, 232.

On *Salix caprea*, *aurita*, *cinerea*, *reticulata*.

June to November. The teleutospores germinate in March and April.

BIOLOGY.—Rostrup and Nielsen state that the æcidiospores of this species are *Cæoma euonymi*. In 1884, I applied the germinating teleutospores from *Salix caprea* to a small plant of *Euonymus europæus* on three separate occasions (March 17, April 6, and May 14), but no result followed. In 1885, I repeated the experiment with the same want of success. Lest there should have been any error in the determination of the teleutospores, I tried them upon a plant of *Ribes rubrum*, but with no result, the above-named botanists having stated that an allied Melampsora (*M. hartigii*) had for its æcidiospores, *Cæoma ribesii*.

## **Melampsora epitea.** (Kze. and Schm.)

*Uredospores*—Sori hypophyllous, rarely epiphyllous, generally very minute, but sometimes larger, at length pulverulent, orange. Spores globose, rarely subelliptical, echinulate, pale yellow, 20μ in diameter. Paraphyses very numerous, clavate, almost globose, hyaline, base acute, smooth, 40 × 22μ.

*Teleutospores*—Hypophyllous, very small, at first brown, then almost black, subverrucæform, hemispherical, crowded. Spores cylindrical, pale brown, 30–34 × 12–14μ. Epispore smooth, 2μ thick.

*Synonyms.*

*Melampsora epitea.* Thüm., "Monograph. der Weidenrust.," p. 17.

*Uredo epitea.* Kze. and Schm., "Mycol. Heft.," i. p. 68.

*Æcidium salicis.* Sow., t. 398, fig. 1.

*Lecythea epitea*, Lév. Cooke, "Micro. Fungi," 4th edit., p. 221.

On *Salix viminalis*.
July to November.

BIOLOGY.—Rostrup and Nielsen, according to Schröter ("Krypt. Flor. Schl.," vol. iii. p. 361), refer *Cæoma ribesii* to this species; but, as I understand the Danish botanists, they refer *C. ribesii* to *M. hartigii*, which is not a British species.

## **Melampsora mixta.** (Schlecht.)

*Uredospores*—Sori on the leaves, hypophyllous, orange, crowded, at length pulverulent, on the young branches large, erumpent, and confluent, also on the inflorescence. Spores elliptical or ovate, orange, echinulate, 14–18 × 12–15μ. Paraphyses very abundant, as much as 50μ long, upper end swollen, capitate, 15–28μ wide, hyaline.

*Teleutospores*—In small blackish crusts, hypophyllous.

*Synonyms.*

Thüm., "Monograph.," p. 20.

*Cæoma mixtum.* Schlecht., "Flor. Berol.," vol. ii. p. 124.

*Lecythea mixta*, Lév. Berk., *Ann. Nat. Hist.*, No. 478. Cooke, "Hdbk.," p. 531; "Micro. Fungi," 4th edit., p. 221.

*Exsiccati.*

Berk., 120.

On *Salix purpurea.*
May to November.

### Melampsora vitellinæ. (D. C.)

*Uredospores*—Sori small, round, soon pulverulent, golden-yellow, amphigenous. Spores elliptical or ovate, echinulate, orange, 25–28 × 15–20μ. Paraphyses fairly abundant, hyaline, smooth, globose, with a short attenuated base, 30–36 × 30μ (Thüm.).

*Teleutospores*—Hypophyllous in small crusts (Schröt.).

*Synonyms.*

Thüm., "Monograph," p. 21. Schröt., *loc. cit.*, p. 361.

*Uredo vitellinæ.* D. C., "Flore franç.," vol. ii. p. 231. Grev., "Flor. Edin.," p. 437. Purton, "Midl. Flor.," vol. iii. p. 298. Johnst., "Flor. Berw.," vol. ii. p. 200.

*Uredo saliceti*, Schlecht. Berk., "Eng. Flor.," vol. v. p. 385.

*Lecythea saliceti*, Lév. Berk., "Outl.," p. 334. Cooke, "Hdbk.," p. 532; "Micro. Fungi," 4th edit., p. 221.

*Exsiccati.*

Cooke, i. 316; ii. 63.

On *Salix pentandra, fragilis, triandra viminalis, vitellina.*

### Melampsora tremulæ. Tul.

*Uredospores*—Sori small, generally hypophyllous, orange-yellow, on the twigs larger, erumpent, pulverulent. Spores subglobose or ovate, echinulate, orange, 15–20 × 14–16μ. Paraphyses numerous, clavate, 40–50 × 10–15μ.

*Teleutospores*—Sori hypophyllous, abundant, reddish brown, then black. Spores elongated, compressed, slightly attenuated downwards, 45–50 × 10–12μ.

*Synonym.*

*Melampsora tremulæ*, Tul. *Ann. Sc. Nat.*, 2nd series, p. 95. Cooke, "Hdbk.," p. 522; "Micro. Fungi," 4th edit., p. 219. Schröt., *loc. cit.*, p. 362.

*Exsiccati.*

Cooke, i. 85; ii. 154; "L. F.," 48. Vize, "Micro. Fungi Brit.," 44.

On *Populus tremula.*
June to November.

BIOLOGY.—Rostrup states that the æcidiospores of this species are *Cæoma mercurialis* and *C. pinitorquum*. The last-named species is not British. I have applied the germinating teleutospores from *P. tremula* to *Mercurialis perennis* in 1883 and 1884, but could obtain no result (Exp. 277–350). Although *C. pinitorquum* is not a British species, yet I applied the germinating teleutospores of this Melampsora to *Pinus sylvestris* (Exp. 351), but with no result. I have also tried the germinating teleutospores on *Orchis maculata* and *latifolia*, but with no result (Exp. 650-652, 654, 655).

Hartig ("Allgem. Forst-und Jagdztg.," 1885, pp. 325–327) states that the æcidiospores of *M. tremulæ* are *Cæoma laricis*, and that by applying the spores of *C. laricis* to the leaves of *Populus tremula* he produced the Melampsora, and adds that *C. laricis* and *C. pinitorquum* may either be two distinct species, or the same species modified by the difference of the host-plant. I have repeated Hartig's experiments on three occasions (Exp. 710, 768, 769), by placing the germinating teleutospores of *M. tremulæ* on young larch-trees, but without result. Professor Trail, however, finds these two species growing in company near Aberdeen.

Probably more than one species of Melampsora occur on *Populus tremula.*

## Melampsora æcidioides. (D. C.)

*Uredospores*—Sori small, roundish, surrounded by a white wreath of large crowded paraphyses. Spores spherical, elliptical, or ovate, 17–24 × 15–17$\mu$. Spore-walls colourless, echinulate, contents orange-red. Paraphyses clavate, 40–60 × 15–20$\mu$.

*Teleutospores*—Forming small brown crusts. Spores cylindrical, brown, cohering laterally, apex truncate, 50 × 10$\mu$.

*Synonyms.*

Schröt., *loc. cit.*, p. 362.
*Uredo æcidioides.* D. C., "Flore franç.," vol. ii. p. 236.
*Cæoma ægirinum.* Schlecht., "Flor. Berol.," vol. ii. p. 124.

On *Populus alba.*

BIOLOGY.—The æcidiospores of this species are *Cæoma mercurialis.* After performing many cultures with the Melampsoræ, I at last placed the germinating teleutospores of this species, obtained from Dinmore Hill, Herefordshire, in October, 1887, on a plant of *M. perennis* (Exp. 711); the spermogonia appeared in about ten days. A duplicate culture on *Clematis vitalba* was without effect.

### **Melampsora populina.** (Jacq.)

*Uredospores*—Sori hypogenous, roundish, at first covered by the epidermis, orange. Spores elongate, elliptical, or ovate, echinulate, orange-yellow, 28–40 × 15–20$\mu$. Paraphyses more or less abundant, capitate or ovate, 45–55 × 20$\mu$.

*Teleutospores*—Sori hypophyllous, flat, generally crowded, often confluent, forming reddish brown, then blackish crusts. Spores cylindrico-prismatic, in section polygonal, pale brown, larger above, 40–50 × 10–15$\mu$.

*Synonyms.*

*Lycoperdon populina.* Jacq., "Collect.," Sup., t. ix. figs. 2, 3.

*Uredo cylindrica*, Strauss. Berk., "Eng. Flor.," vol. v. p. 385, in part.

*Lecythea populina*, Lév. Cooke, "Micro. Fungi," 4th edit., p. 221.

*Melampsora populina*, Lév. Cooke, "Hdbk.," p. 523; "Micro. Fungi," 4th edit., p. 219, t. ix. figs. 195, 196. Schröt., *loc. cit.*, p. 362.

*Uredo populina.* Grev., "Flor. Edin.," p. 442. Johnst., "Flor. Berw.," vol. ii. p. 200. Purton, "Midl. Flor.," vol. iii. t. 27, No. 1542.

*Exsiccati.*

Cooke, i. 83; ii. 61, 135; "L. F.," 47. Vize, "Fungi Brit.," 135, 231; "Micro. Fungi Brit.," 326, 443.

On *Populus nigra*, *balsamifera*, *italica*.
June to November.

BIOLOGY.—Ráthay states that the æcidiospores of this species are *Æc. clematidis.* I have applied the germinating teleutospores from *P. pyramidalis* in four cultures to *Clematis vitalba*, but could obtain no result. Schröter thinks that *Cæoma alliorum*, in part at least, belongs to this Melampsora; I have tested this statement by culture on *Allium ursinum*, but without result.

### **Melampsora hypericorum.** (D. C.)

*Uredospores*—Sori orange, small, pulverulent, scattered, mostly hypophyllous. Spores globose or elliptical, finely echinulate, orange-yellow, 14–21 × 12–17μ. Paraphyses absent.

*Teleutospores*—Dark brown, small, flat. Spores subcylindrical, section polygonal, brown, 25 × 15μ.

*Synonym.*

Winter, *loc. cit.*, p. 241. Schröt., *loc. cit.*, p. 363.

*Uredo hypericorum.* D. C., "Flore franç.," vol. vi. p. 81. Berk., "Eng. Flor.," vol. v. p. 380. Cooke, "Hdbk.," p. 526; "Micro. Fungi," 4th edit., p. 215, t. viii. figs. 174, 175.

*Exsiccati.*

Baxt., 42. Cooke, i. 118; ii. 321. Vize, "Fungi Brit.," 56; "Micro. Fungi Brit.," 127.

On *Hypericum perforatum*, *androsæmum*, *pulchrum*.

### **Melampsora betulina.** (Pers.)

*Uredospores*—Sori pale orange, small, roundish, numerous, flat. Pseudoperidia persistent. Spores ovate or oblong, finely echinulate, orange-yellow, 25–40 × 10–20μ. Without paraphyses.

*Teleutospores*—Sori at first yellow, then brown, at length black, generally hypophyllous, flat, roundish. Spores cylindrical, section polygonal, pale yellow-brown, 50 × 16μ.

*Synonyms.*

Winter, *loc. cit.*, p. 238.

*Uredo populina*, var. *betulina*. Pers., "Syn.," p. 219.

*Uredo cylindrica*, Strauss. Berk., "Eng. Flor.," vol. v. p. 385, in part.

*Uredo ovata.* Grev., "Flor. Edin.," p. 442. Johnst., "Flor. Berw.," vol. ii. p. 198.

*Lecythea betulina*, Lév.

*Melampsora betulina*, Desm. Cooke, "Hdbk.," p. 522; "Micro. Fungi," 4th edit., p. 219, t. ix. figs. 189, 190.

*Exsiccati.*

Cooke, i. 124; ii. 62. Vize, "Micro. Fungi," 140; "Micro. Fungi Brit.," 230.

On *Betula alba*.
May to November.

BIOLOGY.—In the presence of so many assertions that the æcidiospores of the Melampsoræ are to be found amongst the Cæomata, and from the fact that the only station in which I find *C. orchidis* is under a birch-tree that is annually affected with *M. betulinæ*, I have tried several cultures by placing the germinating teleutospores on *Orchis maculata* and *latifolia* (Exp. 256, 284, 343, 672, 674, 675), but always without success. As a plant of *Lonicera periclymenum* grows not very far off, I tried the germinating teleutospores on Lonicera (Exp. 113, 113*a*, 114, 136), but with no result. I have also applied the germinating teleutospores to young healthy birch plants, but of course without result.*

II. PUCCINIASTRUM. Otth.

Teleutospores formed outside the epidermal cells, becoming longitudinally or obliquely divided into from two to four cells.

**Melampsora pustulata.** (Pers.)

*Uredospores*—Sori orange, pustular, often widely confluent, small. Pseudoperidia hemispherical, spores subglobose or ovate, echinulate, orange-yellow, 14–20 × 10–15$\mu$.

* Since Rostrup pointed out, in 1883, the connection between the Melampsoræ and the Cæomata, I have made upwards of forty experimental cultures with these species. The results which I have obtained differ so materially from those of other botanists, that it is evident several species of Melampsora are at present confounded with one another. The satisfactory recognition of these can only be accomplished by careful biological research.

*Teleutospores*—Sori at first reddish brown, then blackish, at length confluent. Spores intercellular, subcylindrical, cells rarely one, generally three or four, superimposed or lateral, 25–30μ long.

*Synonym.*

Schröt., *loc. cit.*, p. 364. Winter, *loc. cit.*, p. 243.

*Uredo pustulata.* Pers., "Syn.," p. 219. Berk., "Eng. Flor.," vol. v. p. 381. Grev., "Flor. Edin.," p. 440. Cooke, "Hdbk.," p. 526; "Micro. Fungi," 4th edit., p. 215.

*Exsiccati.*

Cooke, i. 210; ii. 322; "L. F.," 29.

On *Epilobium palustre, augustifolium.*

**Melampsora circææ.** (Schum.)

*Uredospores*—Sori brownish yellow, pustular, small, scattered. Pseudoperidia hemispherical. Spores round or ovate, pale yellow, echinulate, 15–20 × 10–14μ.

*Teleutospores*—Sori pustular, then forming brownish, waxy crusts. Spores 1–4 cellular, cylindrico-prismatic, brown, intercellular, component cells lateral, prismatic cuneiform, 25–30 × 24–28μ.

*Synonyms.*

Winter, *loc. cit.*, p. 243. Schröt., *loc. cit.*. p. 364.

*Uredo circææ.* Schum., "Enum. Plant. Sæll.," vol. ii. p. 228. Cooke, "Micro. Fungi," 4th edit., p. 217, t. vii. figs. 135, 136.

*Puccinia circææ*, Pers. Cooke, "Hdbk.," p. 507, in part.

*Exsiccati.*

Berk., 342. Cooke, i. 62; ii. 74; "L. F.," 31.

On *Circæa lutetiana, intermedia.*

III. THECOPSORA. Magnus.

Teleutospores formed in the epidermal cells, becoming confluent into irregular but circumscribed masses. Uredospores in pustular heaps.

### Melampsora padi. (Kze. and Schm.)

*Uredospores*—Sori hypophyllous, small, pustular, whitish. Pseudoperidia on violet or brownish polygonal spots, hemispherical, persistent, opening with an apical perforation. Spores elongate or subglobose, often angular from mutual pressure, echinulate, white or yellowish, 15–20 × 10–15$\mu$.

*Teleutospores*—Sori epiphyllous, small, round, or angular, often forming black-brown crusts. Spores divided longitudinally, into cylindrico-prismatic cells, from two to four in number, intracellular, brown, 18–23$\mu$ long.

*Synonyms.*

*Uredo padi.* Kze. and Schm., Exs., 187. Cooke, "Hdbk.," p. 527.

*Uredo porphyrogenita*, Kze. Cooke, "Mico. Fungi," 4th edit., p. 216.

*Melampsora padi.* Cooke, "Hdbk.," p. 523.

*Exsiccati.*

Cooke, i. 536. Vize, "Fungi Brit.," 146.

On *Prunus padus.*
July to November.

### Melampsora vacciniorum. (Link.)

*Uredospores*—Sori yellow, small, pustular, scattered or crowded. Pseudoperidia hemispherical, perforate above. Spores subglobose or ovate, fine, echinulate, orange-yellow, 18–24 × 12–18$\mu$.

*Teleutospores*—Sori forming very small, pale brown, irregular crusts, hypophyllous, on the fallen leaves. Spores formed in the epidermal cells (intracellular), which they generally fill, septate longitudinally, brown, 14–18$\mu$.

*Synonym.*

Schröt., *loc. cit.*, p. 365. Winter, *loc. cit.*, p. 244.

*Uredo vacciniorum.* Link., "Sp. Plant.," vol. ii. p. 15. Berk., "Eng. Flor.," vol. v. p. 378. Cooke, "Hdbk.," p. 527; "Micro. Fungi," 4th edit., p. 216. Johnst., "Flor. Berw.," vol. ii. p. 199.

*Exsiccati*

Cooke, i. 119. Vize, "Fungi Brit.," 59; "Micro. Fungi Brit.," 226.

On *Vaccinium myrtillus.*
June to November.

**Melampsora pyrolæ.** (Gmelin.)

*Uredospores*—Sori small, round, globose, yellow, solitary or aggregate on yellowish or brownish spots. Pseudoperidium at length perforate above. Spores elliptical or subpyriform, echinulate, orange-yellow, 26–33 × 13–15μ.

*Synonyms.*

Schröt., *loc. cit.*, p. 366.
*Æcidium pyrolæ.* Gmelin., "Linné Syst. Nat.," vol. ii. p. 1473.
*Uredo pyrolæ.* Grev., "Flor. Edin.," p. 440. Berk., "Eng. Flor.," vol. v. p. 378. Johnst., "Flor. Berw.," vol. ii. p. 199.
*Trichobasis pyrolæ.* Berk., "Outl.," p. 332. Cooke, "Hdbk.," p. 529; "Micro. Fungi," 4th edit., p. 223.

*Exsiccati.*

Cooke, i. 438.

On *Pyrola rotundifolia* and *minor.*
August to October.

Biology.—Only the uredospores have at present been observed, but from their general appearance this fungus can hardly be other than a Melampsora (Schröter).

IV. MELAMPSORELLA. Schröt.

Teleutospores undivided, formed inside the epidermal cells (intracellular), hyaline, confluent in wide-spreading masses. Promycelial spores hyaline. Uredospores echinulate, enclosed in a pseudoperidium.

**Melampsora cerastii.** (Pers.)

*Uredospores*—Sori pustulate, scattered, small, round. Pseudoperidium hemispherical, persistent, apically perforate. Spores subglobose, ovate or clavate, echinulate, yellow, 20–25 × 14–20μ.

*Teleutospores*—Sori vernal on last year's living leaves, whitish or pinkish, often covering the whole leaf-surface. Spores spherical or shortly prismatic, in section polygonal, membrane thick, smooth, colourless, contents very pale reddish, simple, intercellular, 13–15$\mu$ in diameter. Promycelial spores globose, 8–10$\mu$, colourless.

*Synonyms.*

Schröt., *loc. cit.*, p. 366. Winter, *loc. cit.*, p. 242.

*Uredo pustulata*, var. *cerastii.* Pers., "Syn.," p. 219.

*Uredo caryophyllacearum.* Johnst., "Flor. Berw.," vol. ii. p. 199. Berk., "Eng. Flor.," vol. v. p. 381. Cooke, "Hdbk.," p. 526; "Micro. Fungi," 4th edit., p. 216.

*Exsiccati.*

Cooke, i. 6; ii. 75; "L. F.," 30. Vize, "Fungi Brit.," 145.

On *Cerastium arvense*, *Stellaria graminea.*

BIOLOGY.—We are indebted to Schröter for the life-history of this fungus. The uredospores first make their appearance in June, and are found throughout the summer until October. The teleutospores are found in the following year in May and June, in colourless or pale flesh-coloured sori on those living leaves which have survived the winter.

## COLEOSPORIUM. Lév.

Teleutospores composed of several superimposed cells enclosed in a thick transparent membrane, confluent in flat waxy masses. Each cell germinates by a single unseptate promycelial tube, which produces at its end a single promycelial spore. Uredospores formed in basipetal chains.

### I. EUCOLEOSPORIUM. Winter.

Heterœcious, having spermogonia, æcidiospores, uredospores, and teleutospores.

### Coleosporium senecionis. (Pers.)

*Æcidiospores*—Of two kinds. (1) On the fir needles, scattered or in small groups. Pseudoperidia cylindrical or laterally compressed, mouth torn irregularly, 2–2·5 mm. high. (2) On the young branches, producing fusiform swellings from the perennial mycelium. Pseudoperidia larger, crowded, whitish, saccate, mouth patent, widely torn, 5–6 mm. broad, 2·5–3 mm. high. Spores orange, spherical or ovate or angular, coarsely and thickly verrucose, 30–40 × 18–30μ.

*Uredospores*—Sori orange, small, scattered, soon pulverulent Spores in short chains, elliptical, ovate, or almost cylindrical, epispore densely verrucose, 20–30 × 15–25μ.

*Teleutospores*—Sori at first orange, then forming red crusts. Spores cylindrical or clavate, consisting of about four superimposed cells, dark orange-red, 90–110 × 17–30μ.

*Synonyms.*

Winter, *loc. cit.*, p. 248. Schröt., *loc. cit.*, p. 367.

*Uredo farinosa*, var. *senecionis*. Pers.. "Syn.," p. 218.

*Æcidium pini*, Pers. Berk., "Eng. Flor.," vol. v. p. 374. Grev., t. 7; "Flor. Edin.," p. 444.

*Peridermium pini*, Chev. Cooke, "Hdbk.," p. 535; "Micro. Fungi," 4th edit., p. 191, t. ii. figs. 27, 28.

*Peridermium acicolum*, Link. Cooke, "Micro. Fungi," 4th edit., p. 191.

*Puccinia glomerata* (uredospores). Cooke, "Hdbk.," p. 500.

*Uredo senecionis*, Schlecht. Berk., "Eng. Flor.," vol. v. p. 379. Grev., "Flor. Edin.," p. 438. Johnst., "Flor. Berw.," vol. ii. p. 198.

*Trichobasis senecionis*. Berk., "Outl.," p. 322. Cooke, "Micro. Fungi," 4th edit., t. vii. figs. 145, 146.

*Coleosporium senecionis*, Fries. Cooke, "Micro. Fungi," 4th edit., p. 218, t. vii. figs. 145, 146.

*Exsiccati.*

Cooke, i. 66; ii. 53. Vize, "Micro. Fungi Brit.," 20.

Æcidiospores on the leaves and on the young branches of *Pinus sylvestris*, *austriaca*, May and June.

Uredospores and teleutospores on *Senecio vulgaris*, *viscosus*, *sylvaticus*, and *jacobæa*, all the year.

Biology.—The connection between *Coleosporium senecionis* and *Peridermium pini* was first demonstrated by Wolff in 1872. He first found that the sowing of the æcidiospores of *Æc. pini*, both from the leaves and also from the young branches, on *Senecio viscosus, sylvaticus, vernalis, jacobæa,* and *vulgaris*, gave rise to the uredospores of the Coleosporium. In 1882, I repeated Wolff's culture on *S. vulgaris* with the æcidiospores from the leaves with success. In 1883, the Rev. Dr. Keith sent me from Forres a specimen of *Æc. pini* on the bark of a young fir-branch, the spores from which I used for infecting two plants of *S. vulgaris*, but without success. Too much importance must not be attached to this failure, considering the distance from which the æcidiospores came. I have had, however, so many failures in infecting *S. vulgaris* with the æcidiospores from *Æc. pini*, var. *acicola*, that I think there must be more than one species included under this name. My friend M. Max. Cornu informs me that in France he has succeeded in producing *Cronartium asclepiadeum* by sowing the æcidiospores of *Æc. pini*, var. *acicola* on *Vincetoxicum officinale.* As neither the Cronartium nor its host-plant occur in Britain, we must conclude that the æcidiospores which M. Cornu employed belong to a distinct species.

### II. HEMICOLEOSPORIUM. Winter.

Having uredospores and teleutospores, which occur on the same host-plant.

### Coleosporium sonchi. (Pers.)

*Uredospores*—Sori rounded, soon pulverulent and scattered, orange. Spores in short chains, rounded, oblong, or sub-cylindrical, coarsely and densely verrucose, orange-yellow, 20–35 × 15–20$\mu$.

*Teleutospores*—Sori at first orange, then red, flat, often confluent, forming crusts. Spores cylindrical or cylindrico-clavate, generally four-celled, 60–70 × 15–25$\mu$, sometimes as much as 130–140$\mu$ long.

*Synonyms.*

Schröt., *loc. cit.*, p. 368. Winter, *loc. cit.*, p. 247.

*Uredo sonchi-arvensis.* Pers., "Syn.," p. 217.

*Uredo compransor*, Schlecht. Berk., "Eng. Flor.," vol. v. p. 379.

*Uredo sonchi*, Pers. Purton, "Midl. Flor.," vol. iii., No. 1547. Johnst., "Flor. Berw.," vol. ii. p. 198. Grev., "Flor. Edin.," p. 441.

*Uredo tussilaginis*, Pers. Grev., "Flor. Edin.," p. 438. Johnst., "Flor. Berw.," vol. ii. p. 198.

*Uredo petasites*. Grev., "Flor. Edin.," p. 441. Johnst., "Flor. Berw.," vol. ii. p. 198.

*Coleosporium sonchi-arvensis*, Lév. Cooke, "Hdbk.," p. 521; "Micro. Fungi," 4th edit., p. 218, t. viii. figs. 178, 179.

*Coleosporium cacaliæ*, D. C. Cooke, "Micro. Fungi," 4th edit., p. 218.

*Coleosporium tussilaginis*, Lév. Cooke, "Hdbk.," p. 520; "Micro. Fungi," 4th edit., p. 217, t. viii. figs. 180, 181.

*Coleosporium petasitis*, Lév. Cooke, "Hdbk.," p. 521; "Micro. Fungi," 4th edit., p. 217.

*Exsiccati.*

Cooke, i. 32, 80, 81; ii. 44, 151, 152; "L. F.," 45. Vize, "Fungi Brit.," 45, 48; "Micro. Fungi Brit.," 49.

On *Sonchus arvensis*, *oleraceus*, *Tussilago farfara*, *Petasites officinalis*, *Cacalia hastata*.

May to October.

BIOLOGY.—I have followed Winter and Schröter in uniting these species, but further biological observations will probably show some of them to be distinct.

### Coleosporium campanulæ. (Pers.)

*Uredospores*—Sori yellow, small, rounded, soon pulverulent and scattered. Spores in short chains, rounded or elongated, verrucose, orange-yellow, 17–25 × 13–20$\mu$.

*Teleutospores*—Sori at first orange, then red, at length brownish, flat, confluent. Spores cylindrical or clavate, generally four-celled, 90–100 × 18–22$\mu$.

*Synonyms.*

Winter, *loc. cit.*, p. 246. Schröt., *loc. cit.*, p. 369.

*Uredo campanulæ*. Pers., "Syn.," p. 217. Grev., "Flor. Edin.," p. 440. Johnst., "Flor. Berw.," vol. ii. p. 200.

*Coleosporium campanulæ*, Lév. Cooke, "Hdbk.," p. 521; "Micro. Fungi," 4th edit., p. 218.

*Uredo crustacea*. Berk., "Eng. Flor.," vol. v. p. 378.

*Exsiccati.*

Baxt., 41. Berk., 336. Cooke, i. 81; ii. 421. Vize, "Fungi Brit.," 46.

On *Campanula rotundifolia, trachelium.*
May to October.

**Coleosporium euphrasiæ.** (Schum.)

*Uredospores*—Sori orange, irregular, soon pulverulent and scattered. Spores concatenate, subglobose or elongate, often irregular, verrucose, orange-yellow, 20–30 × 15–20$\mu$.

*Teleutospores*—Sori at first orange, then red. Spores orange, cylindrical or clavate, 80–100 × 20–25$\mu$.

*Synonyms.*

Winter, *loc. cit.*, p. 246. Schröt., *loc. cit.*, p. 370.

*Uredo euphrasiæ.* Schum., "Enum. Plant. Sæll.," vol. ii. p. 230.

*Uredo rhinanthacearum*, D. C. Berk., "Eng. Flor.," vol. v. p. 377. Grev., "Flor. Edin.," p. 439. Johnst., "Flor. Berw.," vol. ii. p. 200.

*Coleosporium rhinanthacearum*, Lév. Cooke, "Hdbk.," p. 521; "Micro. Fungi," 4th edit., p. 218, t. viii. figs. 176, 177.

*Exsiccati.*

Cooke, i. 299; ii. 156. Vize, "Fungi Brit.," 47.

On *Melampyrum arvense, Bartsia odontites, Euphrasia officinalis, Rhinanthus crista-galli.*

**CHRYSOMYXA.** Unger.

Teleutospores formed of a series of superimposed cells, of which the lower are sterile, forming flat or slightly elevated orange or reddish waxy crusts. Uredospores in linear series. Germination of the teleutospores by a multicellular promycelium from each cell, which produces from three to four, mostly four, promycelial spores.

HEMICHRYSOMYXA. Winter.

Having only uredospores and teleutospores, which occur on the same host-plant.

### Chrysomyxa pyrolæ. (D. C.)

*Uredospores*—Sori hypophyllous, scattered, generally covering the under surface of the leaves, roundish, pulverulent, yellow. Spores ovate or elliptical, often angular, coarsely verrucose, yellowish orange, 20–23 × 15–22μ.

*Teleutospores*—Sori orange, then red, flat, waxy. Spores, 110–120 × 8–10μ.

*Synonyms.*

Schröt., *loc. cit.*, 372. Winter, *loc. cit.*, p. 250.

*Æcidium pyrolæ.* D. C., "Flore franç.," vol. vi. p. 99.

*Uredo pyrolæ.* Grev., "Flor. Edin.," p. 440. Johnst., "Flor. Berw.," vol. ii. p. 198.

*Trichobasis pyrolæ.* Berk., "Outl.," p. 332. Cooke, "Hdbk.," p. 529; "Micro. Fungi," 4th edit., p. 223, in part.

*Exsiccati.*

Vize, "Micro. Fungi Brit.," 52

On *Pyrola rotundifolia.*

### Chrysomyxa empetri. (Pers.)

*Uredospores*—Sori orange, roundish or elongate, at first covered by the epidermis, then surrounded by it. Spores ovate, elliptical, or pyriform, verrucose, orange-yellow, 25–30 × 17–25μ.

*Synonyms.*

Schröt., *loc. cit.*, p. 372.

*Uredo empetri*, Pers. Moug. and Nest., Exs., No. 391. Cooke, *Seem. Jour. Bot.*, vol. iv. p. 99; "Hdbk.," p. 527; "Micro. Fungi," 4th edit., p. 216.

On *Empetrum nigrum.*
May to September.

## CRONARTIUM. Fries.

Teleutospores unicellular, compacted into a cylindrical body, germinating at maturity on the living host-plant. Promycelial spores almost spherical, pale. Uredospores enclosed in a pseudoperidium, which surrounds the base of the cylinder of teleutospores.

### **Cronartium flaccidum.** Alb. and Schw.

*Uredospores*—Sori small. Pseudoperidia broadly conical, dark yellow, persistent. Spores elliptical or ovate, echinulate, 20–30 × 13–20μ.

*Teleutospores*—Sori cylindrical, often curved, about 2 mm. high. Spores oblong or cylindrical, brown, 10–12μ across.

*Synonyms.*

*Sphæria flaccida.* Alb. and Schw., "Consp.," p. 31.

*Cronartium pæoniæ*, Cast. Cooke, "Micro. Fungi," 4th edit., p. 215.

*Cronartium flaccidum* (Alb. and Schw.). Winter, *loc. cit.*, p. 236.

On *Pæonia officinalis.* In gardens. North Wootton; C. B. P. South of England; Miss Jelly.

---

# APPENDIX.

IMPERFECT FORMS, OF WHICH THE FULL LIFE-HISTORY IS UNKNOWN.

## UREDO.

Spores produced singly on the terminal ends of mycelial hyphæ (basidia). Germination by a germ-tube which does not produce promycelial spores, but enters the host-plant through its stomata.

**Uredo agrimoniæ.** D. C.

Sori pustular, hypophyllous, roundish or irregular, often confluent and wide-spreading, covered by rather thin pseudoperidia. Spores subglobose or ovate, verrucose, orange, 17–23 × 14–17μ.

*Synonyms.*

Winter in Rabh., "Krypt. Flor.," p. 252

*Uredo potentillarum*, var. *agrimoniæ*. D. C., "Flore franç.," vol. vi. p. 81.

*Uredo potentillarum*, D. C. Berk., "Eng. Flor.," vol. v. p. 382. Grev., "Flor. Edin.," p. 438. Johnst., "Flor. Berw.," vol. ii. p. 199. Cooke, "Hdbk.," p. 525; "Micro. Fungi," 4th edit., p. 215.

*Coleosporium ochraceum*, Fckl. Cooke, "Micro. Fungi," 4th edit., p. 218.

*Exsiccati.*

Cooke, i. 635; ii. 149. Vize, "Fungi Brit.," 57; "Micro. Fungi Brit.," 325.

On *Agrimonia eupatoria*.

**Uredo symphyti.** D. C.

Sori minute, very numerous, hypophyllous, yellow. Mycelium perennial. Spores subglobose, ovate, finely echinulate, yellow, 25–35 × 16–24μ.

*Synonyms.*

Winter, *loc. cit.*, p. 254.

*Uredo symphyti*, D. C. "Encyc.," vol. viii. p. 232.

*Trichobasis symphyti*, Lév. Cooke, "Hdbk.," p. 529. Berk., "Outl.," p. 332.

*Coleosporium symphyti*, Fckl. Cooke, "Micro. Fungi," 4th edit., p. 218.

*Exsiccati.*

Berk., 320.

On *Symphytum officinale*.

May to September.

**Uredo polypodii.** Pers.

Sori orange-yellow, rounded or irregular, scattered, often confluent, covered by the epidermis. Spores of two kinds: (*a*) ovate or elliptical, echinulate above, smooth below, orange-yellow, 24–30 × 12–15$\mu$; (*b*) ovate or elliptico-polygonal, orange, with a very thick epispore (5$\mu$) quite smooth, 40–50 × 25–35$\mu$.

*Synonyms.*

Schröt., "Krypt. Flor. Schl.," vol. iii. p. 374.

*Uredo linearis*, var. *polypodii.* Pers., "Syn.," p. 217.

*Uredo filicum*, Desm. Berk., "Eng. Flor.," vol. v. p. 383. Sow., 320. Cooke, "Hdbk.," p. 526; "Micro. Fungi," 4th edit., p. 215.

*Exsiccati.*

Berk., 339. Cooke, i. 633; ii. 73. Vize, "Micro. Fungi," 143.

On *Polypodium dryopteris*, *Cystopteris fragilis*, *Adiantum capillus-veneris.*

**Uredo scolopendri.** Fckl.

Sori hypophyllous, small, pustulate, pale yellow, on brownish irregular spots. Pseudoperidium hemispherical, opening by a small aperture above, through which the spores escape. Spores subglobose or clavate, echinulate, 30–40 × 15–20$\mu$.

*Synonyms.*

Schröt., *loc. cit.*, p. 374.

*Uredo scolopendri*, Fckl.

*Milesia polypodii*, B. White (?). "Scot. Nat.," vol. iv. p. 162, t. ii. fig. 5.

On *Blechnum spicant*, *Scolopendrium vulgare.*

June to October.

**Uredo mulleri.** Schröt.

Spermogonia epiphyllous, on circular yellow spots. Sori amphigenous, roundish, golden-yellow, often circinate. Spores globose, echinulate, orange, 18–20$\mu$ in diameter.

*Synonyms.*

*Uredo mülleri.* Schröt., "Krypt. Flor. Schl.," vol. iii. p. 375.

*Uredo æcidioides.* I. Müller, *Rostp. der Rosa und Rubusarten*, 1886, p. 24, t. ii. fig. 12.

On *Rubus fruticosus.* Haywood Forest, Hereford; C. B. P. Ireland; Mr. Greenwood Pim.

October to November.

Biology.—Müller suggests that this may be connected with *Chrysomyxa albida.*

**Uredo quercus.** Brond.

Sori hypophyllous, yellow, then orange, orbicular, minute, scattered or in clusters. Spores subglobose, echinulate, orange-yellow, 15–25 × 12–15μ.

*Synonym.*

Winter, *loc. cit.*, p. 254.

*Uredo quercus*, Brond. Duby, "Bot. Gall.," vol. ii. p. 893. Berk., "Eng. Flor.," vol. v. p. 383. Cooke, "Hdbk.," p. 526; "Micro. Fungi," 4th edit., p. 216.

*Exsiccati.*

Berk., 239. Cooke, i. 281; ii. 76. Vize, "Fungi Brit.," 58; "Micro. Fungi Brit.," 129.

On *Quercus pedunculata.*

**Uredo iridis.** (Thüm.)

Sori oval, linear, sometimes confluent, amphigenous, at first covered, then naked, chestnut-brown. Spores profuse, mostly globose, rarely oval or pyriform, nucleate, with two germ-pores, brown, echinulate, 30–35 × 20–25μ.

*Synonym.*

*Uromyces iridis* (Lév.). Thüm., "Myco. Univ.," No. 2045.

On *Iris flavissima, spuria, ensata, decora, kingii, pumila, filifolia, caucasica, iberica, tolmicana.* In gardens.

June to September.

BIOLOGY.—This species occurs upon various species of cultivated iris. To Dr. M. Foster, F.R.S., I am indebted for numerous specimens. Repeated examinations at various periods of the year, both on the part of Dr. Foster and myself, have failed to show teleutospores. When applied to *I. pseudo-acorus* and *fœtidissima*, I have almost always failed in obtaining any results. Thümen's specimen in my copy consists only of uredospores, and is on *I. flavissima* from Siberia. Rabenhorst's *Uredo iridis* on *I. pumila* ("Fung. Europ.," No. 1674) seems to be the same plant. Dr. M. Foster says, "It came out with me first on *I. flavissima*, then on *I. tolmieana*, then on *I. iberica*, and all the irises from Central Asia; also on *I. spuria*, *ensata;* finally on *I. decora*, *kingii*, *pumila*, *filifolia*, and *caucasica*. It does not readily attack the broad-leaved Mediterranean forms, such as *I. germanica*, *pallida*, etc.; but I am inclined to think that almost every species would take it. I have not yet, however, seen it on *I. fœtidissima* or *pseudo-acorus*."

### **Uredo phillyreæ.** Cooke.

Sori yellow, rounded or elliptical, on yellowish spots, solitary or in clusters. Spores subglobose, ovate, finely echinulate, orange-yellow, 15–18 × 12–15$\mu$.

*Synonym.*

*Uredo phillyreæ.* Cooke, Exs.

*Exsiccati.*

Cooke, i. 592. Vize, "Fungi Brit.," 142.

On *Phillyrea media.*

### **Uredo tropæoli.** Desm.

Hypogenous spots, pale yellow. Sori minute, roundish, scattered or confluent. Spores ovoid or subglobose, orange.

*Synonym.*

*Uredo tropæoli.* Desm., *Ann. Nat. Sc.*, 1836, vol. vi. p. 243. Cooke, *Seem. Jour. Bot.*, vol. iv. p. 97; "Hdbk.," p. 528; "Micro. Fungi," 4th edit., p. 216.

*Exsiccati.*

Desm., i. 837.

On *Tropæolum aduncum.*

October.

This species, which can hardly be regarded as British, was found by Dr. Capron, in 1865, near Guildford.

**Uredo lynchii.** (B. and Br.)

Spots small, pallid, scattered, rarely confluent. Spores yellow, obovate, beautifully echinulate, with short pedicels, 20–30 × 28–35$\mu$.

*Synonym.*

*Trichobasis lynchii.* B. and Br., *Ann. Nat. Hist.*, No. 1706. W. G. Smith, *Gard. Chron.*, May 30, 1885, p. 693, fig. 154.

On exotic *Spiranthes.* Kew Gardens.

Not truly a British species.

**Uredo plantaginis.** B. and Br.

Spots pallid, pustular, minute, soon ruptured at the apex. Spores elliptical, yellow.

*Synonym.*

*Uredo plantaginis.* B. and Br., *Ann. Nat. Hist.*, No. 1905.

On *Plantago lanceolata.* Wood Newton, Dolgelly.

There is some doubt as to whether this is not a *Synchytrium.*

## CÆOMA.

Spores without a pseudoperidium, accompanied by spermogonia, with or without paraphyses, produced in chains. Germination as in Uredo.

**Cæoma saxifragæ.** (Strauss.)

Sori rounded or oval, flat, orange, amphigenous, scattered. Spores subglobose, finely verrucose, orange-yellow, 17–30$\mu$ in diameter.

*Synonyms.*

Winter, *loc. cit.*, p. 258.

*Uredo polymorpha*, var. *saxifragæ.* Strauss in Wetter., "Ann.," vol. ii. p. 87.

*Uredo saxifragarum*, D.C. Berk., "Eng. Flor.," vol. v. p. 381. Cooke, "Hdbk.," p. 525; "Micro. Fungi," 4th edit., p. 215. Grev., "Flor. Edin.," p. 440.

On *Saxifraga granulata.*
June to September.

### Cæoma euonymi. (Gmelin.)

Spermogonia flat, circinating in small clusters. Sori rounded, often confluent, on the leaves and young branches. Spores round, elliptical, or subpyriform, finely verrucose, pale yellow, 13–25 × 25–30μ.

*Synonyms.*

Winter, *loc. cit.*, p. 259.

*Æcidium euonymi.* Gmelin, "Linné Syst. Nat.," vol. ii. p. 1473.

*Uredo euonymi*, Mart. Cooke, "Micro. Fungi," 4th edit., p. 216.

On *Euonymus europæus.*
August to September.

### Cæoma mercurialis. (Pers.)

Sori flat, roundish or irregular, often circinating and confluent on large yellow spots, orange-yellow. Spores elliptical or subpyriform, finely verrucose, orange-yellow, 17–25 × 10–15μ.

*Synonyms.*

Winter, *loc. cit.*, p. 257.

*Uredo confluens*, var. *mercurialis.* Pers., "Syn.," p. 214.

*Uredo confluens*, D. C. Berk., "Eng. Flor.," vol. v. p. 383. Cooke, "Hdbk.," p. 527; "Micro. Fungi," 4th edit., p. 216, t. vii. figs. 133, 134. Grev., "Flor. Edin.," p. 438. Sow., t. 397, fig. 6.

*Exsiccati.*

Cooke, i. 117; ii. 426. Vize, "Fungi Brit.," 144.

On *Mercurialis perennis.*
May and June.

Biology.—This is stated by Rostrup and Nielsen to be the Æcidium of *Melampsora tremulæ* from *Populus tremula* and *alba*. I repeated their experiments by placing the teleutospores from *P. tremula* on *Mercurialis perennis* on several occasions, but was unable to confirm their statement. The *Cæoma pinitorquum*, which these authors also state to be connected with a form of *M. tremulæ*, has not yet occurred in Britain.

In October, 1887, however, during the annual excursions of the Woolhope Club, I gathered a number of leaves of *Populus alba*, on Dinmore Hill, with a Melampsora on them. In April and May, 1888, I infected with this Melampsora some plants of *Mercurialis perennis*, which had been growing for several years in my garden at King's Lynn, and produced first the spermogonia, and then the perfect Cæoma.

### **Cæoma alliorum.** Link.

Spermogonia flat, yellowish, in circular groups. Sori amphigenous in circular or elliptical clusters. Spores subglobose or elliptical, punctate, yellow, $17–25 \times 10–20\mu$.

*Synonyms.*

*Cæoma alliatum.* Link (?), "Sp. Plant.," vol. ii. p. 43.

*Uredo alliorum*, D. C. Cooke, "Micro. Fungi," 4th edit., p. 217, in part.

On *Allium ursinum, oleraceum, vineale.*

Biology.—Schröter thinks this species, or at least the form of it that occurs on *A. oleraceum*, is connected with *Melampsora populina* "Krypt. Flor. Schl.," vol. iii. p. 363).

### **Cæoma orchidis.** Alb. and Schw.

Spermogonia on yellowish spots. Sori generally irregular, flat, often confluent. Spores subglobose or ovate, punctate, yellow. $15–25 \times 10–20\mu$.

*Synonyms.*

Winter, *loc. cit.*, p. 256.

*Uredo confluens*, var. *orchidis.* Alb. and Schw., "Consp.," p. 122.

*Uredo orchidis*, Mart. Cooke, "Hdbk.," p. 527; "Micro. Fungi," 4th edit., p. 216.

*Exsiccati.*

Cooke, i. 61; ii. 323.

On *Orchis maculata, latifolia, Listera ovata.*

May and June.

BIOLOGY.—In view of so many statements having been made that the Cæomata are the æcidiospores of the Melampsoræ, I have made several cultures with the teleutospores of *M. betulinæ* on *Orchis maculata* and *latifolia*, but always without result (Exp. 256, 284, 343). I was led to do so from finding *C. orchidis* under a birch-tree affected with the Melampsora, but I feel sure there is no connection between these two fungi.

**Cæoma laricis.** (Westd.).

Sori minute, on yellow spots, surrounded by the ruptured epidermis and a number of barren cells. Spores subglobose or elliptical, finely echinulate, orange-yellow, 15–25 × 12–18$\mu$.

*Synonym.*

Winter, *loc. cit.*, p. 256.

*Uredo laricis.* Westd., *Bullet. Acad. Roy. Belgique*, 2nd sér., tome xi., No. 34.

On *Pinus larix.*

May and June.

BIOLOGY.—This species is not uncommon early in the year on larch foliage, but is very inconspicuous and easily overlooked. From certain reasons I was led to apply the spores to a plant of *Senecio jacobæa*, but they produced no result. The investment of barren cells shows its relationship to the Æcidia. Hartig has more recently asserted that this species is connected with one form of *Melampsora tremulæ*, and Professor Trail has found them growing in company near Aberdeen (see p. 241).

## ÆCIDIUM.

Spores surrounded by a pseudoperidium; produced in basipetal series. Germination as in Uredo.

* Mycelium localized, short-lived.

**Æcidium aquilegiæ.** Pers.

Pseudoperidia crowded upon round, thickened, yellow spots on the leaves, or upon elongated swellings on the stems, cylindrical, with white, torn, everted edges, Spores subglobose, finely verrucose, orange, 16–30 × 14–20μ.

*Synonym.*

Winter, *loc. cit.*, p. 268.

*Æcidium aquilegiæ.* Pers, "Ic. Pict.," p. 58, t. 23, fig. 4. Cooke, "Hdbk.," p. 539.

*Exsiccati.*

Vize, "Micro. Fungi Brit.," 58.

On *Aquilegia vulgaris.*
May and June.

**Æcidium grossulariæ.** (Gmelin).

Pseudoperidia on concave thickened spots, which are reddish above and yellow below, roundish on the leaves, and sometimes covering the whole surface of the young fruit, and even on the seeds, shortly cylindrical, hypophyllous, with much-torn white edges. Spores subglobose or polygonal, yellow, finely echinulate, 15–24 × 12–18μ.

*Synonyms.*

Winter, *loc. cit.*, p. 198.

*Æcidium rubellum, β grossulariæ*, Gmelin. "Linné Syst. Nat.," vol. ii. p. 1473.

*Æcidium grossulariæ*, D. C. Grev., t. 62. Berk., "Eng. Flor.," vol. v. p. 372. Cooke, "Hdbk.," p. 541; "Micro. Fungi," 4th edit., p. 197.

*Exsiccati.*

Cooke, i. 10. Vize, "Fungi Brit.," 135; "Micro. Fungi Brit.," 328.

On *Ribes grossularia.*
May and June.

BIOLOGY.—This species varies very much in frequency in different seasons, sometimes being very abundant; at other times scarcely a specimen can be found.

### **Æcidium periclymeni.** Schum.

Pseudoperidia on roundish, pallid, yellow spots on the leaves, hypogenous, cylindrical, with white, irregularly torn edges. Spores subglobose or polygonal, finely verrucose, yellow, 15–27 × 15–25μ.

*Synonym.*

*Æcidium periclymeni.* Schum., "Enum. Plant. Sæll.," vol. ii. p. 225. Cooke, "Hdbk.," p. 539; "Micro. Fungi," 4th edit., p. 196. Winter, *loc. cit.*, p. 264.

*Exsiccati.*

Cooke, i. 102; ii. 96; "L. F.," 54. Vize, "Fungi Brit.," 78.

On *Lonicera periclymenum.*

BIOLOGY.—I have made a great number of experimental cultures with this species during the past five years, but hitherto entirely without success.

### **Æcidium prunellæ.** Winter.

Pseudoperidia on yellowish brown or violet-edged spots, generally arranged concentrically, cup-shaped, white, with whitish, rather torn and everted edges. Spores colourless, polygonal, finely echinulate, 16–21μ in diameter.

*Synonym.*

*Æcidium prunellæ.* Winter in Rabh. (sub. *Uromyces*), "Krypt. Flor.," vol. i. p. 164.

On *Prunella vulgaris.*
Forres; Rev. Dr. Keith.

### **Æcidium convallariæ.** Schum.

Pseudoperidia on pale or yellowish spots, which are roundish on the leaves and elongate on the stems, flat, with torn white edges. Spores subglobose, finely verrucose, yellow, 15–30 × 14–22μ.

*Synonym.*

*Æcidium convallariæ.* Schum., "Enum. Plant. Sæll.," vol. ii. p. 224. Winter, *loc. cit.*, p. 259.

On *Convallaria majalis.* Mr. Birks.
May to July.

**Æcidium barbareæ.** D. C.

Pseudoperidia amphigenous on rubescent spots, in irregular clusters, large, cup-shaped, rather flat, edges whitish, torn or crenulate. Spores orange, subglobose, finely verrucose, 15–25μ in diameter.

*Synonym.*

*Æcidium barbareæ.* D. C., "Flore franç.," vol. ii. p. 244. Duby, "Bot. Gall.," vol. ii. p. 905. Cooke, "Grevillea," vol. x. p. 115. Winter, *loc. cit.*, p. 267.

On *Barbarea præcox.* Plymouth ; Mr. Varenne.

BIOLOGY.—Sometimes the pseudoperidia occupy the whole leaf-surface, causing considerable distortion.

**Æcidium clematidis.** D. C.

Pseudoperidia in roundish clusters on the leaves, often upon swellings on the pedicels and young shoots, causing considerable distortion, shortly cylindrical or flat, with white torn edges. Spores subglobose, minutely verrucose, yellow, 20–27μ in diameter.

*Synonym.*

Winter, *loc. cit.*, p. 270.

*Æcidium ranunculacearum*, var. *clematidis.* D. C., "Flore franç.," vol. ii. p. 243. Cooke, "Hdbk.," p. 539.

On *Clematis vitalba.*
May to July.

BIOLOGY.—Ráthay states that this is connected with *Melampsora populina*, Tul. ("Über auto. und heterœcische Uredineen," *Verhand. zool. bot. Gesell.*, bd. xxxi. p. 13. Wein : 1881.) I have tried this culture on four occasions in 1886, with *M. populina*, from *P. pyramidalis*, but always without success; and in 1888, from *P. alba*, with the same result.

### Æcidium sonchi. Johnst.

"Minute, scattered, whitish or cream-coloured, prominent. Spores ovate, rather large."

*Synonym.*

Johnst., "Flor. Berw.," vol. ii. p. 205.

On the under surface of the leaves of *Sonchus arvensis*, in autumn.

"The cover does not split so regularly, and is not so decidedly cupped, as is common in this genus."

### Æcidium ranunculacearum. D. C.

Pseudoperidia hypophyllous in roundish or elongated clusters of various sizes, cup-shaped, whitish, margin brittle. Spores polygonal, orange-yellow, 17–28 × 14–20μ.

*Synonym.*

Winter, *loc. cit.*, p. 269. Grev., "Flor. Edin.," p. 446. D. C., "Flore franç.," vol. vi. p. 97.

On *Ranunculus lingua*. Duddingston Loch; Dr. Greville.

### Æcidium dracontii. Schw.

Pseudoperidia on extensive pallid spots on the leaves, sometimes almost covering them, arranged without order, elongate. Spores orange.

*Synonym.*

*Æcidium dracontii.* Schw., *Trans. Am. Phil. Soc.*, 1834. Cooke, "Hdbk.," p. 538; "Micro. Fungi," 4th edit., p. 200.

On *Arum triphyllum*. In gardens, Melbury, 1863; M. J. B. This can hardly be considered a British species.

### Æcidium strobilinum. (Alb. and Schw.)

Pseudoperidia numerous, in clusters on the inner surface of the scales of the fallen cones, hemispherical or polygonal from mutual pressure, about 1 mm. wide, mostly opening obliquely. Spores irregularly rounded or sub-elliptical, yellow, 18–34 × 16–23μ.

*Synonyms.*

Winter, *loc. cit.*, p. 260.

*Licea strobilina.* Alb. and Schw., "Consp.," p. 109, t. vi. fig. 5.

*Phelonitis strobilina*, Pers. Cooke, "Hdbk.," p. 409.

*Perichæna strobilina.* Fries, "Syst. Myc.," vol. iii. p. 190. Berk., "Eng. Flor.," vol. v. p. 321. Grev., t. 275.

*Exsiccati.*

Berk., 292. Cooke, i. 522.

On the scales of the cones of *Pinus abies.*

**Æcidium incarceratum.** B. and Br.

"Sori minute, crowded in irregular spots. Peridia included in the parenchyma of the leaf. Pseudospores pallid."

*Synonym.*

*Æcidium incarceratum.* B. and Br. in Rabh., "Fung. Europ.," No. 1492.

On *Sagittaria sagittifolia.*

I can detect but little difference between this and *Doassansia sagittariæ.*

**Æcidium phillyreæ.** D. C.

Pseudoperidia mostly thickly crowded, in large numbers, with almost entire, slightly inverted edges. Spores very variously shaped, roundish, oval, oblong, or pyriform, verrucose, orange-yellow, 18–35 × 15–20μ.

*Synonym.*

*Æcidium phillyreæ.* D. C., "Flore franç.," vol. vi. p. 96. Cooke, "Hdbk.," p. 539. Winter, *loc. cit.*, p. 263.

*Exsiccati.*

Thüm., "Myc. Univ.," 1717. Vize, "Micro. Fungi Brit.," 236.

On *Phillyrea media.*

BIOLOGY.—This is a doubtful Æcidium, being little more than a condition of *Cæoma phillyreæ*, as far as I have observed it.

**Æcidium glaucis.** Dozy and Molk.

Pseudoperidia irregularly scattered, sometimes circinating, in various-sized groups, shortly cylindrical or flattish, with white torn edges. Spores polygonal, finely granular, colourless, 16–25 × 15–20$\mu$.

*Synonym.*

*Æcidium glaucis.* Dozy. and Molk., *Tijdschr. v. Natur. Ges.*, vol. xii. p. 16. Cooke, "Grevillea," vol. xv. p. 29. Winter, *loc. cit.*, p. 262.

On *Glaux maritima.*

**Æcidium poterii.** Cooke.

Pseudoperidia hypogenous in subrotund or elongate clusters on the leaves, also upon the petioles, scattered or circinate, immersed, edges torn into minute fugacious teeth. Spores yellowish, oval.

*Synonym.*

*Æcidium poterii.* Cooke, *Seem. Jour. Bot.*, vol. ii. p. 39, t. xiv. fig. 3; "Hdbk.," p. 540; "Micro. Fungi.," 4th edit., p. 198.

On *Poterium sanguisorba.*

May and June.

Probably this is the species referred to by Mr. Berkeley in the "Eng. Flor.," vol. v. p. 373, as having been found by Dr. Greville.

** Mycelium perennial, diffused in the host-plant.

**Æcidium punctatum.** Pers.

Pseudoperidia uniformly scattered over the whole of the foliage, hypophyllous, flat, semi-immersed, with torn yellowish edges. Spores subglobose, pale yellowish-brown, 15–24$\mu$ in diameter. Spermogonia scattered, blackish, punctiform.

*Synonyms.*

Winter, *loc. cit.*, p. 269.

*Æcidium punctatum*, Pers. *Ann. Bot.*, 1796, vol. xx. p. 135.

*Æcidium quadrifidum*. D. C., "Flor. franc.," vol. vi. p. 190. Berk., "Eng. Flor.," vol. v. p. 371. Cooke, "Hdbk.," p. 536; "Micro. Fungi," 4th edit., p. 194.

*Exsiccati.*

Berk., 227. Cooke, i. 101; ii. 310. Vize, "Fungi Brit.," 73.

On *Anemone ranunculoides* (*coronaria*).
April and May.

BIOLOGY.—The affected leaves are thicker, and borne on longer stems, than the healthy ones.

**Æcidium leucospermum.** D. C.

*Ecidiospores*—Spermogonia scattered equally over the whole surface of the leaves, and generally upon all the leaves of the affected plants, which are altered in habit thereby. Pseudoperidia shortly cylindrical, white, with torn edges. Spores polygonal, colourless, smooth, 15–22$\mu$ in diameter, sometimes 26$\mu$ long. Mycelium perennial.

*Synonyms.*

*Æcidium leucospermum*, D. C. Berk., "Eng. Flor.," vol. v. p. 371. Cooke, "Hdbk.," p. 536; "Micro. Fungi," 4th edit., p. 194, t. i. figs. 4–6.

*Puccinia fusca* (Relh.). Winter in Rabh., "Krypt. Flor.," vol. i. p. 199, in part. Schröt., "Krypt. Flor.," vol. iii. p. 343, in part.

*Lycoperdon innatum*. Withering, vol. iv. p. 352.

*Exsiccati.*

Baxter, 89. Berk., 220. Cooke, i. 3; ii. 77. Vize, "Fungi Brit.," 74.

On *Anemone nemorosa*.
April and May.

BIOLOGY.—Although Winter and Schröter consider this species to be the Æcidium of *Puccinia fusca*, yet, in the absence of direct biological evidence, I venture to differ from them. The Puccinia in question is very common, but the Æcidium is rare in Britain.

### **Æcidium bunii.** D. C.

*Æcidiospores*—Pseudoperidia generally on swellings on the stems, shortly cylindrical, with whitish, torn, everted edges. Spores polygonal, finely verrucose, orange-yellow, 15–20$\mu$ in diameter, sometimes 25$\mu$ long.

*Synonym.*

*Æcidium bunii*, D. C. "Syn.," p. 51. Berk., "Eng. Flor.," vol. v. p. 370. Cooke, "Hdbk.," p. 540; "Micro. Fungi," 4th edit., p. 196. Grev., "Flor. Edin.," p. 445.

On *Conopodium denudatum* (*Bunium flexuosum*).

BIOLOGY.—This Æcidium is comparatively rare in Britain, and I have never been able to obtain any evidence that it is connected with the Puccinia which occurs on the same host-plant.

### **Æcidium euphorbiæ.** Gmelin.

Pseudoperidia equally scattered over the whole leaf-surface, at first conical, then cup-shaped, with whitish torn edges. Spores polygonal or subglobose, finely verrucose, orange-yellow, 19–26 × 30–35$\mu$.

*Synonym.*

Winter, *loc. cit.*, p. 261.

*Æcidium euphorbiæ.* Gmelin, "Linné Syst. Nat.," vol. ii. p. 1473, in part. Purton, "Midl. Flor.," vol. iii. No. 1537 (?).

On *Euphorbia exigua.*
Hampshire; Mr. Hill.

### **Æcidium elatinum.** Alb. and Schw.

Mycelium perennial in the affected branches, causing fusiform swellings, from which are given off deformed shoots (*Hexenbesen*), which bear pale green, short, swollen leaves on all sides. Spermogonia immersed, conical, honey-coloured. Pseudoperidia white, opening irregularly. Spores elliptical or polygonal, coarsely verrucose, 16–30 × 15–17μ.

*Synonyms.*

*Æcidium elatinum.* Alb. and Schw., "Consp.," p. 121.

*Peridermium elatinum*, Alb. and Schw. Cooke. *Seem. Jour. Bot.*, vol. ii. p. 34; "Hdbk.," p. 535; "Micro. Fungi," 4th edit., p. 194.

On *Pinus picea.*

### Æcidium pseudo-columnare. J. Kühn.

Pseudoperidia in two rows on the under side of the affected leaves, which are not altered in length or breadth, but are paler than the healthy ones. Pseudoperidia spherical, ovate, or elongate, ·5–2 mm. high, edges irregularly torn. Spores white, finely verrucose, ovate, elongate elliptical, or irregular, often triangular in section, 33–37 × 18–25μ.

*Synonyms.*

J. Kühn, "Hedwigia," 1884, p. 168.

*Peridermium columnare*, Alb. and Schw. Cooke, "Hdbk.," p. 535; "Micro. Fungi," 4th edit., p. 194, t. ii. figs. 27, 28.

*Exsiccati.*

Vize, "Micro. Fungi Brit.," 60. Cooke, i. 214.

On *Abies pectinata, nordmaniana, amabilis, cephalonica.*
Lyme Regis, Mr. Munro.

Biology.—*Æcidium columnare* was found by De Bary to be the Æcidium of *Melampsora goeppertiana*, Kühn; but the teleutospores have not up to the present time been found in England, although their host-plant, *Vaccinium vitis-idæa*, is by no means rare. I therefore submitted specimens of this species to Professor Kühn, who found them to be his *Æc. pseudo-columnare*, the life-history of which is distinct from *Æc. columnare.*

# DESCRIPTIONS OF THE BRITISH USTILAGINEÆ.

## **USTILAGINEÆ.** Schröter.

Fungi parasitic in the tissues of living plants. Mycelium widely spreading, but soon vanishing. Teleutospores produced in the interior of mycelial branches, which often become gelatinized. Germination of the spores by a promycelium and promycelial spores (sporidia), or by continuous sprouting, after the manner of yeast-spores.

### **Ustilago.** Pers.

Teleutospores simple, produced in the interior of much gelatinized swollen hyphæ; when mature forming pulverulent masses. Germination by a septate promycelium, which produces promycelial spores (sporidia) both laterally and apically, but principally the former.

* Teleutospores smooth or obscurely punctate.

### **Ustilago longissima.** (Sow.)

Forming long parallel lines, mostly on the upper side of the leaves, soon pulverulent, olive-brown. Teleutospores globose, sometimes discoid, often irregular, pale olive-brown, smooth, 3·5–4$\mu$ in diameter, 6–7$\mu$ long. Promycelia very short.

*Synonyms.*

*Ustilago longissima.* Winter in Rabh., "Krypt. Flor.," vol. i. p. 85.

*Ustilago longissima*, Tul. Cooke, "Hdbk.," p. 512; "Micro. Fungi," 4th edit., p. 229, t. v. figs. 105–107.

*Uredo longissima.* Sow., t. 139. Berk., "Eng. Flor.," vol. v. pt. ii. p. 375.

*Exsiccati.*

Baxter, 230. Cooke, i. 55; ii. 71. Vize, "Fungi Brit.," 33.

On *Glyceria aquatica* and *fluitans*.

Throughout the summer and autumn.

BIOLOGY.—The affected plants seldom if ever develop their inflorescence. Intermixed with the teleutospores are a number of pale cells, the function of which is unknown. For an account of the germination of the teleutospores, see p. 81.

### Ustilago hypodytes. (Schlecht.)

Produced on the culms, beneath the leaf-sheaths, soon exposed, black, with a tinge of olive, pulverulent. Teleutospores subglobose, oblong, or angular, pale yellowish brown, subtransparent, smooth, 3–6 × 3–4·5$\mu$, mixed with others much larger.

*Synonyms.*

*Cæoma hypodytes.* Schlecht., "Flor. Berol.," vol. ii. p. 129.

*Ustilago hypodytes*, Fries. Winter, *loc. cit.*, p. 87. Cooke, "Hdbk.," p. 513; "Micro. Fungi," 4th edit., p. 229, t. v. figs. 100, 101.

*Exsiccati.*

Cooke, i. 56; ii. 433. Vize, "Fungi Brit.," 35.

On *Triticum repens*, *junceum*, *Elymus arenarius*, *Bromus erectus*, *Psamma arenaria*.

June to November.

BIOLOGY.—See p. 80.

### Ustilago segetum. (Bull.)

Produced in the receptacle and rachis, destroying the whole inflorescence, dusty, black, with an olive-brown lustre. Teleutospores globose or irregular, often flattened on one side, pale yellowish-brown, subtransparent, smooth or obscurely punctate or granular, 5–8$\mu$ in diameter.

*Synonyms.*

*Reticularia segetum.* Bull., "Champ.," vol. i. p. 90, t. 472, fig. 2. With., vol. iv. p. 356. Purton, vols. ii., iii. No. 1079. Johnst., "Flor. Berw.," vol. ii. p. 203. Grev., "Flor. Edin.," p. 442.

*Uredo segetum.* Pers., "Disp. Meth.," p. 56. Berk., "Eng. Flor.," vol. v. p. 314.

*Ustilago carbo,* Tul. Cooke, "Hdbk.," p. 512.

*Ustilago segetum* (Bull.). Winter, *loc. cit.*, p. 90. Cooke, "Micro. Fungi," 4th edit., p. 229, t. v. figs. 98, 99.

*Exsiccati.*

Cooke, i. 54; ii. 428, 430, 432. Baxter, 43. Vize, "Fungi Brit.," 31.

On *Avena elatior, sativa, Triticum vulgare, Hordeum vulgare, distichum, hexactichum.*

May and June.

BIOLOGY.—See p. 74, and p. 101. Mr. J. L. Jensen, of Copenhagen, has recently published his observations on the biology of this species. He has succeeded in infecting oats, barley, and wheat, by removing the external envelopes of the seeds and applying the spores to the bare kernels. He finds that the varieties of *U. segetum* which occur upon the above plants are biologically distinct, the Ustilago from the one cereal being incapable of infecting the others. Upon barley he finds two well-marked forms, var. *tecta* and *nuda.* Of these *nuda* is by far the most common. The spores have a tinge of yellow when seen *en masse*; they are rather smaller than those of *tecta.* The affected ears are completely destroyed, excepting the rachis, and the spores soon scattered. *Tecta*, on the other hand, has somewhat larger spores, which, when seen *en masse*, are jet-black. The affected ears are nearly twice as broad as the healthy ones, and do not emerge from the top of the leaf-sheath, but burst through laterally. Each of the affected kernels retains its shape for a considerable time, being enclosed in a membranous investment. The spores escape through minute fissures which appear in this membrane. *Tecta* has been found in the island of Iona, but it doubtless occurs all over Britain.

### Ustilago grandis. Fries.

On the caulms beneath the leaf-sheath, often occupying whole internodes, black, dusty. Teleutospores globose, elongated or irregular, pale brown, smooth, 7–12 × 6–8μ.

*Synonyms.*

*Ustilago grandis.* Fries, "Syst. Mycol.," vol. iii. p. 518. Schröt., "Krypt. Flor. Schl.," vol. iii. p. 268. Winter, *loc. cit.*, 87.

*Ustilago typhoides.* B. and Br., *Ann. Nat. Hist.*, No. 480. Cooke, "Hdbk.," p. 513; "Micro. Fungi," 4th edit., p. 229, t. vi. figs. 128, 129.

On *Phragmites communis.*

July to November.

BIOLOGY.—See p. 83.

### Ustilago grammica. B. and Br.

Forming little transverse bands, consisting of short, parallel black lines, 2 mm. or more long. Teleutospores globose, very minute.

*Synonym.*

*Ustilago grammica.* B. and Br., *Ann. Nat. Hist.*, No. 483. Cooke, "Hdbk.," p. 514; "Micro. Fungi," 4th edit., p. 229, t. vi. figs. 120, 121.

On *Aira cæspitosa* and *Glyceria aquatica.*

I have never met with this species; and am unable to give any details concerning its spore-germination.

### Ustilago marina. Durieu.

Spore-masses blackish brown, forming swellings on the roots. Teleutospores of two forms: (*a*) globose or obtusely ovoid or elongated, 10–13μ; (*b*) irregularly ovoid elongate, 16 × 10–13μ, olive-brown, smooth.

*Synonym.*

F. v. Waldh., "Syst. des Ust.," p. 17.

*Ustilago marina.* Cooke, "Grevillea," vol. xiv. p. 90.

On *Scirpus parvulus.*

**Ustilago hypogæa.** Tul.

Spore-masses black, compact, formed round the root-stock, intersected by white fibres. Teleutospores rounded or rounded polygonal, dark brown, scarcely transparent, smooth, contents very oleaginous, 20–24 × 14–20μ.

*Synonym.*

*Ustilago hypogæa.* Tulasne, "Fung. Hypogæi," 1862, p. 196. F. v. Waldh., "Monograph," p. 18. "Grevillea," vol. xiii. p. 52.

On *Linaria spuria.* Isle of Wight.

**Ustilago caricis.** (Pers.)

Spore-masses black, produced within the glumes, forming a firm, black, globose body. Teleutospores very irregular in form, spherical, elliptical, discoid or angular. Epispore dark brown, opaque, obscurely punctate, 12–24 × 7–20μ.

*Synonyms.*

Winter, *loc. cit.*, p. 92. Schröt., *loc. cit.*, p. 270.

*Uredo caricis.* Pers., "Syn.," p. 225.

*Uredo urceolorum*, D. C. Berk., "Eng. Flor.," vol. v. p. 375. Grev., "Flor. Edin.," p. 443. Johnst., "Flor. Berw.," vol. ii. p. 204.

*Ustilago urceolorum*, Tul. Cooke, "Hdbk.," p. 512. "Micro. Fungi," 4th edit., p. 229, t. vi. figs. 109–111.

*Ustilago montagnei.* B. and Br., No. 479. Cooke, "Hdbk.," p. 513; "Micro. Fungi," 4th edit., p. 229, t. v. figs. 96, 97.

*Farinaria carbonaria.* Sow., t. 396, fig. 4.

*Exsiccati.*

Cooke, i. 541. Berk., 114. Vize, "Micro. Fungi," 131.

On *Carex præcox, stellulata, recurva, glauca, dioica, vulgaris, panicea, pseudocyperus, hirta, Rhyncospora alba.*

June to August.

### Ustilago bistortarum. (D. C.)

Produced in bullate masses in the leaves, at first hemispherical, then bursting, black. Teleutospores subglobose, dark violet, obscurely punctate, 10–20 × 12–15$\mu$.

*Synonyms.*

Schröt., *loc. cit.*, p. 271. Winter, *loc. cit.*, p. 95.

*Uredo bistortarum*, var. D. C., "Flore franç.," vol. vi. p. 76.

*Tilletia bullata*, Fckl. Cooke, "Micro Fungi," 4th edit., p. 233.

On *Polygonum bistorta* and *Rumex sp.* (?)
July and August.

### Ustilago olivacea. (D. C.)

Produced in the fructification, at length pulverulent and mixed with filaments, dark olive-brown. Teleutospores variable in form and size, roundish, angular, ovate, elongate, or subcylindrical and curved, pale olive-brown, smooth or finely punctate. The globose spores are about 5$\mu$ in diameter, the elongate 5–15 × 4–5$\mu$.

*Synonyms.*

Winter, *loc. cit.*, p. 91. Schröt., *loc. cit.*, p. 269.

*Uredo olivacea.* D. C., "Flore franç.," vol. vi. p. 78. Berk., "Eng. Flor.," vol. v. p. 376.

*Ustilago olivacea*, Tul. Cooke, "Hdbk.," p. 513; "Micro. Fungi," 4th edit., p. 230, t. vi. figs. 126, 127.

*Exsiccati.*

Cooke, i. 298; ii. 435. Vize, "Fungi Brit.," 32.

On *Carex riparia.*
June and July.

BIOLOGY.—The function and origin of the filaments are unknown. See p. 84.

**Ustilago bromivora.** (Tul.)

Produced in the inflorescence, soon pulverulent, black. Teleutospores spherical or oval, dark brown, obscurely punctate, 7–12 × 7–10$\mu$.

*Synonyms.*

*Ustilago carbo*, var. *bromivora*. Tul., *Ann. Sc. Nat.*, ser. iii., vol. vii. p. 81.

*Ustilago bromivora*, Waldh. Cooke, "Micro. Fungi," 4th edit., p. 230. Winter, *loc. cit.*, p. 91. Schröt., *loc. cit.*, p. 260.

On *Bromus secalinus* and *mollis*.
May and June.

Biology.—See p. 83.

** Teleutospores echinulate or verrucose.

**Ustilago maydis.** (D. C.)

Olive-brown, dusty, in the ovule and female inflorescence, and on the stems and leaves. Teleutospores subglobose, rarely elongate, pale brown, closely echinulate, 10–13 × 8–10$\mu$.

*Synonym.*

Schröt., *loc. cit.*, p. 271. Winter, *loc. cit.*, p. 97.

*Ustilago maydis*, Corda. Cooke, "Hdbk.," p. 513; "Micro. Fungi," 4th edit., p. 230, t. v. fig. 108.

*Exsiccati.*

Cooke, i. 433; ii. 431.

On *Zea mäys*.
September and October.

Biology.—See p. 79.

**Ustilago vinosa.** (Berk.)

Spore-masses black, produced in the swollen flower-buds. Teleutospores spherical, very pale violet, transparent, thickly covered with large hemispherical warts, 10–12 × 7–10$\mu$.

*Synonyms.*

*Uredo vinosa.* Berk. in litt.

*Ustilago vinosa*, Tul. Cooke, "Hdbk.," p. 514; "Micro. Fungi," 4th edit., p. 230.

On *Oxyria reniformis.*

*** Teleutospores reticulated.

**Ustilago scabiosæ.** (Sow.)

Produced in the anthers and filling the whole of the florets, pale flesh-colour. Teleutospores subglobose, nearly colourless, finely reticulated with low ridges, 10–12 × 8–10$\mu$.

*Synonyms.*

Winter, *loc. cit.*, p. 99. Schröt., *loc. cit.*, p. 272.

*Farinaria scabiosæ.* Sow., "Eng. Fung.," t. 396, fig. 2.

*Ustilago flosculorum*, Tul. Cooke, "Hdbk.," p. 515; "Micro. Fungi," 4th edit., p. 231, t. vi. figs. 123–125, in part.

On *Scabiosa arvensis.*
June to September.

BIOLOGY.—See p. 79.

**Ustilago flosculorum.** (D. C.).

Produced in the anthers, white, pale brownish, dirty violet or violet. Teleutospores transparent, seldom quite spherical, irregularly round, rarely sub-elliptical, ridges low, forming close reticulations, 10–18 × 10–20$\mu$.

*Synonyms.*

Winter, *loc. cit.*, p. 99.

*Uredo flosculorum.* D. C., "Flore franç.," vol. vi. p. 79.

*Ustilago intermedia*, Schröt. Rabh., "Fung. Europ.," 1696.

*Ustilago succisæ.* Magnus, "Hedwigia," 1875, p. 17. Cooke. "Micro. Fungi," 4th edit., p. 230.

On *Scabiosa columbaria*, *arvensis*, and *succisa.*
July to October.

BIOLOGY.—See p. 79.

**Ustilago utriculosa.** (Nees.)

Spore-masses dark violet, produced within the perianth, causing the blossoms to become swollen. Teleutospores globose, transparent, violet, reticulated with very high ridges which form a wide network, 9–12$\mu$ in diameter.

*Synonyms.*

Winter, *loc. cit.*, p. 100. Schröt., *loc. cit.*, p. 273.

*Cæoma utriculosum.* Nees, "System," p. 14, t. i., fig. 6.

*Uredo utriculosa*, D. C. Berk., "Eng. Flor.," vol. v. p. 377.

*Ustilago utriculosa*, Tul. Cooke, "Hdbk.," p. 514; "Micro. Fungi," 4th edit., p. 230, t. vi. figs. 112–116.

*Exsiccati.*

Vize, "Micro. Fungi," 132.

On *Polygonum lapathifolium*, *persicaria*, *convolvulus*, and *hydropiper.*

June to September.

Biology.—See p. 85.

**Ustilago violacea.** (Pers.)

Spore-masses violet, pulverulent, produced in the anthers. Teleutospores rounded, violet, transparent, epispore covered by ridges which form a close network, 6–9$\mu$ in diameter.

*Synonyms.*

Winter, *loc. cit.*, p. 98. Schröt., *loc. cit.*, p. 273.

*Uredo violacea.* Pers., "Disp. Meth.," p. 57.

*Farinaria stellariæ.* Sow., "Eng. Fung.," t. 396, fig. 1.

*Uredo antherarum*, D. C. Grev., "Flor. Edin.," p. 443. Berk., "Eng. Flor.," vol. v. p. 381.

*Ustilago antherarum*, Fries. Cooke, "Hdbk.," p. 515; "Micro. Fungi," 4th edit., p. 230, t. v. figs. 102–104.

*Exsiccati.*

Cooke, ii. 427. Vize, "Fungi Brit.," 84.

On *Silene inflata*, *maritima*, *nutans*, *Cerastium viscosum*

*Stellaria graminea, holostea, Lychnis flos-cuculi, vespertina*, and *diurna.*

June to October.

BIOLOGY.—See p. 78.

### **Ustilago major.** Schröt.

Spore-masses blackish violet, produced in the anthers, pulverulent. Teleutospores spherical or elongated, violet, reticulated with ridges about $1\mu$ high and $1\mu$ distant from each other, $7-13 \times 7-9\mu$.

*Synonym.*

*Ustilago major.* Schröt. in Cohn's "Krypt. Flor. Schl.," vol. iii. p. 273.

On *Silene otites.* Near Brandon; Mr. Frank Norgate.

July and August.

BIOLOGY.—See p. 84.

### **Ustilago kühneana.** Wolff.

Produced on the stems, flowers, and on the leaves in the form of spots or striæ, rusty violet. Teleutospores globose, violet, rather transparent, reticulated with very slightly elevated ridges, which form a close network on the epispore, $10-18\mu$ in diameter.

*Synonym.*

Winter, *loc. cit.*, p. 98.

*Ustilago kühneana.* Wolff in *Bot. Zeit.*, 1874, p. 814. Cooke, "Micro. Fungi," 4th edit., p. 231.

On *Rumex acetosa* and *acetosella.*

June and July.

BIOLOGY.—See p. 80.

### **Ustilago tragopogi.** (Pers.)

Spore-masses blackish violet, destroying the whole of the flowers, enclosed in the involucre. Teleutospores globose, generally irregularly rounded, opaque, dark violet, with very fine reticulations, $13-17 \times 10-13\mu$.

*Synonyms.*

Schröt., *loc. cit.*, p. 274. Winter, *loc. cit.*, p. 101.
*Uredo tragopogi.* Pers., "Disp. Meth.," p. 57.
*Ustilago receptaculorum*, Fries. Cooke, "Hdbk.," p. 515; "Micro. Fungi," 4th edit., p. 230, t. v. figs. 92–95.

*Exsiccati.*

Cooke, i. 59; ii. 434. Vize, "Micro. Fungi," 134.

On *Tragopogon pratensis.*
May to July.

BIOLOGY.—See p. 80.

**Ustilago cardui.** F. v. Waldh.

Spore-masses brown, produced in the florets. Teleutospores globose or shortly elliptical, violet, with very high reticulating ridges, 15–20 × 10–15μ.

*Synonym.*

*Ustilago cardui*, Waldh. *Bull. Sc. Nat. de Moscou*, 1867, vol. i. Cooke, "Micro. Fungi," 4th edit., p. 231. Schröt., *loc. cit.*, p. 274. Winter, *loc. cit.*, p. 101.

On *Carduus acanthoides.*
July and August.

BIOLOGY.—See p. 78.

**SPHACELOTHECA.** De Bary.

Teleutospores contained in a compact receptacle, formed of barren cells provided with a columella, and open above. Teleutospores simple. Germination as in Ustilago.

**Sphacelotheca hydropiperis.** (Schum.)

Spore-masses black, elongated, projecting from the perianth, formed in the ovary, at length opening at the top and allowing the dusty teleutospores to escape. Teleutospores globose or elongated, dark violet, transparent, smooth, 9–12 × 8–11μ.

*Synonyms.*

Schröt. in Cohn's "Krypt. Flor. Schl.," vol. iii. p. 275.

*Uredo hydropiperis.* Schum., "Enum. Plant. Sæll.," vol. ii. p. 234.

*Ustilago candollei,* Tul. Cooke, "Micro. Fungi," 4th edit., p. 229.

*Exsiccati.*

Cooke, i. 58; ii. 72. Vize, "Micro. Fungi Brit.," 221.

On *Polygonum hydropiper.*
August to October.

BIOLOGY.—See p. 85.

## TILLETIACEI. Schröt.

Promycelium producing spores only at its apex. Conidia either produced on the host-plant or from mycelium.

### Tilletia. Tulasne.

Teleutospores simple, produced separately as outgrowths from the gelatinized mycelium, when mature, pulverulent. Conidia produced from mycelium in nährlösung.

### Tilletia tritici. (Bjerk.)

Spore-mass dusty, olive-black, within the ovary, when rubbed emitting an offensive fishy odour. Teleutospores spherical, olive-brown. Epispore reticulated with ridges, from 1 to 1·5$\mu$ high and about 4$\mu$ apart, 16–20$\mu$ in diameter.

*Synonyms.*

Winter in Rabh., "Krypt. Flor.," vol. i. p. 110. Schröt., *loc. cit.*, p. 277.

*Lycoperdon tritici.* Bjerk., *Act. Succ.*, 1775, p. 326.

*Uredo caries,* D. C. Grev., "Flor. Edin.," p. 443. Berk., "Eng. Flor.," vol. v. p. 373. Johnst., "Flor. Berw.," vol. ii. p. 204.

*Tilletia caries,* Tul. Cooke, "Hdbk.," p. 511; "Micro. Fungi," 4th edit., p. 223, t. v. figs. 84–91.

*Exsiccati.*

Baxter, 113. Cooke, i. 53. Vize, "Micro. Fungi," 130.

On *Triticum vulgare.*

BIOLOGY.—See p. 86.

**Tilletia decipiens.** (Pers.)

Spore-masses black, compact, produced within the ovary, fetid. Teleutospores globose, reticulated, ridges 2·5–3μ high and about 4μ apart, dark brown, 24–28μ in diameter.

*Synonyms.*

Schröt., *loc. cit.*, p. 278. Winter, *loc. cit.*, p. 111.

*Uredo segetum*, var. *decipiens.* Pers., "Syn.," p. 225.

*Tilletia sphærococca*, F. v. Waldh. Cooke, "Grevillea," vol. xii. p. 99.

On *Agrostis vulgaris.*

BIOLOGY.—Professor J. W. H. Trail states that the presence of the fungus in *Agrostis vulgaris* causes it to assume the form called *pumila*, usually regarded as a distinct variety of this grass.

**Tilletia striæformis.** (Westd.)

Spore-masses black, in parallel lines on the leaves, leaf-sheaths, and stems. Teleutospores globose or irregularly rounded, thickly covered with minute spicules, which towards the base of the spores tend to form reticulations, olive-brown, 10–15 × 9–12μ.

*Synonyms.*

Schröt., *loc. cit.*, p. 278. Winter, *loc. cit.*, p. 108.

*Uredo striæformis.* Westd., *Bull. Acad. de Brux.*, 1851, p. 406.

*Ustilago salveii.* B. and Br., *Ann. Nat. Hist.*, No. 482. Cooke, "Hdbk.," p. 514; "Micro. Fungi," 4th edit., p. 230, t. vi. figs. 117–119.

*Exsiccati.*

Cooke, i. 57. Vize, "Micro. Fungi," 133; "Micro. Fungi Brit.," 222.

On *Dactylis glomerata*, *Triticum repens*, *Holchus lanatus*. June to September.

BIOLOGY.—Fischer v. Waldheim says that the germination is similar to that of *T. tritici*.

## UROCYSTIS. Rab.

Pulverulent; spore-balls consisting of a variable number of dark, smooth, large teleutospores placed centrally, surrounded by a number of smaller, paler pseudospores, which do not germinate. Germination of the central teleutospores as in Tilletia.

### Urocystis occulta. (Wallr.)

Spore-masses black, forming long lines on the leaves, stems, and leaf-sheaths. Spore-balls round or elliptical, consisting of 1 to 3 spores, surrounded by a variable number of pseudospores, 16–25 × 15–20$\mu$. Teleutospores irregularly rounded, contiguous surfaces flattened, smooth, opaque, dark brown, 10–18$\mu$ in diameter. Pseudospores irregularly but generally distributed around the circumference of the spore-ball, subhemispherical, pale brown, 4–6$\mu$ high. Mycelium annual.

*Synonyms.*

Schröt., *loc. cit.*, p. 279. Winter, *loc. cit.*, p. 119.

*Erysibe occulta.* Wallr., "Flor. Crypt. Germ.," vol. ii. p. 212.

*Urocystis occulta*, Preuss. Cooke, "Hdbk.," p. 517.

*Urocystis parallela*, B. and Br. Cooke, "Micro. Fungi," 4th edit., p. 232, t. ix. figs. 187, 188.

*Uredo parallela*, Sow. Berk., "Eng. Flor.," vol. v. p. 375, in part.

On *Secale cereale.*

BIOLOGY.—See p. 93.

### Urocystis agropyri. (Preuss.)

Produced in parallel lines on the leaves, black. Spore-balls subglobose, 20–26 × 15–20$\mu$. Teleutospores from 1 to 3, surrounded by a single layer of pseudospores, 5–9$\mu$ in diameter. Mycelium perennial.

*Synonyms.*

Schröt., *loc. cit.*, p. 279.

*Uredo agropyri*, Preuss. Sturm's "Deut. Flor.," 3rd edit., vol. vi. p. 1, t. i.

*Urocystis parallela*, B. and Br. Cooke, "Micro. Fungi," 4th edit., p. 232, in part.

On *Triticum repens* and *Avena elatior*.
June to October.

**Urocystis fischeri.** Körn.

Spore-masses black, in parallel lines on the leaves. Spore-balls with 1 or 2, rarely 3 teleutospores, 30–45$\mu$ in diameter. Teleutospores rather larger as a rule than in *U. occulta*, generally 18–20$\mu$ in diameter. Pseudospores firmly attached, numerous, completely enclosing the teleutospores, dark brown.

*Synonym.*

Winter, *loc. cit.*, p. 120.
*Urocystis fischeri.* Körn., "Hedwigia," 1877, p. 34.

On *Carex glauca*.
July to September. Mr. Soppitt.

BIOLOGY.—See p. 93.

**Urocystis colchici.** (Schlecht.)

Spore-masses black, forming in the leaves swollen lines and patches, which soon burst and become pulverulent. Spore-balls rounded or irregular, composed of from two to four roundish, polygonal, smooth, brown teleutospores, which are surrounded by a great number of yellowish-brown pseudospores. Spore-balls 20–25 × 15–20$\mu$; teleutospores 12 × 15$\mu$ in diameter; pseudospores 8–11$\mu$ in diameter.

*Synonyms.*

Winter, *loc. cit.*, p. 120. Schröt., *loc. cit.*, p. 280.

*Cæoma colchici.* Schlecht., "Linnæa," vol. i. p. 241.
*Urocystis colchici*, Tul. Cooke, "Hdbk.," p. 517; "Micro. Fungi," 4th edit., p. 232. B. and Br., No. 485.

*Exsiccati.*

Berk., 309.

On *Colchicum autumnale.*
April to July.

BIOLOGY.—See p. 68.

### Urocystis sorosporioides. Körn.

Spore-masses black, in thick, flat pustules on the leaves, and in fusiform swellings on the stems. Spore-balls rounded, compact, opaque, 25–45 × 15–25μ. Teleutospores from three to six in number, roundish, compressed, dark brown, 12–16μ in diameter. Pseudospores hemispherical, pale brown, equally distributed around the teleutospores, 7–10μ in diameter.

*Synonyms.*

Schröt., *loc. cit.*, p. 280. Winter, *loc. cit.*, p. 124.

*Urocystis sorosporioides*, Körn. Fckl., "Symb. Nach.," vol. iii. p. 10. Cooke, "Micro. Fungi," 4th edit., p. 232.

On *Thalictrum minus*, and its var. *maritima.*
June to August.

### Urocystis gladioli. (Requien.)

Spore-masses black within the corms. Spore-balls roundish. 40–50μ in diameter. Teleutospores rounded on the outer side, compressed on the inner, brown, 4–6μ in diameter. Pseudospores very numerous, pale, brown, evenly distributed.

*Synonyms.*

Winter, *loc. cit.*, p. 121.

*Uredo gladioli*, Requien. Duby. "Bot. Gall.," vol. ii. p. 901.
*Urocystis gladioli*, Smith. Cooke, "Micro. Fungi," 4th edit., p. 232.

On *Gladiolus communis.*

**Urocystis anemones.** (Pers.)

Producing roundish, elongate, or irregular swellings on the stems and midribs. Spore-masses black. Spore-balls variable in size, form, and composition, 35 × 25μ. Teleutospores one or two in number, rarely more, subglobose or polygonal, opaque, dark brown, obscurely punctate, 10–15μ in diameter. Pseudospores not very numerous, globose, or semiovate, pale brown, seldom completely surrounding the teleutospores, 8–10μ in diameter.

*Synonyms.*

Schröt., *loc. cit.*, p. 280. Winter, *loc. cit.*, p. 123.

*Uredo anemones.* Pers., "Disp. Meth.," p. 56.

*Urocystis pompholygodes*, Lév. Cooke, "Hdbk.," p. 517; "Micro. Fungi," 4th edit., p. 232, t. ix. figs. 183, 184.

*Exsiccati.*

Berk., 236. Cooke, i. 79; ii. 148. Vize, "Micro. Fungi," 36; "Micro. Fungi Brit.," 4.

On *Anemone nemorosa*, *Ranunculus repens* and *bulbosus*.
May to October.

BIOLOGY.—See p. 94.

**Urocystis violæ.** (Sow.)

Spore-masses black, in swellings on the stems and midribs. Spore-balls irregularly rounded, 35–45μ in diameter, containing from one to six or more teleutospores, which are roundish or polygonal, dark brown, 10–15μ in diameter. Pseudospores distinctly hemispherical, pale brown, 6–10μ in diameter.

*Synonyms.*

Schröt., *loc. cit.*, p. 280. Winter, *loc. cit.*, p. 122.

*Granularia violæ.* Sow., "Eng. Fung.," t. 440.

*Urocystis violæ*, B. and Br. Cooke, "Hdbk.," p. 517; "Micro. Fungi," 4th edit., p. 232, t. ix. figs. 185, 186.

*Exsiccati.*

Cooke, i. 78. Vize, "Micro. Fungi," 137.

On *Viola odorata, sylvatica.*
May to September.

BIOLOGY.—See p. 94.

### Urocystis primulicola. Magnus.

Spore-masses brownish black, produced in the ovary, pulverulent. Spore-balls roundish or irregular, 25–50μ in diameter. Teleutospores subglobose, ovate, or polygonal, dark brown, smooth, 10–15μ in diameter. Pseudospores much resembling the teleutospores, but rather smaller and not quite so dark in colour, numerous. Promycelial spores ovate or elongate, elliptical, 10–12 × 4–5μ

*Synonym.*

*Urocystis primulicola.* Magnus, *Hedwigia*, 1879, p. 19.

On *Primula farinosa*, Rev. C. Wolley Dod; on *P. vulgaris*, Prof. J. W. H. Trail.
July and August.

BIOLOGY.—See p. 95.

## ENTYLOMA. De Bary.

Mycelium intercellular, not gelatinized. Teleutospores formed within the hyphæ, intercalated, never becoming pulverulent, in circumscribed clusters. Promycelial spores produced on the end of the promycelium. Conidia in some species produced on the living host-plant from the mycelium.

* Conidia formed on the living host-plant.

### Entyloma fergussoni. (B. and Br.)

Spots circular, 1–3mm. in diameter, becoming whitish from the development of the conidia. Teleutospores globose or irregularly rounded. Epispore rather thin, pale brown, smooth, 11–13μ in diameter.

*Synonyms.*

*Entyloma canescens*, Schröt. Cohn's "Beitr.," vol. ii. p. 372. Trail, *Scott. Nat.*, 1884, pp. 128–181. Winter, *loc. cit.*, p. 113.

*Protomyces fergussoni.* B. and Br., *Ann. Nat. Hist.*, No. 1473. Cooke, "Grevillea," vol. xii. p. 99; "Micro. Fungi," 4th edit., p. 227.

On *Myosotis arvensis*, *palustris*, *cæspitosa*.
June to October.

**Entyloma bicolor.** Zopf.

Spots roundish, yellow. Teleutospores globose or polygonal. Epispore gelatinous, of variable thickness, at first colourless, then brown, 20–23 × 15–18$\mu$. Conidia cylindrical, curved, simple or septate, rounded above, attenuated towards the base, 10–20 × 3$\mu$.

*Synonym.*

*Entyloma bicolor.* Zopf in Rabh., "Fung. Europ.," No. 2496. Winter, *loc. cit.*, p. 112.

On *Papaver rhœas*.
June and July.

**Entyloma ranunculi.** (Bon.)

Spots circular, 2–5mm. across, at first whitish, then yellow or brown. Conidia produced freely, fusiform or filiform, 40 × 2$\mu$. Teleutospores globose. Epispore smooth, pale brown, 10–14$\mu$ in diameter.

*Synonyms.*

Winter, *loc. cit.*, p. 112. Schröt., *loc. cit.*, p. 282.

*Fusidium ranunculi.* Bon., "Handb. d. Mycol.," p. 43.

*Protomyces ficariæ*, Cornu and Roze.

*Entyloma ficariæ*, F. v. Waldh.

*Glœosporium ficariæ*, Berk. Cooke, "Hdbk.," p. 475. Trail, *Scott Nat.*, 1884, p. 228.

*Exsiccati.*

Cooke, i. 533. Berk., 212.

On *Ranunculus ficaria* and *sceleratus.*
May and September.

BIOLOGY.—See p. 91.

### Entyloma matricariæ. Trail.

Spots on stems (irregularly rounded or oval), or more often on leaves, affecting the segments on all their surfaces, small, nearly white while conidia are being formed, but becoming brown and dry; conidia (produced freely on conidiophores pushed out in clusters from the stomata) fusiform or filiform, 15–20 × 1·5–2$\mu$, pale yellowish, pluri-guttulate or faintly 3-4 septate. Teleutospores abundantly formed in the tissues of the host, round or polygonal from pressure, 10–12$\mu$ in diameter. Epispore smooth, about 1·5$\mu$ thick, at first hyaline, becoming brown.

On *Matricaria inodora.*
At Finstown, in Orkney, in August, and plentifully near Aberdeen in September; J. W. H. Trail.

** Conidia not formed on the living host-plant.

### Entyloma chrysosplenii. (B. and Br.)

Spots whitish, thickened, roundish, 2–6 mm. across. Teleutospores globose or shortly elliptical, smooth, colourless, 10–12$\mu$ in diameter.

*Synonym.*

Schröt., *loc. cit.*, p. 283.
*Protomyces chrysosplenii*, B. and Br. Cooke, "Micro. Fungi," 4th edit., p. 227.

On *Chrysosplenium oppositifolium.*
June to September.

### Entyloma microsporum. (Ung.)

In round or fusiform swellings, which become yellowish brown. Teleutospores globose or irregular, with a thick compound epispore, pale yellowish brown, 15–25 × 12–15$\mu$.

*Synonyms.*

Schröt., *loc. cit.*, p. 284.

*Protomyces microsporus.* Ung., "Exanth. Plant.," p. 343.

*Entyloma ungeriaum*, De Bary. Cooke, "Micro. Fungi," 4th edit., p. 233. Trail, *Scott. Nat.*, 1884, p. 288.

On *Ranunculus repens* and *acris.*
June to November.

BIOLOGY.—The promycelial spores are cylindrico-fusiform, and conjugate in pairs.

**Entyloma calendulæ.** (Oud.)

Spots round, at first whitish, then becoming brown. Teleutospores globose or polygonal, smooth, nearly colourless or pale yellowish brown, 10–16$\mu$ in diameter.

*Synonyms.*

Schröt., *loc. cit.*, p. 283.

*Protomyces calendulæ.* Oud., "Flor. Myc. Neerland," vol. ii. p. 42.

*Protomyces hieracii.* Berk. in Herb., No. 5248. Cooke, "Grevillea," vol. xii. p. 99.

*Entyloma calendulæ.* Trail, *Scott. Nat.*, January, 1884, p. 124.

On *Calendula officinalis*, *Hieracium vulgatum*, *murorum.*
June to September.

## MELANOTÆNIUM. De Bary.

Spores simple, in indeterminate groups, produced within the hyphæ, not pulverulent, germination as in Entyloma.

**Melanotænium endogenum.** De Bary.

Mycelium intercellular, permeating the whole plant, sending tufted haustoria into the adjacent cells. Teleutospores forming black masses under the cuticle, rounded, elliptical, or polygonal from mutual pressure, dark brown, smooth, 15–20 × 12–20$\mu$. Epispore thick, opaque, dark brown. Endospore thin, colourless.

*Synonym.*

Schröt., *loc. cit.*, p. 283.

*Melanotænium endogenum.* De Bary, *Bot. Zeit.*, 1874, p. 108. Trail., *Scott. Nat.*, 1884, p. 243.

On *Galium verum.*

Biology.—The affected plants are deformed by the presence of the parasite, being shorter and stiffer, seldom blossoming, and resembling an Equisetum in habit (see p. 95).

## TUBERCINIA. Fries.

Spore-balls consisting of numerous teleutospores, at length pulverulent. Germination as in Entyloma. Conidia white, ovate, produced from mycelium on the living host-plant.

### Tubercinia trientalis. B. and Br.

Spore-masses black, in bullate swellings on the leaves and stems. Spore-balls irregularly rounded or elongated, containing a large number of teleutospores (50 to 100), from 20–100$\mu$ in diameter. Teleutospores rounded or polygonal, dark brown, smooth, firmly connected, 15–30 × 10–20$\mu$. Conidia in white, shining, extended patches on the under side of the leaves, ovate or pyriform, attenuated above, hyaline.

*Synonyms.*

Schröt., *loc. cit.*, p. 285.

*Ascomyces trientalis.* Berk., "Outl.," p. 376. Cooke, "Hdbk.," p. 737 (conidial stage).

*Tubercinia trientalis*, B. and Br., No. 488. Cooke, "Hdbk.," p. 516.

*Sorosporium trientalis*, Woron. Cooke, "Micro. Fungi.," 4th edit., p. 231.

*Sorosporium paridis* (Ung.). Winter, *loc. cit.*, p. 102.

*Exsiccati.*

Vize, "Micro. Fungi," 137.

On *Trientalis europæa.*
June to October.

BIOLOGY.—The affected plants are taller and more slender in their form. In May and June they produce conidia on the under sides of their leaves. The stems are blackened by the presence of spores beneath the bark (see also p. 96).

**Tubercinia scabies.** Berk.

" Spores subglobose, composed of minute cells, forming together a hollow globe, with one or more lacunæ, generally attached laterally by a slender thread, olive."—*Berkeley.*

*Synonyms.*

*Erysibe subterranea,* Wallr. Regens., *Bot. Zeit.*, 1846, p. 119.

*Protomyces tuberum solani.* Martius's *Die Kartoffel Epid.*, 1842, p. 28, t. ii. figs. 9–13 ; t. iii. figs. 36-38.

*Tubercinia scabies.* Berk., *Jour. Royal Hort. Soc.*, 1846, vol. i. p. 33, figs. 30, 31 ; *Ann. Nat. Hist.*, No. 489. Cooke, "Hdbk.," p. 516 ; "Micro. Fungi," 4th edit., p. 231, t. iii. fig. 54.

*Exsiccati.*

Cooke, i. 445.

On potatoes (*Solanum tuberosum*).

I am unacquainted with this species. In my copy of Cooke Exs. I can find no spores.

**DOASSANSIA.** Cornu.

Spores in dense masses, surrounded by an investment of sterile cells, not pulverulent, germination as in Entyloma.

**Doassansia alismatis.** (Nees.)

Spots rounded, yellowish, crowded with brown dots, on the leaves and stems. Spore-balls very numerous, 3 mm. across. Teleutospores roundish or polygonal, pale brown, smooth, 10–15 × 8–10$\mu$. Promycelial spores abundant, cylindrical.

*Synonym.*

*Doassansia alismatis.* Cornu, *Ann. Sc. Nat.*, 6th series, vol. xv. p. 285, t. 16, figs. 1–4. Trail., *Scot. Nat.*, January, 1884, p. 124. Schröt., *loc. cit.*, p. 286.

On *Alisma plantago.*
July to October.

BIOLOGY.—See p. 92.

**Doassansia sagittariæ.** (Fckl.)

Spots yellowish, pustulate, rounded, ·5–1 cm. across, hypophyllous. Spore-balls roundish, 50–60$\mu$ in diameter. Teleutospores firmly cohering, angular, pale brown-yellow, smooth, 8–11$\mu$ in diameter. Promycelial spores numerous, cylindrical.

*Synonym.*

Schröt., *loc. cit.*, p. 286.
*Protomyces sagittariæ*, Fckl. Cooke, "Micro. Fungi," 4th edit., p. 227.

*Exsiccati.*

On *Sagittaria sagittifolia.*
May to September.

BIOLOGY.—See p. 92.

## THECAPHOREI. Schröt.

Spore-forming hyphæ not gelatinized. Teleutospores compound. Promycelium sometimes branched. Promycelial spores produced in concatenate chains from the end of the promycelium.

### Thecaphora. Fingerhuth.

Spore-balls composed of large teleutospores, firmly adherent, convex externally. Promycelium filiform, sometimes with lateral branches.

**Thecaphora hyalina.** Fing.

Spore-masses rich chestnut-brown, pulverulent, produced in the seeds. Spore-balls irregularly rounded, very variable in size, brownish yellow, 50–60 × 20–50$\mu$. Teleutospores from two to ten in number, firmly cohering, broadly and shortly cuneiform internally, internal flat surfaces smooth, external rounded surface rough with large rounded tubercles, brown, 15–20$\mu$ in diameter.

*Synonyms.*

Schröt., *loc. cit.*, p. 288. Winter, *loc. cit.*, p. 105.

*Thecaphora hyalina.* Fing., "Linnæa," vol. x. p. 230. B. and Br., *Ann. Nat. Hist.*, No. 1148. Cooke, "Hdbk.," p. 515; "Micro. Fungi," 4th edit., p. 231.

*Exsiccati.*

Cooke, i. 313. Vize, "Micro. Fungi Brit.," 45.

On *Convolvulus sepium* and *soldanella.*
September and October.

BIOLOGY.—See p. 86.

**Thecaphora trailii.** Cooke.

Produced in the florets. Teleutospores purple-brown, globose, usually in fours, rarely two or three, compressed on the inner face. Epispore finely verrucose, 12–14$\mu$ in diameter.

*Synonyms.*

*Thecaphora trailii.* Cooke, "Grevillea," vol. xi. p. 155.

*Thecaphora cirsii.* Boudier, *Bull. Soc. Mycol. de France*, 1887, vol. iii. p. 149, pl. xv. fig. 1.

On *Carduus heterophyllus.* Prof. J. W. H. Trail.
August.

BIOLOGY.—Pulverulent, having very much the habit of *Ustilago cardui*, but differs in being a true Thecaphora, and in the epispore being verrucose and not reticulate. M. C. C.

## SOROSPORIUM. Rudolphi.

Teleutospores very numerous, small, in large spore-balls, but easily separable, produced from intertwining gelatinized hyphæ, having at first a gelatinous investment. Promycelium filiform.

**Sorosporium saponariæ.** Rudolphi.

Produced in the inflorescence, destroying the reproductive organs. Spore-masses rusty brown. Spore-balls roundish, composed of numerous loosely connected teleutospores, 50–100$\mu$ in diameter. Teleutospores roundish, shortly elliptical or polygonal from mutual pressure, yellowish brown, transparent, rough, with minute tubercles and toothed ridges, 12–18 × 10–12$\mu$.

*Synonym.*

*Sorosporium saponariæ.* Rudol., "Linnæa," vol. iv. p. 116. Schröt., *loc. cit.*, p. 288. Winter, *loc. cit.*, p. 104.

On *Dianthus deltoides.* Norwich, in gardens.

BIOLOGY.—See p. 85.

---

# SUPPLEMENT.

## ALLIED AND ASSOCIATED SPECIES.

Doubtful Ustilaginei (Graphiola, Entorrhiza, Tuberculina). The descriptions of the British species of Protomyces are also given, both because they are apt, on cursory examination to be confounded with the Entylomata, and also because they have been described with them.*

### GRAPHIOLA. Poiteau.

Mycelium in the tissues of the living plant, forming small conceptacles, which burst through the cuticle of the plant. Peridia roundish, the outer hard, formed by the intertwining of the mycelial hyphæ, the inner peridium thin, enclosed by the outer, filled with hyphæ, sterile and spore-bearing. Spore-forming hyphæ at the base of the conceptacle, yellow, filamentous, crowded, becoming septate above into short joints, of which the uppermost gradually mature. Spores formed from cells, which are given off laterally from the joints, and which become abjointed

* Professor Marshall Ward considers that the *Schinzia leguminosarum* of Frank, which causes the tubercular swellings on the roots of the Leguminosæ, is allied to the Ustilagineæ, but the evidence (*Phil. Trans.*, 1887, pp. 539–562) adduced by him seems hardly to be conclusive.

by transverse septa. Spores spherical or elliptical. Between the spore-forming hyphæ arise bundles of sterile hyphæ, which grow up out of the peridium and carry up with them the spores. The spores germinate by the development of a filamentous mycelium, or by the formation of fusiform sporidia. From the investigations of E. Fischer,* it would appear that this fungus is related to the Ustilaginei both by its spore-formation and by the manner in which its spores germinate.

### **Graphiola phœnicis.** (Moug.)

Conceptacles erumpent on the leaves of the host-plant, 1–1·5 mm. wide, ·5 mm. high, opening above and allowing the sterile hyphæ to protrude. Outer peridium black and corneous; inner peridium delicate, colourless. Sterile hyphæ yellow, protruding 2 mm. or more. Spores *en masse* yellow, globose or elliptical, 3–6$\mu$ across. Membrane thick, colourless, smooth.

*Synonyms.*

Schröt. in Cohn, "Krypt. Flor.," vol. iii. p. 289.

*Graphiola phœnicis*, Moug. Schröt., "Krypt. Flor. Schl.," vol. iii. p. 289.

*Corda*, "Anleitung," t. C. 26, Nos. 5–8. Cooke, "Hdbk.," p. 546.

On the leaves of *Phœnix dactylifera* in greenhouses.

## **ENTORRHIZA.** C. Weber.

Mycelium parasitic in the tissues of living plants, producing large spores at the ends of lateral branches. Spores simple, single or numerous in the cells of the host-plant. Epispore thick, germinating by one or more thin germ-tubes, which are sometimes slightly branched, and which develop, both at their ends and lower down, small sickle-shaped promycelial spores †

* E. Fischer, *Beitrag. zur Kenntniss der Gattung Graphiola. Bot. Zeit.*, 1883.

† C. Weber, *Ueber den Pilze der Wurzelanschwellungen von Juncus bufonius. Bot. Zeit.*, 1884.

### **Entorrhiza cypericola.** Magnus.

In the cells of the periblem of the living root, causing swellings from 3 mm. thick to 10 mm. long. Spores elliptical, sometimes pointed at the end, 17–20 × 15–17$\mu$. Epispore thick, yellow or brown, rough with large hemispherical or irregular warts. Promycelial spores crescentic, whorled and very small (Schröt., *loc. cit.*, p. 290).

On the roots of *Juncus bufonius* and *lamprocarpus*. The Links, Old Aberdeen; Prof. J. W. H. Trail.

Mr. P. Cameron records it on *Juncus uliginosus* and *squarrosus*.

## **TUBERCULINA.** Sacc.

Mycelium parasitic on the hyphæ and spore-beds of the Uredineæ. Spore-beds flat, formed by erect hyphæ, which bear on their summits conidia. Conidia spherical, smooth, pulverulent, germinating by erect branched promycelia, which bear sickle-shaped promycelial spores. Older spore-beds forming sclerotia (Schröt., *loc. cit.*).

### **Tuberculina persicina.** Ditm.

Parasitic upon æcidiospores and uredospores, at first flat, pale violet or dirty red, spore-bearing hyphæ 30–60 × 2–3$\mu$. Conidia globose or shortly elliptical, 7–14$\mu$ wide. Epispore smooth, almost colourless, or very pale violet. Sclerotia convex, sometimes globose, smooth externally, violet, internally white.*

*Synonyms.*

Schröt., *loc. cit.*, p. 291.

*Tuberculina persicina*, Ditm. Schröt., "Krypt. Flor. Schl.," vol. iii. p. 291.

*Tuberculina vinosa*, Sacc.

On *Æcidium asperifolii*, *tussilaginis*, *Ræstelia lacerata*.

* C. Gobi, *Über Tubercularia persicina Mém. de l'Académie Impériale des Sciences de St. Pétersbourg*, s. vii., t. xxxii., 1884.

## PROTOMYCES. Unger.

Mycelium intercellular, parasitic in the tissues of living plants. Spores formed in the continuity of the mycelial hyphæ, inside the tissues of the living host-plant, causing indurated swellings on the host-plant. Germination by the development of numerous minute sporidia inside the resting spores. (Plate VIII. figs. 14–20.)

### Protomyces macrosporus. Unger.

Tumefactions at first translucent, pale yellow, then white, at length brownish, 1–4 mm. long, 2 mm. wide and thick, firm, at first closed, then open. Spores irregularly spherical or elliptical, 40–80 × 35–60$\mu$. Epispore as much as 5$\mu$ thick, pale yellow, contents colourless, sporidia cylindrical, 2–2·2 × 1$\mu$.

*Synonym.*

*Protomyces macrosporus*, Unger. Schröt., *loc. cit.*, p. 259. Cooke, "Micro. Fungi," 4th edit., p. 227.

On *Ægopodium podagraria*, *Helosciadium nodiflorum*, *Heracleum sphondylium*, *Angelica sylvestris*, *Anthriscus sylvestris*, *Œnanthe crocata*.

May to October.

### Protomyces rhizobius. Trail.

Spores in the cortex of the roots in groups of from two to eight, spherical, nearly smooth, with very thick walls, pale brown or nearly colourless, 30–33$\mu$ in diameter.

*Synonym.*

*Protomyces rhizobius.* Trail, *Scott. Nat.*, January, 1884, p. 125.

On the roots of *Poa annua.* Old Aberdeen, May, 1883.

### Protomyces pachydermus. Thüm.

Forming elongate or confluent swellings in the leaf-stalk or mid-rib of the leaves. Spores scattered, intercellular, subglobose or elliptical, thick-walled. Epispore smooth, pale brown, 15–20$\mu$ in diameter.

*Synonym.*

*Protomyces pachydermus.* Thüm., "Hedwigia," 1874, p. 97. Trail, *Scott. Nat.*, July, 1883, p. 33.

On *Taraxacum officinale.* Aberdeen, Prof. J. W. H. Trail.

**Protomyces menyanthis.** De Bary.

Spores aggregated in roundish or confluent patches, immersed in the substance of the leaves, purplish on the surface. Spores brownish, subglobose.

*Synonym.*

*Protomyces menyanthis,* De Bary. Cooke, "Micro. Fungi," 4th edit., p. 228.

*Exsiccati.*

Vize, "Micro. Fungi," 151.

On the leaves of *Menyanthes trifoliata* and *Comarum palustre.*

Biology.—This species Professor Trail finds commonly in Aberdeenshire, on *Menyanthes*, but not upon *Comarum.*

**Protomyces ari.** Cooke.

Spots aggregated in elongated patches, immersed in the substance of the leaves and petioles, always covered, globose, simple, brown. Endochrome granular. Episporc smooth.

*Synonym.*

*Protomyces ari.* Cooke, "Micro. Fungi," 4th edit., p. 227.

On the leaves and petioles of *Arum maculatum.*

## THE BARBERRY LAW OF MASSACHUSETTS.

Anno Regni Regis Georgii II. Vicesimo Octavo, chap. x. (published January 13, 1755).

*An Act to prevent Damage to English Grain arising from Barberry Bushes.*

Whereas it has been found by experience, that the Blasting of Wheat and other English Grain is often occasioned by Barberry Bushes, to the great loss and damage of the inhabitants of this province :—

Be it therefore enacted by the Governour, Council, and House of Representatives, that whoever, whether community or private person, hath any Barberry Bushes standing or growing in his or their Land, within any of the Towns in this Province, he or they shall cause the same to be extirpated or destroyed on or before the thirteenth Day of June Anno Domini One Thousand Seven Hundred and Sixty.

Be it further enacted that if there shall be any Barberry Bushes standing or growing in any land within this Province, after the said 10th day of June, it shall be lawful, by Virtue of this Act, for any Person whosoever to enter the Lands wherein such Barberry Bushes are, first giving one month's notice of his intention to do so to the Owner or Occupant thereof, and to cut them down, or pull them up by the root, and then to present a fair account of his labour and charge therein to the owner or occupant of the said land; and if such owner or occupant shall neglect or refuse by the space of two months next after the presenting said account, to make to such person reasonable payment as aforesaid, then the person who cut down or pulled up such bushes, may bring the action against such owner or occupant, owners or occupants, before any Justice of the Peace, if under forty shillings, or otherwise before the Inferior Court of Common Pleas in the County where such Bushes grew, who upon proof of the cutting down or pulling up of such bushes by the person who brings the action, or such as were employed by him, shall and is hereby respectively empowered to enter up judgment for him to recover double the value of the reasonable expense and labour in such service and award execution accordingly.

Be it further enacted that if the lands on which such Barberry Bushes grew are common and undivided lands, that then an action may be brought as aforesaid, against any one of the proprietors in such manner as the Laws of this Province provide, in such cases where Proprietors may be sued.

Be it further enacted, that the Surveyors of the Highways, whether public or private, be and hereby are empowered and required *ex officio* to destroy and extirpate all such Barberry Bushes as are or shall be in the Highways in their respective Wards or Districts, and if any such shall remain after the aforesaid tenth Day of June, Anno Domini One Thousand Seven Hundred and Sixty, that then the Town or District in which such bushes are shall pay a Fine of two shillings for every bush standing or growing in such Highway, to be recovered by Bill Plaint, Information, or on the Presentment of a Grand Jury, and to be paid one Half to the Informer and the other Half to the Treasury of the County in which such bushes grew, for the use of the County.

Be it further enacted, That if any Barberry Bush stand or grow n any Stone Wall or other Fence, either pointing on Highway, or lividing between one Propriety and another, that an Action may ›e brought as aforesaid against the Owner of the said Fence or he Person occupying the Land to which such Fence belongs; ınd if the Fence in which such Bushes grew is a Divisional Fence ›etween the Lands of one Person or Community and another, and uch fence hath not been divided, by which means the particular hare of each Person or Community is not known, then an Action nay be brought as aforesaid against either of the Owners or Occupants of the said Land.

Be it further enacted, That where the Occupant of any Land hall eradicate and destroy any Barberry Bush growing thereon, or in any of the Fences belonging to the same (which such Occupant is hereby authorized to do, and every action to be brought against him for so doing, shall be utterly barred), or shall be obliged, pursuant to this Act, to pay for pulling them up or cutting them down, that then the owner or proprietor of such Land shall pay the said Occupant the full value of his Labour and Cost in destroying them himself, or what he is obliged to pay to others as aforesaid; and if such Owner or Owners shall refuse so to do, then it shall be lawful for the said Occupant or Occupants to withhold so much of the Rents or Income of the said Land as shall be sufficient to pay or reimburse his cost and charge arising as aforesaid.

This Act to continue in Force until the Tenth Day of June, One Thousand Seven Hundred and Sixty-four.

*From the Province Laws of Massachusetts*, 1736–1761, p. 153.

# GLOSSARY.

*Abjoint.* A spore is said to be abjointed when it is developed from the end of a mycelial hypha or from a cell, from which it is first cut off by a septum ; the lower part of the cell is then termed the stalk-cell.

*Abstriction*, the separation of a spore by the contraction of the spore-forming hypha below it without the previous production of a septum, as the promycelial spores, Uredine spermatia.

*Acropetal*, proceeding in the direction of the apex from below upwards.

*Æcidiospore*, those spores of the Uredineæ which arise from the promycelium produced by a promycelial spore, in basipetal series.

*Æcidium*, the generic name formerly applied to the æcidiospores.

*Amphigenous*, growing on both the upper and lower surfaces of the leaves.

*Anastomosing*, uniting into a network by cross branches.

*Autœcious.* Those parasitic fungi which pass the whole of their life-cycle upon the same species of host-plant are said to be autœcious.

*Base*, the lower end of a spore, that which is nearest its attachment.

*Basidium*, the cell or hypha from the apex of which spores are abjointed or abstricted.

*Basipetal*, proceeding in the direction of the base from above downwards.

*Bullate*, swollen in the form of a blister.

*Capitate*, having a head, as when the end of an erect hypha is abruptly enlarged in a spherical manner. *See* Hooded.

*Carpogonium*, the female cell, which is fertilized by the pollinodium.

*Circinate*, arranged in a circular manner.

*Clavate*, club-shaped.

*Concatenate*, united in a continuous series, like the links of a chain or beads of a necklace.

*Conidiophore*, the cell which produces conidia.

*Conidium*, an asexual spore.

X

*Deciduous*, falling off.

*Discrete*, separate, distinct, not confluent.

*Echinulate*, covered with short sharp spines.

*Elliptical*, having the form of an ellipse or oval with the ends rounded.

*Endochrome*, the protoplasmic contents of spores or hyphæ, usually applied to the coloured cell-contents of spores.

*Epiphyllous*, growing on the upper surface of the leaves.

*Erumpent*, bursting through the surface of the host-plant or matrix.

*Fimbriate*, torn into a fringe.

*Foramen*, a small hole or perforation.

*Germ-pore*, the opening in the walls of a spore through which the germ-tube is protruded.

*Germ-tube*, the tube emitted from a spore in germination, which may become a promycelium or may develop into a mycelium.

*Globose*, spherical, the shape of a globe.

*Glomerulus*, a small round head.

*Haustorium*, a short lateral branch of a mycelial hypha, which enters a cell of the host-plant and acts the part of a sucker.

*Heterœcious*. Those parasites which pass one part of their lives upon one host-plant, and the other part upon another of a different species, are said to be heterœcious.

*Hooded*. Those spores which have a much thickened membrane at their upper ends are said to be hooded. Sometimes they are called capitate.

*Hymenium*, the base of an æcidial cup formed by those hyphæ which produce the æcidiospores.

*Hypha*, a branch of mycelium, consisting of an elongated cell.

*Hypogenous*, *Hypophyllous*, growing on the lower side of the leaves.

*Intercellular*, between the cells of the host-plant.

*Intracellular*, inside the cells of the host-plant.

*Isodiametric*, having the transverse, longitudinal, and perpendicular diameters the same length.

*Laciniate*, cut or torn into segments or laciniæ.

*Ligulate*, tongue-shaped.

*Lumen*, the calibre of a tube.

*Mesospore*, a unicellular teleutospore occurring in a Puccinia spore-bed; a middle spore-form between Uromyces and Puccinia.

*Metœcious*, the same as heterœcious.

*Mycelium*, the vegetative part of parasitic fungi, which consists of an assemblage of hyphæ or tubes containing protoplasm.

*Oblong*, in the sense employed in the description of the spores of the Uredineæ, it means an oval body with very blunt ends, or an oblong body with rounded angles.

*Obovate*, the reverse of ovate, egg-shaped, with the base narrower than the apex.

*Oval*, a rounded figure with one diameter longer than the other, like an oblique section of a cylinder; almost synonymous with elliptical.

*Ovate, Ovoid*, the shape of an egg—that is, oval, with the base somewhat broader than the apex.

*Papillate*, covered with, or terminating in, a papilla.

*Paraphyses*, barren hyphæ, which may be filiform and dark-coloured or variously swollen and hyaline.

*Parasite*, an organism that can exist only upon a living plant or animal, from which it takes its food.

*Parenchyma*, the cellular tissue of the host-plant.

*Periblem*, in the sense used in the text, may be taken to mean that part of the root of the host-plant which lies beneath the cortex.

*Peridium*, the case or envelope that encloses the spores in some groups of fungi.

*Perithecium*, the flask-shaped receptacle which encloses the ascigerous fructification of the Sphæriæ, etc.

*Persistent*, lasting, not soon falling off.

*Pollinodium*, the cell which acts the part of the male organ and fertilizes the carpogonium.

*Promycelial spore*, a spore produced from a promycelium.

*Promycelium*, the germ-tube of a teleutospore, which is of limited length, and produces a few spores, unlike the teleutospore.

*Protoplasm*, the living contents of cells, or hyphæ, consisting of albumenoid substances.

*Pseudoperidium*, the outer investing case of the æcidiospores, composed of sterile cells.

*Pseudospores*, the barren spores which invest the fertile spores in Urocystis. The term has been applied by some authors to the uredospores and æcidiospores.

*Pulverulent*, dusty.

*Pulvinate*, cushion-shaped.

*Pyriform*, pear-shaped.

*Receptacle*, (1) in fungi, the structure which encloses the spores; (2) in flowering plants, that part of the flower-stalk to which the flowers are attached.

*Reniform*, kidney-shaped.

*Reticulate*, covered by lines or ridges which cross each other and form a network.

*Saprophyte*, as opposed to parasite, a fungus which lives upon dead organic matter only.

*Scattered*, spread about, not localized in distinct groups or clusters.

*Sclerotium*, an old genus of fungi, comprising hard compact bodies, which are now known to be the resting condition of the mycelium.

*Separate*, as applied to the teleutospores, distinct from one another, not compacted into a solid mass.

*Septate*, partitioned off into compartments.

*Sorus*, a heap or cluster of spores, a spore-bed.

*Spermatium*, a minute cell abstricted from a hypha. These hyphæ are contained in a receptacle called a spermogonium. By some botanists a male sexual function is attributed to the spermatia of the Uredineæ.

*Spermogonium*, the receptacle which encloses the spermatia.

*Spherical*, globose, in the form of a sphere.

*Spore*, a cell which, on becoming free, has the faculty of reproducing the fungus.

*Sporophore*, a mycelial hypha that bears spores.

*Sterigma*, the cell from which spores are abjointed or abstricted ; the same as basidium.

*Sub*, a prefix implying a slight degree of anything.

*Teleutospore*, the last-formed spore in the life-cycle of the Uredineæ and Ustilagineæ, which germinates by the production of a promycelium and promycelial spores.

*Trabecula*, literally a little beam ; a small band of tissue joining others together, often transverse in direction.

*Trichogyne*, the threadlike filament in lichens which, becoming impregnated above by the spermatia, gives rise below to the carpogonium or perithecium.

*Truncate*, cut off abruptly.

*Urceolate*, shaped like a pitcher with a contracted mouth.

*Uredo*, the old generic name applied to the uredospores.

*Uredospore*, a spore which is abjointed from the apex of a mycelial hypha, endowed with a limited period of vitality, and germinating by the emission of a germ-tube, which develops into a mycelium.

*Verrucæform*, shaped like a wart.

*Verrucose*, covered with small warts.

*Yeast-spore*, a spore produced by budding from another spore. Generally these spores occur in large numbers, and are then called yeast-colonies.

# AUTHORS QUOTED.


*Albertini, J. B.*, and *Schweinitz, L. D.* Conspectus Fungorum in Lusatiæ superioris agro Niskiensi crescentium. 1805.
*Annales des Sciences naturelles* (Botanique). 6 séries. Paris: 1824–88

*Bachmann, E.* Spectroskopische Untersuchungen von Pilzfarbstoffen. 1886.
*Bagnis, C.* Osservazioni sulla vita e morfologia di alcuni funghi Uredinei. 1875.
*Barclay, A.* On *Æcidium urticæ*, Schum., var. *Himalayense*. Scientific Memoirs by Medical Officers of the Army in India. 1887.
*Banks, J.* Annals of Agriculture, vol. xliii. 1805.
*Berkeley, M. J.*, in Sir J. E. Smith's English Flora, vol. v. pt. ii. 1837.
—— in Hooker's Flora Antarctica. 1844–47.
—— Transactions of the Royal Horticultural Society. 1847.
—— Outlines of British Fungology. 1865.
—— and *Broome, C. E.*, in Annals and Magazine of Natural History. 1838–85.
*Bonorden, H. F.* Handbuch der allgemeinen Mykologie. 1852.
*Botanische Zeitung.* 1843–88.
*Brefeld, O.* Botanische Untersuchungen über die Schimmelpilze. 1881.
—— Botanische Untersuchungen über die Hefenpilze. 1883.
*Bulletin* de la Société Mycologique de la France. 1887.
*Bulliard, P.* Histoire des champignons de la France. Paris: 1791–98.
*Byerkander*, in Abhandl. d. Schwedischen Akadem. 1775.

*Castagne, L.* Observations sur quelques plantes acotyledonées recueillies dans la départment des Bouches du Rhone. 1842–43.
—— Catalogue des plantes que croisent naturellement aux environs de Marseille. 1845–51.
*Chevallier, F.* Flora générale des environs de Paris. 1826–36.
*Cooke, M. C.* Handbook of British Fungi. 1871.
—— Introduction to the Study of Microscopic Fungi. 4th edit. 1878.
—— On *Uromyces*. Grevillea, vol. vii.
—— Circumnutation in Fungi. *Quekett Journal*, 1884, vol. i., 2nd series.
*Corda, A.* Icones Fungorum. 6 fasc. 1838–56.
*Cornu, M.* Bulletin de la Société Botanique de France, 1876, vol. xxiii.
—— Comptes rendus, January 21, 1875.

*Cornu, M.* Contributiones à l'étude des Ustilaginées, in Annales des Sciences naturelles. 6me série (Bot.), tome xv. 1883.

*De Bary, A.* Untersuchungen über die Brandpilze. 1853.
—— Neue Untersuchungen über Uredineen, in Monatsber. der Berliner Akademie. 1863.
—— Recherches sur la développement de quelques champignons parasites in Annales des Sciences naturelles. 4me série (Bot.), tome xx. 1863.
—— Morphologie und Physiologie der Pilze, Flechten, und Myxomyceten. 1866.
—— Neue Untersuchungen über Uredineen, in Monatsberichten der K. Akademie der Wissenschaften zu Berlin. 1865–66.
—— *Protomyces microsporus* und seine Verwandten. in Botan. Zeitung. 1874.
—— Vergleichende Morphologie und Biologie der Pilze, Mycetozoen, und Bacterien. 1884.
—— Comparative Morphology and Biology of the Fungi, Mycetozoa, and Bacteria, translated by Henry E. F. Garnsey and Isaac B. Balfour. 1887.
—— and *Woronin.* Beiträge zur Morphologie und Biologie der Pilze. 1864–82.
*De Candolle, A. P.*, in Lamarck's Encyclopédie méthodique botanique. 1783–1817.
—— Flora française. 1805.
—— Physiologie végétale. 3 vols. 1832.
—— and *Lamarck, J. B.* Synopsis Plantarum. 1806.
*Desvaux, A. N.* Journal de Botanique, rédigé par une société de botanistes, tome ii. 1809.
*De Vaureal*, in Schutzenberger on Fermentation. International Scientific Series, vol. xx. 1876.
*Dickson, J.* Fasciculi plantarum cryptogamicarum Britanniæ. 1785–1801.
*Dietel, P.* Beiträge zur Morphologie und Biologie der Uredineen. 1887.
*Duby, J. S.* Botanicon Gallicum. 1822–30.

*Eysenhardt*, in Linnæa, bd. iii. 1828.

*Fingerhuth, K. A.* Mykologische Beiträge. Linnæa, bd. x.
*Fische, C.* Entwickelungsgeschichte von *Doassansia Sagittariæ.* Berichte der deutschen botanischen Gesellschaft. 1882.
*Fischer, Ed.* Beiträge zur Kenntniss der Gattung *Graphiola.* Inaugural Dissertation, in Botan. Zeitung, Nos. 45–48. 1883.
*Fontana Felice.* Osservazioni sopra la Ruggine del Grano. 1767.
*Fries, E.* Observationes Mycologicæ. 1815–18.
—— Systema Mycologicum. 3 vols. 1821–30.
*Fuckel, L.* Symbolæ Mycologicæ. 1869–75.

*Gardener's Chronicle.* 1841–88.
*Gardiner, W.* Flora of Forfarshire. 1848.
*Gmelin, J. F.*, in Linné Systema Vegetabilium, vol. ii. 1796.
*Gobi, C.* Ueber den *Tubercularia persicina*, Ditm. in Mémoires de l'Aca-

démie Impériale des Sciences de St. Pétersbourg. 7me série, tome xxxii., No. 14. 1885.
*Greville, R. K.* Scottish Cryptogamic Flora. 1823–28.
—— Flora Edinensis. 1824.
*Grevillea.* A Quarterly Record of Cryptogamic Botany. London: 1872–88.

*Hansen, A.* Der Chlorophyllfarbstoff, in Arbeiten des. botan. Institutes zu Würzburg, bd. iii.
*Hartig, R.* Wichtige Krankheiten der Waldbäume. 1874.
—— Lehrbuch der Baumkrankheiten. 1882.
*Hartsen, M.* Comptes rendus. 1874.
*Hedwigia.* Notizblatt für Kryptogamische Studien. 1852–88.
*Hoffmann, H.* Ueber der Flugbrand. 1866.

*Ihne, E.* Studien zur Pflanzengeographische Geschichte der Einwanderung von *Puccinia malvacearum* und *Elodea canadensis*. 1880.

*Jacquin, J. F.* Collectanea ad Botanicam, Chemicam, et Historiam Naturalem Spectantia. 1786–96.
*Johnston, G.* Flora of Berwick-on-Tweed. 2 vols. 1829.

*Körnicke, F.* Mykologische Beiträge, in Hedwigia. 1876–77.
—— in Fuckel's Symbolæ Mycologicæ, Nachtrag 3. 1875.
*Kühn, J.* Krankheiten der Kulturgewächse. 1859.
—— in Rabenhorst's Fungi Europæi, Cent. xviii.
—— Uber die Entwickelungsformen des Getreidebrandes. 1874.
—— *Calyptospora.* Nov. gen. Uredinearum in Hedwigia. 1869.
*Kühne, W.* Ueber lichtbestandige Farben der Netzhaut. Untersuchungen aus der Physiolog. Inst. d. Univ. Heidelburg, bd. i.
*Kunze* and *Schmidt*. See *Schmidt* and *Kunze*.

*Léveillé, J. H.* Novus genus Uredinorum (*Uredo sempervivi*), in Bullet. Sc. Nat., tome vi. 1825.
—— Sur la disposition des Urédinées. Annales des Sciences naturelles (Bot.). 3me série, tome viii. 1847.
—— in Annales des Sciences naturelles. 3me série, tome ix.
—— in D'Orbigny's Dictionnaire universelle de l'histoire naturelle. 13 vols. 1841, etc.
*Link, H. F.*, in Linneus' Species Plantarum exhibentes Plantas rite cognitas. 4th edit., vol. vi. 1825.
*Linnæa.* Journal fur die Botanik. 1826–88.

*Magnus, P.* Ueber *Æcidium urticæ* und *Puccinia caricis*, in Verhandl. des botan. Vereins der Provinz Bradenburg. 1872.
—— Mykologische Bemerkungen. in Hedwigia. 1873.
—— Ueber die Familie der Melampsoreen. *Ibid.* 1875.
—— Bemerkungen über einige Uredineen. *Ibid.* 1877.
*Marshall, W.* Rural Economy of Norfolk. 2nd edit. 1795.

*Marshall, W.* Rural Economy of the Midland Counties. 1790.
*Martius, H.* Prodromus floræ Mosquensis. 1812.
—— *C. F. P.* Die Kartoffel-Epidemie. Munich: 1842.
*Meyen, F. J.* Pflanzen-pathologie. 1841.
*Montagne, J. F. C.*, in Gay's Historia fisica y politica de Chile. 1845.
*Mougeot, J. B.*, and *Nestler, C.* Stirpibus cryptogamis Vogeso-Rhenanis. 15 fasc. 4to. 1810-64.
*Müller, J.* Die Rostpilze der Rosa und Rubusarten und die auf ihnen vorkommenden Parasiten in Landwirthschaft Jahrbüche. 1886.

*Nees von Essenbeck, C. G.* Das System der Pilze. 1817.
*Nielsen, P.* Bemerkungen über einige Rostarten (*Puccinia poarum*) in Botanik Tidskr. 1877.
—— Om Schoeler in Ugeskrift fur Landmæna. 1884.

*Oersted, A. S.* Om Sygdome hos planterne. 1863.
—— Podisoma und Rœstelia, R. Danske vidensk. Selskab. Skrifter. 5th series, vol. vii., 1863, in Bulletin de l'Acad. Roy. des Sc. de Copenhagen. 1866-67.
*Otth, H.*, in Mittheil. Naturfors. Gesellschaft. in Bern. 1861.
*Oudemans, C. A. J. A.* Matériaux pour la flore mycologique de la Néerlande. 1873.

*Persoon, C. H.* In Romer's Neues Magazin fur die Botanik.
—— Observat. in Usteri annalen des Botanik, vol. xv.
—— Tentamen Dispositionis Methodicæ Fungorum. 1797.
—— Synopsis Methodica Fungorum. 1797.
—— Icones pictæ rariorum fungorum. 1803-6.
*Pirotta, R.* Nuova Giornale Bot. Ital., vol. xiii., 1881.
*Plowright, C. B.* Mimicry in Fungi. Grevillea, vol. x.
—— Germination of the Uredines. Grevillea, vol. ix.
—— Reproduction of the Heterœcious Uredines. *Journal of the Linnean Society*, vol. xxi.
—— *Mahonia aquifolia* as a nurse of the wheat mildew (*Puccinia graminis*). *Proceedings of the Royal Society*, No. 228. 1883.
—— On the Life History of the Dock Æcidium. *Ibid.*, No. 228. 1883.
—— The *Ranunculi æcidia* and *Puccinia schœleriana*. *Quarterly Journal of Microscopical Science*, vol. xxv., new series.
—— On the Life History of *Æcidium bellidis*. *Journal of the Linnean Society*, vol. xx. 1884.
—— On Certain British Heterœcious Uredines. *Ibid.*, vol. xxiv. 1887.
*Pointeau*, in Annales des Sciences naturelles. 1re. série, tome iii. 1824.
*Prevost, B.* Memoire sur la cause immédiate de la carie. 1807.
*Prillieux, E.* Sur la formation et la germination des spores des Urocystis, in Annales des Sciences naturelles. 6me séries (Bot.), tome x.
*Purton, T.* Flora of the Midland Counties. 1817-21.

*Rabenhorst, L.* Herbarium vivum Mycologicum. 1855-60.
—— Fungi Europæi Exsiccati. 1861-86.

*Rabenhorst, L.* Union. itin., 1866 (quoted by Winter).
*Räthay, E.* Uber das Eindringen der *Puccinia malvacearum*, in Verhandl. d K.K. Zool. bot. Ges., bd. xxxi. 1881.
—— Über einige autoecische und heterœcische Uredineen. 1881.
—— Untersuchungen über die spermogonien der Rostpilze. 1882.
*Rebentisch, J.* Prodromus floræ Neomarchicæ. 1804.
*Rees, M.* Die Rostpilze der deutschen Coniferen. 1869.
*Relhan, R.* Flora Cantabrigiensis. 1785–93.
*Rostrup, E.* Ueber eine genetische verbindung zwischen *Puccinia moliniæ* und *Æcidium orchidearum*, in Botan. Tidskr. Kjöbenhaven: 1874.
—— Heterœciske Uredineen, in K. D. vidensk Selsk. Forhandl. 1884.
*Rudolphi, F.* Plantarum vel novarum vel minus cognitarum descriptiones in Linnæa, vol. iv.

*Saccardo, P. A.*, in Michelia, ii. 1880.
*Schindler, F.* Ueber den Einfluss verschiedener Temperaturen auf die Keimfähigkeit der Stienbrandsporen, in Forchungen auf Geb. der Agrikulturphysik, bd. iii. 1880.
*Schlechtendal, D.* Flora berolensis. 1823–24.
——, *F. L.* Fungorum novorum et descriptorum illustrationes. Linnæa, vol. i. 1826.
*Schmidt, J. C.* and *Kunze, G.* Deutschlands Schwämme, in getrockneten Exemplaren. 1816–18.
*Schneider, W. G.* Herbar. Schlesien.
*Schøler.* Berberissens skudelige Indflydelse paa Sæden Lund. 1818.
*Schöpf, J. D.* Reise durch die mittleren und sudlichen vereinigten Nordamerikanischen Staaten. 1788.
*Schrank, F.* Baierische Flora. 1789.
—— in Hoppe's Botanisches Taschenbuch. 1793.
*Schröter, J.* Die Brand und Rostpilze Schlesiens in Abhandlung der Schles. Gesellsch. 1869–72.
—— Mittheilungen über einige Schlesische Uredineen, in Cohn's Beiträge zur Biologie der Planzen, bd. i.
—— *Melampsorella*, eine neue Uredineen Gattung, in Hedwigia. 1873.
—— Beobachtungen über der Zusammengehörigkeit von *Æcidium euphorbiæ* und *Uromyces pisi*, in Hedwigia. 1875.
—— Bemerkungen und Beobachtungen über einige Ustilagineen, in Cohn's Beiträge zur Biol., bd. ii. 1877.
—— Entwickelungsgeschichte einiger Rostpilze. *Ibid.*, bd. iii. 1879.
—— in Cohn's Kryptogamen Flora von Schlesien, bd. iii. 1885–88.
*Schultz, C. F.* Prodromus Floræ Stargardiensis. 1819.
*Schumacher, C. F.* Ennumeratio plantarum in partibus Sællandiæ septentrionalis et orientalis crescentium. 1801–3.
*Scottish Naturalist.* 8vo. 1872–88.
*Shakespeare, W.* King Lear.
*Smith, W.* Smaller Dictionary of Greek and Roman Antiquities. 4th edit. 1863.
——, *Worthington, G.* Diseases of Field and Garden Crops. 1884.
*Sommerfelt, S. C.* Supplementum floræ lapponicæ. 1826.

*Sowerby, J.* English Fungi. 1797–1809.

*Strauss, F.* Abhandlungen über die Persoon'schen Gattungen *Stilbospora*, *Uredo*, und *Puccinia*, in Annalen der Wetterauischen Gesellschaft für die gesammte Naturkunde. 1811.

*Sturm, J.* Deutschlands Flora, Pilze. 1817–51.

*Thümen, F.* *Melampsora salicina*, in Mittheilungen aus dem forstlichen Versuchswesen Oesterreichs, bd. ii. 1879.

*Trail, J. W. H.* On some Leaf-parasites new or rare in Britain, in *Scottish Naturalist*, new series, vol. i. pp. 124, 180.

—— On the Species of Entyloma parasitic in Species of Ranunculus in Scotland, *loc. cit.*, p. 227.

—— Two new British Ustilagineæ, *loc. cit.*, 241.

*Trelease, W.* Parasitic Fungi of Wisconsin. 1884.

*Tulasne, L. R.* and *C.* Mémoire sur les Ustilaginées comparées aux Urédinées, in Annales des Sciences naturelles. 1847.

—— Second mémoire sur les Urédinées et les Ustilaginées. *Ibid.* 1854.

—— Comptes rendus, tomes xxxii. et xxxvi.

*Unger, F.* Die Exantheme der Pflanzen. 1833.

—— Beiträge zur vergleichenden Pathologie, Sendschreiben an Herrn Prof. Schönlein. 1840.

*Von Liebenberg, A.*, in Oesterreich landw. Wochenblatt. 1879.

*Von Waldheim, A. F.* Sur la structure des spores des ustilaginées, in Bulletin de la Sociéte des Naturalistes de Moscou. 1867.

—— Beiträge zur biologie und entwickelungsgeschichte der Ustilagineen, in Pringsheim's Jahrbucher. 1869.

—— Translation of the above in the *Transactions of the New York Agricultural Society.* 1870.

—— Aperçu systematique des Ustilaginées. 1877.

—— Les Ustilaginées et leurs plantes nourricières, in Annales des Sciences naturelles. 6$^{me}$ séries (Bot.), tome iv. 1877.

*Wallroth, F. G.* Flora Cryptogamica Germaniæ. 1831–33.

*Ward, Marshall.* On *Hemileia vastatrix*, in *Linnean Society's Journal* (Botany), vol. xix., and *Quarterly Journal of Microscopical Science*, new series, vol. xxi.

—— On the Structure and Life History of *Entyloma ranunculi* (Bon.). *Philosophical Transactions of the Royal Society*, vol. 178 (1887), Botany.

—— On the Tubercular Swellings on the Roots of *Vicia faba*. *Philosophical Transactions of the Royal Society*, vol. 178 (1887), Botany.

*Weber, C.* Ueber den Pilze der Wurzelanschwellungen von *Juncus bufonius*, in Botanische Zeitung. 1884.

*Westendorp, G. D.* Herbier Cryptogamique Belge, No. 677, in Bulletin de l'Academ. Roy. de la Belgique. Bruxelles: 1851.

*Winter, G.* Cultur der *Puccinia sessilis* und dessen *Æcidium*, in Sitzungber. der naturf. Gesellsch. zu Leipzig.

—— Ueber der Æcidium der *Puccinia arundinacea*, in Hedwigia. 1875.

*Winter, G.* Ustilagineen, in Flora. 1876.
—— in Rabenhorst's Kryptogamen Flora von Deutschland, Oesterreich, und der Schweiz. 2nd edit. 1884.
*Withering, W.* Botanical Arrangement. 2nd edit., 1787, and 4th edit.
*Wolff, R.* Der Brand des Getreides. 1874.
—— Beitrag zur Kenntniss der Ustilagineen, in Botanische Zeitung. 1873.
—— *Æcidium pini* und sein Zusammenhang mit *Coleosporium senecionis.*
*Woronin, M.* Beitrag zur Kenntniss der Ustilagineen. 1882.

*Young, A.* Annals of Agriculture, vol. xliii. 1805.

# EXSICCATI.

*Baxter, W.* Stirpes Cryptogamicæ Oxoniensis. Oxford: 1825.
*Berkeley, M. J.* British Fungi. 4 vols. 4to. London.
*Cooke, M. C.* Fungi Britannici Exsiccati. 1st series, i–vi. 8vo. London: 1865–71.
—— 2nd series, i–vii. 4to. London: 1875–79.
*Rabenhorst, L.* Fungi Europæi Exsiccati. 4to. Dresden: 1861–85.
*Thümen, F.* Mycotheca Universalis. 23 cent. Bayreuth: 1875–80.
*Vize, J. E.* Fungi Britannici. 2 cent. Forden.
—— Micro Fungi Britannici. 5 cent. Forden.

# DESCRIPTION OF PLATES.

## PLATE I.

Fig. 1.—Mycelium of the æcidiospores of *Puccinia porri* on *Allium schœnoprasum* near the spore-bed, showing the granular protoplasm and nuclei. × 500. C. B. P.

Fig. 2.—Mycelium of the æcidiospores of *Puccinia tragopogonis*. × 500. C. B. P.

Fig. 3.—Three spermogonia of *Æcidium berberidis* in different stages of development, the two younger show the converging fibres which form their necks. These subsequently separate and become paraphyses. × 200. Modified from De Bary.

Fig. 4.—Two spermogonia in a more advanced condition, seen in section. × 200. De Bary.

Fig. 5.—Spermatia being abstricted from their hyphæ. × 200. De Bary.

Fig. 6.—Three spermatia of *Æcidium berberidis*. × 475, and enlarged to 1000. C. B. P.

Fig. 7.—Five spermatia of *Æcidium ranunculi repentis*, April 29, 1883, showing their different forms and somewhat irregular contour in the same spermogonium. × 475, enlarged to 1000. C. B. P.

Fig. 8.—Three of the same spermatia after being twelve hours in sugar and water; they are just beginning to bud. Enlarged to 1000. C. B. P.

Fig. 9.—The same after forty-eight hours. Enlarged to 1000. C. B. P.

Fig. 10.—Spermatia of the æcidiospores of *Puccinia adoxæ* from Mr. A. Lister's figure. × 710, enlarged to 1000.

Fig. 11.—Two spermatia of *Æcidium punctatum* in the act of budding, May 7, 1883. Enlarged to 1000. C. B. P.

Fig. 12.—Spermatia of *Æcidium bellidis*, December, 1883. Enlarged to 1000. C. B. P.

Figs. 13–15.—The same germinating in honey and water, after twelve hours. Enlarged to 1000. C. B. P.

Fig. 16.—The same as they appeared on the sixth day. Enlarged to 1000. C. B. P.

## PLATE II.

Fig. 1.—The early stage of *Æcidium crassum*, showing the basipetal spore series enclosed by the pseudoperidial cells, and covered by the epidermis. × 200. De Bary.

Fig. 2.—*Æcidium grossulariæ* at maturity, after rupture of the epidermis and pseudoperidial cell-layer. × 150. De Bary.

Fig. 3.—Four æcidiospores of *Puccinia poarum* (*Æcidium tussilaginis*). × 475. C. B. P.

Fig. 4.—One of the same beginning to germinate and showing its germ-pores, five of which are visible, the sixth being on the opposite side of the spore. × 475. C. B. P.

Fig. 5.—Æcidiospore of *Phragmidium rubi* in the same condition, and showing five germ-pores. × 475. C. B. P.

Fig. 6.—Æcidiospore of *Puccinia poarum* twelve hours after germination, showing the circumnutation of the germ-tube, and the migration of the endochrome to its growing end. × 475. C. B. P.

Fig. 7.—Æcidiospore of *Gymnosporangium clavariæforme* (*Ræstelia lacerata*), with six germ-pores, from one of which a comparatively short germ-tube has been protruded, which has become branched and full of endochrome from the spore, at its peripheral extremity. × 475. C. B. P.

Fig. 8.—A basipetal series of æcidiospores of *Chrysomyxa rhododendri* showing the alternate abortive cells. × 600. De Bary.

Fig 9.—Æcidiospore-chain of *Ræstelia cancellata*, showing the undeveloped mother-cells between the spores. × 300. De Bary.

Fig. 10.—Æcidiospores of *Puccinia poarum* (*Æcidium tussilaginis*) germinating upon the cuticle of *Poa trivialis*, and entering the stomata. × 475. C. B. P.

Fig. 11.—Part of a leaf of *Vinca major*, with the spermogonia and æcidia of *Puccinia vincæ*. × 2. C. B. P.

Fig. 12.—Section of the same leaf, showing two of the pulvinate æcidia. × 10. C. B. P.

Fig. 13.—Section of a small æcidium. × 475. C. B. P.

Fig. 14.—Two æcidiospores germinating. × 500. C. B. P.

## Plate III.

Fig. 1.—Sorus of the uredospores of *Melampsora farinosa*, showing the paraphyses. × 200. De Bary.

Figs. 2-4.—Uredospores of the same in different stages of development. × 300. De Bary.

Fig. 5.—Uredospores and mycelium of *Puccinia suaveolens* treated with caustic potash, showing the development of the spores. × 300. De Bary.

Figs. 6-8.—Development of the teleutospores of *Uromyces fabæ*. × 300. De Bary.

Fig. 9.—Uredospore of *Puccinia graminis*. × 475. C. B. P.

Fig. 10.—Two uredospores of the same in the initial stage of germination, showing two germ-pores. × 475. C. B. P.

Fig. 11.—Uredospore of the same after five hours and forty minutes' immersion in water; it has emitted a germ-tube from each germ-pore. × 475. C. B. P.

Fig. 12.—The same after twenty-three hours. One germ-tube has ceased to grow and is, like the spore, empty; the other, which is empty and septate below, contains all the orange endochrome, and, having grown to a considerable length, has taken three circumnutatory turns. × 475. C. B. P.

Fig. 13.—Another uredospore of the same, which, after twenty-six hours, has become branched as well as having made several circumnutatory movements. × 475. C. B. P.

Fig. 14.—Abnormal germination of one of the same uredospores (after ninety hours), in which a spore of reserve has been produced at the upper end of the germ-tube. × 475. C. B. P.

Fig. 15.—The germ-tubes of two uredospores of *Puccinia graminis* entering the stomata of a wheat plant. × 475. C. B. P.

Fig. 16.—Two hyaline paraphyses of the uredospores of *Melampsora epitea*. × 200. C. B. P.

Fig. 17.—Two capitate hyaline paraphyses of the uredospores of *Puccinia anthoxanthi*. × 200. C. B. P.

Figs. 18–21.—Development of a teleutospore of *Puccinia graminis*. × 200. De Bary.

Fig. 22.—Teleutospore of *Gymnosporangium clavariæforme* germinating. It has protruded two promycelia, one from each segment near the septum. The endochrome is passing from the spore into the promycelia, leaving the apex of the spores empty. × 475. C. B. P.

Fig. 23.—The same further advanced. × 475. C. B. P.

Fig. 24.—Four promycelial spores of the same, three of which have germinated, and one has formed a spore of reserve. × 475. C. B. P.

## Plate IV.

Fig. 1.—Promycelial spores of *Puccinia phalaridis* piercing by their germ-tubes the epidermal cells of *Arum maculatum*. × 475. C. B. P.

Fig. 2.—Teleutospore of *Uromyces fabæ* germinating. × 200. C. B. P.

Fig. 3.—Teleutospore of *Puccinia coronata* germinating from both cells; both germ-canals are placed laterally. × 250. C. B. P.

Fig. 4.—Teleutospore of *Puccinia arundinacea* germinating. × 250. C. B. P.

Fig. 5.—Teleutospore of *Phragmidium rubi*, three cells of which have germinated. × 300. C. B. P.

Fig. 6.—Teleutospore of *Triphragmium ulmariæ* germinating. × 475. C. B. P.

Fig. 7.—Teleutospore of *Endophyllum euphorbiæ* germinating; two of the promycelial spores have already emitted germ-tubes while still attached to the promycelium. × 475. C. B. P.

Fig. 8.—Three teleutospores of *Melampsora betulina*, two of which have germinated and are emptied of their endochrome. × 475. C. B. P.

Fig. 9.—Three teleutospores of *Coleosporium senecionis*, the upper compartments of two of which have germinated and produced single promycelial spores. × 300. C. B. P.

Fig. 10.—Germination of *Chrysomyxa rhododendri*. After De Bary.

Fig. 11.—*Gymnosporangium sabinæ*, thin-walled teleutospore germinating (12 hours); two germ-tubes have been emitted from each segment of the spore. × 200. C. B. P.

Fig. 12.—A thick-walled teleutospore of the same. × 200. C. B. P.

Fig. 13.—*Gymnosporangium confusum* thick-walled teleutospore germinating. × 200. C. B. P.

Fig. 14.—Thin-walled teleutospore of the same, germination rather further advanced. × 200. C. B. P.

Figs. 15-17.—Three teleutospores of *Puccinia amorphæ*, showing the gelatinous investment. × 400. C. B. P.

## Plate V.

Fig. 1.—Mycelial hypha of *Ustilago segetum* from the base of the stem of *Avena clatior*. × 500. C. B. P.

Fig. 2.—Mycelial hypha from the rachis of the same plant. × 500. C. B. P.

Fig. 3.—Mycelial hypha of *U. longissima*, the extremity of which has become gelatinized. × 500. C. B. P.

Fig. 4.—Two gelatinized hyphæ of the same, one of which (*a*) has become nodose, and in it spore-formation has begun (from *Glyceria aquatica*). × 500. C. B. P.

Fig. 5.—Gelatinized hypha of *U. segetum* from *Avena clatior*. × 500. C. B. P.

Fig. 6.—Mycelial hypha of *Sorosporium saponariæ*, which, after piercing a cell-wall, has become suddenly changed into a spore-forming hypha. × 600. Von Waldheim.

Fig. 7.—Gelatinized hypha of *U. maydis*, showing the contents breaking up for spore-formation. × 900. Von Waldheim.

Fig. 8.—A number of coalesced gelatinized hyphæ of *U. violacea* in which spore-formation is taking place, and the contour of the spores is distinctly observable. × 900. Von Waldheim.

Fig. 9.—A gelatinized hypha of *S. saponariæ*, which is coiled upon itself preparatory to the formation of a spore-ball. × 500. Von Waldheim.

Fig. 10.—A young spore-ball of *S. saponariæ*, with the commencement of four spores seen in section, surrounded by the coalesced gelatinized hyphæ. Treated with iodine. × 500. Von Waldheim.

Fig. 11.—A larger spore-ball seen before the coalescence of the hyphæ. × 300. Von Waldheim.

Fig. 12.—Two spores from the interior of a spore-ball still attached to their spore-forming hypha. × 800. Von Waldheim.

Fig. 13.—The terminal branches of a spore-forming hypha of *Urocystis colchici* becoming curved inwards. × 1000? Winter.

Fig. 14.—The same more advanced; the external branches from which the pseudospores are developed are embracing the young spore-ball. × 1000? Winter.

Fig. 15.—Another view of the same. × 1000? Winter.

## Plate VI.

Fig. 1.—Two mycelial hyphæ of *Tubercinia trientalis* with botryform haustoria. × 520. Woronin.

Fig. 2.—The primary mycelium of *Urocystis occulta* growing across the interspace between the inner surface of the primary embryonic sheath and the outer surface of the embryo of a rye plant. × 375. Wolff.

Fig. 3.—Promycelial spore of *Tilletia levis* entering the primary sheath of a wheat plant and invaginating a sheath of cellulose. × 600. Wolff.

Fig. 4.—Spore-forming hypha of *Tilletia tritici* from wheat, showing the budding of the spores; from an ovary 1 mm. long. × 900. Von Waldheim.

Fig. 5.—The extremity of a spore-forming hypha of the same, showing two spores in different degrees of development. × 500. Von Waldheim.

Fig. 6.—Transparent single-contoured spore of the same, with shrivelled remains of the spore-forming hypha. × 500. Von Waldheim.

Fig. 7.—Teleutospore of *T. tritici* germinating after forty-eight hours in water. × 500. C. B. P.

Fig. 8.—The same more advanced, showing the H-shaped spores (primary promycelial spores conjugated), two of which have already fallen away from the promycelium. × 300. C. B. P.

Figs. 9, 10.—Primary promycelial spores conjugated, empty, and septate, each pair of which has produced a secondary promycelial spore. × 300. C. B. P.

Fig. 11.—Two secondary promycelial spores conjugating. × 350. C. B. P.

Fig. 12.—The same a little later. A germ-tube has been given off from one of the spores. × 350. C. B. P.

Fig. 13.—A single unconjugated secondary promycelial spore giving off a germ-tube. × 350. C. B. P.

Fig. 14.—Mycelium from a single primary promycelial spore in nährlösung producing conidia. × 100. Reduced from Brefeld.

Fig. 15.—A small portion of the same more highly magnified. × 357. After Brefeld.

## PLATE VII.

Fig. 1.—*Ustilago segetum.* Teleutospore germinating in water after sixteen and a half hours. × 500. C. B. P.

Fig. 2.—The same, with a secondary promycelial spore produced from a primary while still attached to the promycelium. × 500. C. B. P.

Fig. 3.—A promycelium of the same forming a buckle joint. × 500. C. B. P.

Fig. 4.—A promycelium of the same forming a bow joint. × 500. After Brefeld.

Fig. 5.—Two promycelia conjugating. × 500. C. B. P.

Fig. 6.—A promycelium growing out into a germ-tube. × 500. After Wolff.

Fig. 7.—Two promycelia given off from one teleutospore. × 500. C. B. P.

Fig. 8.—Yeast-spores produced from a promycelium in nährlösung (three days). × 500. C. B. P.

Fig. 9.—A promycelium with a buckle joint and secondary promycelial spores in exhausted nährlösung (twelve days). × 500. C. B. P.

Fig. 10.—A yeast-spore colony, produced from a single teleutospore. × 500. Brefeld.

Fig. 11.—Promycelial spores conjugating. × 500. C. B. P.

Fig. 12.—Promycelial spore emitting a germ-tube. × 500. C. B. P.

Fig. 13.—Promycelial spore producing a secondary promycelial spore in water. × 500. C. B. P.

Fig. 14.—*Ustilago longissima.* Teleutospore just beginning to germinate. × 500. C. B. P.

Fig. 15.—The same a little later, showing the septation of the promycelium and the separation of the promycelial spore. × 500. C. B. P.

Fig. 16.—The same after two promycelial spores have been thrown off, and a third is almost ready to be so. × 500. C. B. P.

Fig. 17.—*Ustilago kühneana.* Teleutospore germinated, showing the promycelial spores in whorls on the promycelium. × 350. Brefeld.

Fig. 18.—*Ustilago scabiosæ* on *Scabiosa arvensis*, twelve hours after germination in water (September 1, 1884). × 500. C. B. P.

Fig. 19.—*Ustilago major* on *Silene otites*, gathered near Paris, October, 1887; germinated at the beginning of November. Twenty hours in water, showing the promycelium, which has nearly attained its full development. × 500. C. B. P.

Fig. 20.—The same at twenty-four hours, having become septate. × 500. C. B. P.

Fig. 21.—Teleutospore with promycelium still attached (thirty-six hours), which has produced a promycelial spore at its apex. This is quite an exceptional circumstance with this species. × 500. C. B. P.

Fig. 22.—Three promycelia which have fallen off the teleutospores (twenty four hours). × 500. C. B. P.

Figs. 23-25.—Three promycelia which have produced promycelial spores both terminally and laterally, after having fallen away from the teleutospores (forty-eight hours). × 500. C. B. P.

Fig. 26, 27.—Spore-balls of *Urocystis primulicola*, germinating (August 31, 1884) after forty-eight hours in water. They have emitted promycelia, probably from the central spores—but on this point I was unable to satisfy myself,—which have given origin to promycelial spores, one of which (Fig. 27) has produced a secondary promycelial spore. × 475. C. B. P.

Fig. 28.—Two promycelial spores conjugating, after fifty hours in water. × 500. C. B. P.

Fig. 29.—Two promycelial spores which have become vacuolate and septate, after two hundred and sixty-four hours in water. × 500. C. B. P.

Fig. 30.—Promycelial spore emitting a germ-tube, after one hundred and sixty-eight hours in water. × 500. C. B. P.

Fig. 31.—*Urocystis anemones.* Teleutospore germinating in water, November 25, 1884. × 475. C. B. P.

Fig. 32.—Two promycelial spores of the same conjugating. × 500. C. B. P.

Fig. 33.—Two promycelial spores of the same which have become vacuolate and septate. × 500. C. B. P.

Fig. 34.—Spore-ball of *Urocystis fischeri*, one teleutospore of which has germinated. × 500. C. B. P.

Fig. 35.—Three promycelial spores of the same. × 500. C. B. P.

PLATE VIII.

Fig. 1.—*Tubercinia trientalis* conidiophores growing outwards between the epidermal cells and producing conidia. × 320. Woronin.

Fig. 2.—Two conidia of the same germinating. × 320. Woronin.

Fig. 3.—Conidium of the same forcing its germ-tube between two epidermal cells of the host-plant. × 520. Woronin.

Fig. 4.—*Doassansia alismatis.* Section of a leaf of *Alisma plantago*, showing a receptacle of *D. alismatis.* × 20. After Cornu.

Fig. 5.—Teleutospore of the same germinating, showing the septum in the promycelium, and apical cluster of promycelial spores. × 500. Cornu.

Fig. 6.—*Graphiola phœnicis.* Three spores germinating, *a* has just emitted its germ-tube ; in *b* and *c* the germ-tubes have become promycelia, and are each producing a single promycelial spore. × 600. Fischer.

Fig. 7.—Promycelial spore of the above which has from one end emitted a germ-tube, and has become septate and empty. × 600. Fischer.

Fig. 8.—Spore of the same which has produced a well-developed germ-tube. × 600. Fischer.

Fig. 9.—*Tuberculina persicina.* Ripe spore just beginning to germinate in a weak saccharine solution. × 420. Gobi.

Fig. 10.—The same further advanced, and developing a promycelial spore at its distal extremity. × 420. Gobi.

Fig. 11.—Apex of promycelium just before the promycelial spore is abstricted. × 420. Gobi.

Fig. 12.—The extremity of a branched mycelial hypha producing spores from the surface of ripe sclerotium. × 420. Gobi.

Fig. 13.—Three promycelial spores of various sizes. × 420. Gobi.

Fig. 14.—*Protomyces macrosporus.* Mycelial hypha with intercalated spore. × 700. De Bary.

Fig. 15–18.—The same in various stages of maturity and germination. × 200. De Bary.

Fig. 19.—The escaped spores. × 200. De Bary.

Fig. 20.—The same conjugating. × 300. De Bary.

Plate I

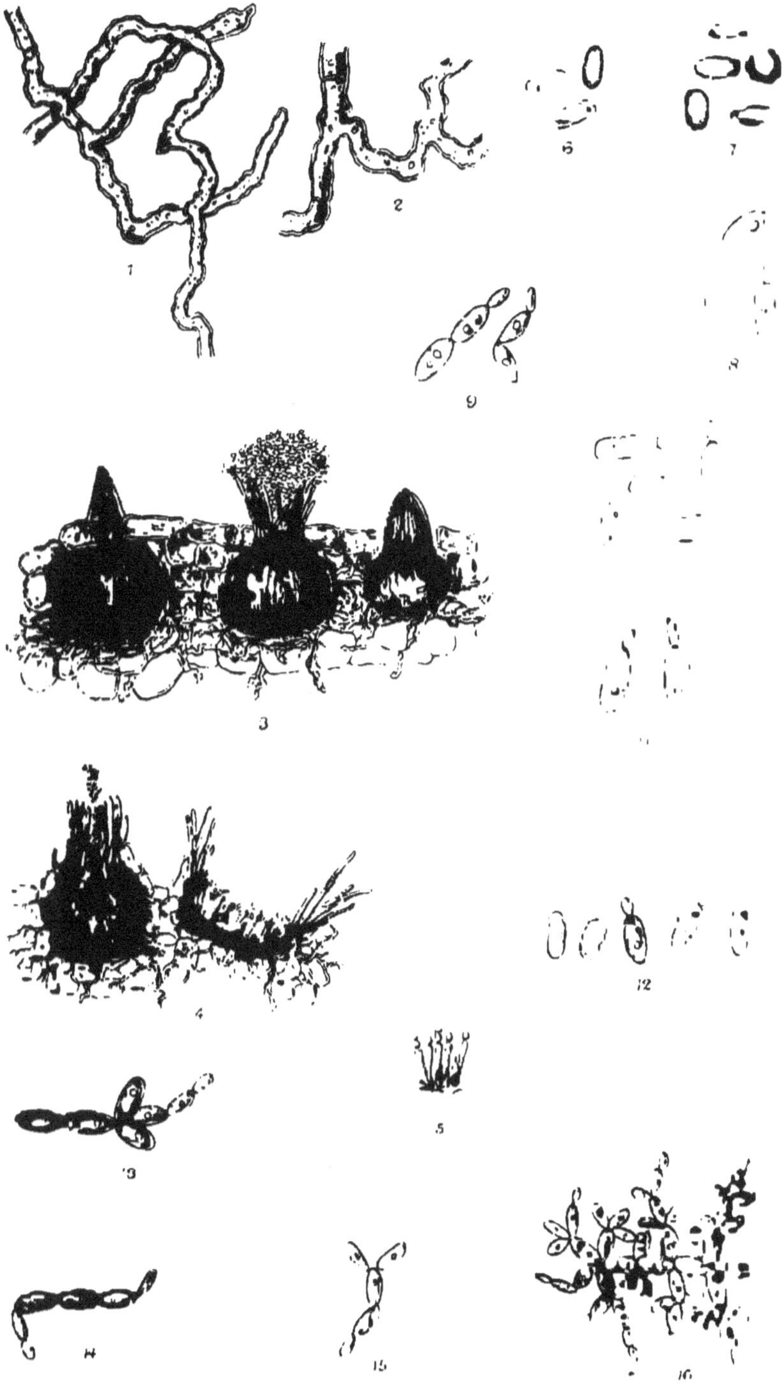


P del.

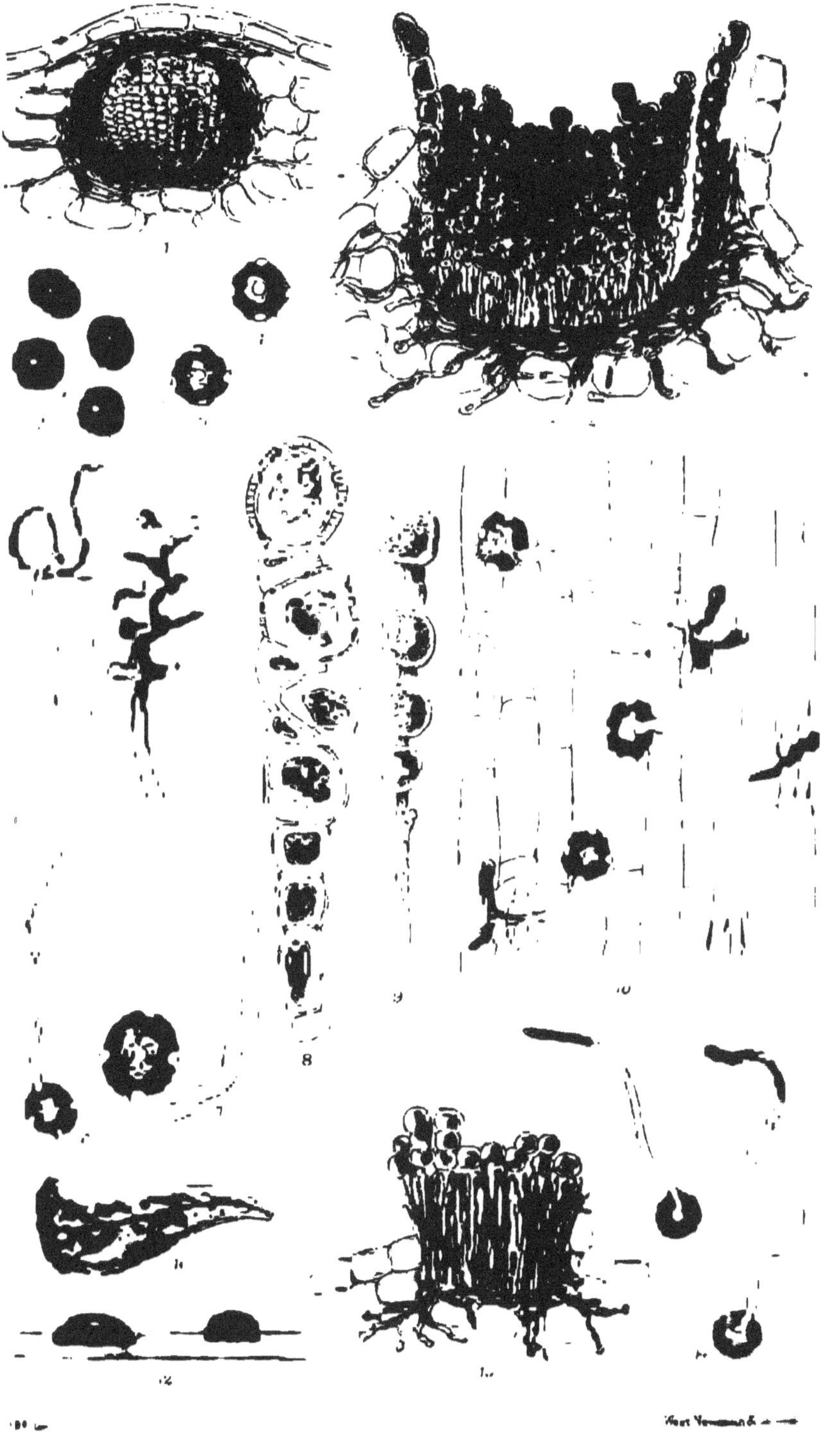

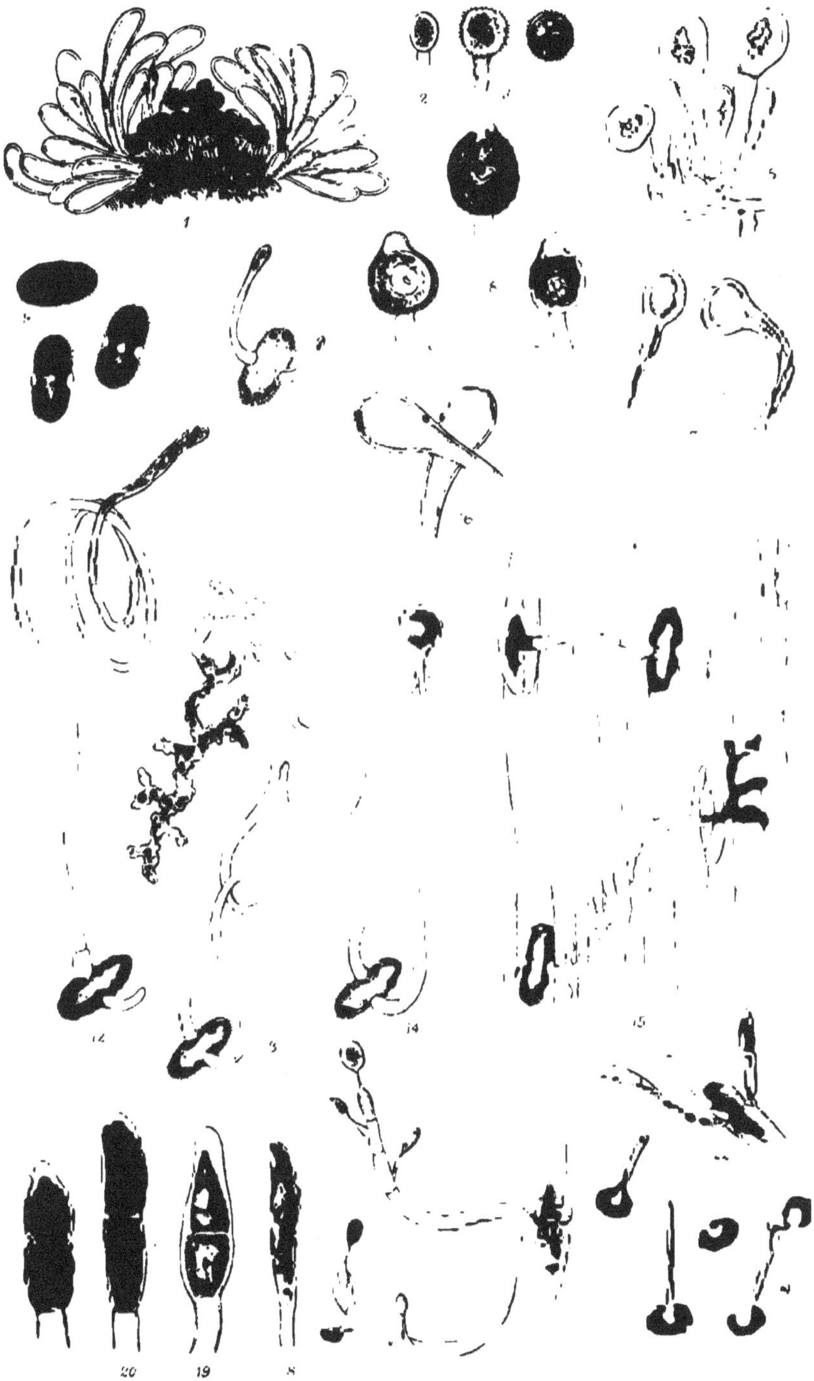

del West Newman & Co.

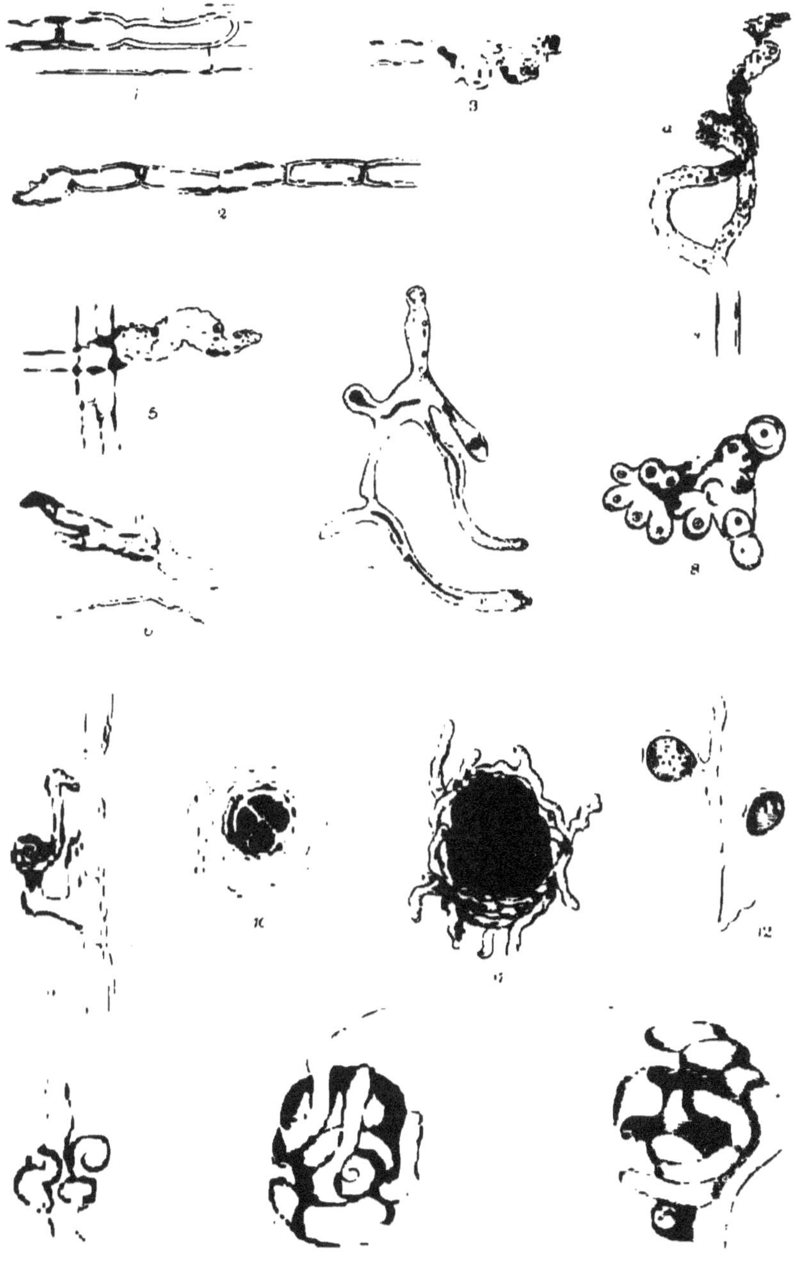
3
2
5
8
12

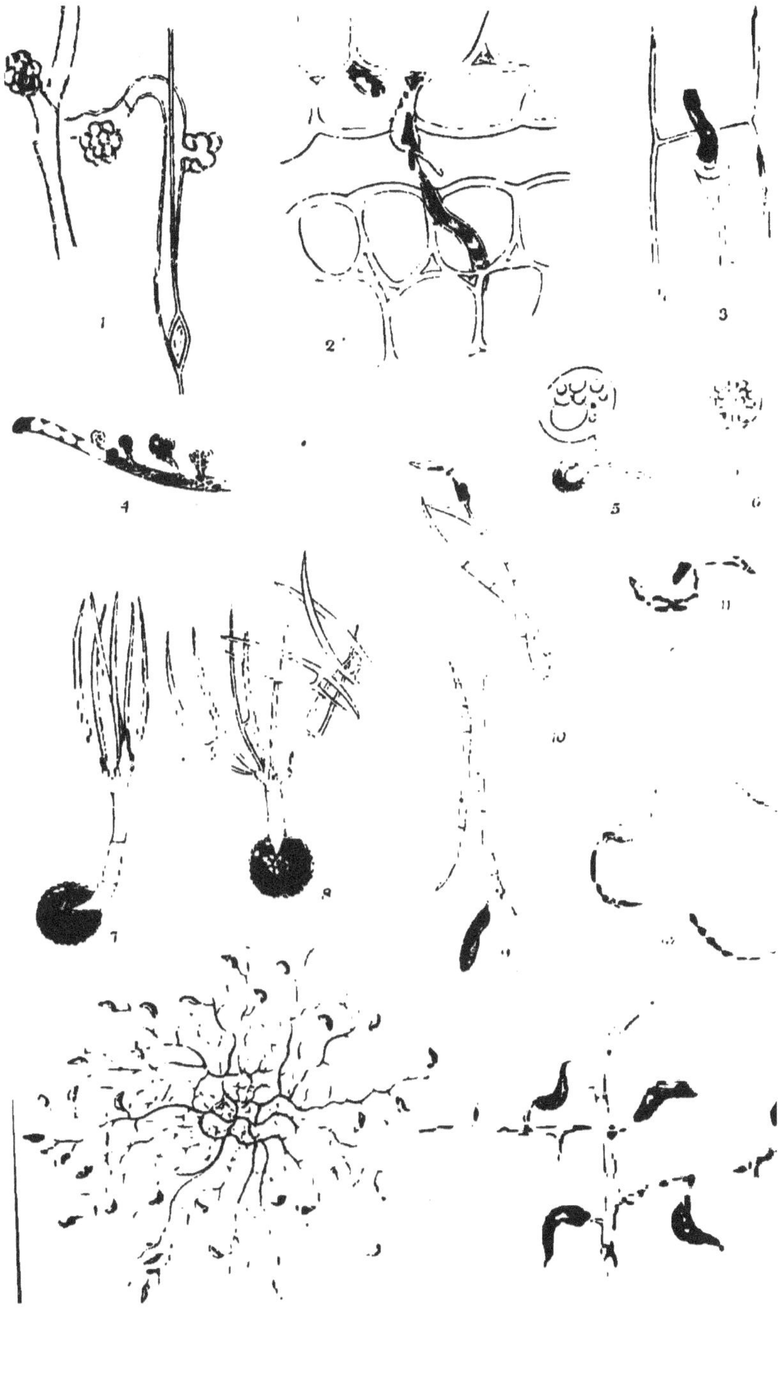

1
2
3
4
5
6
7
8
10
11

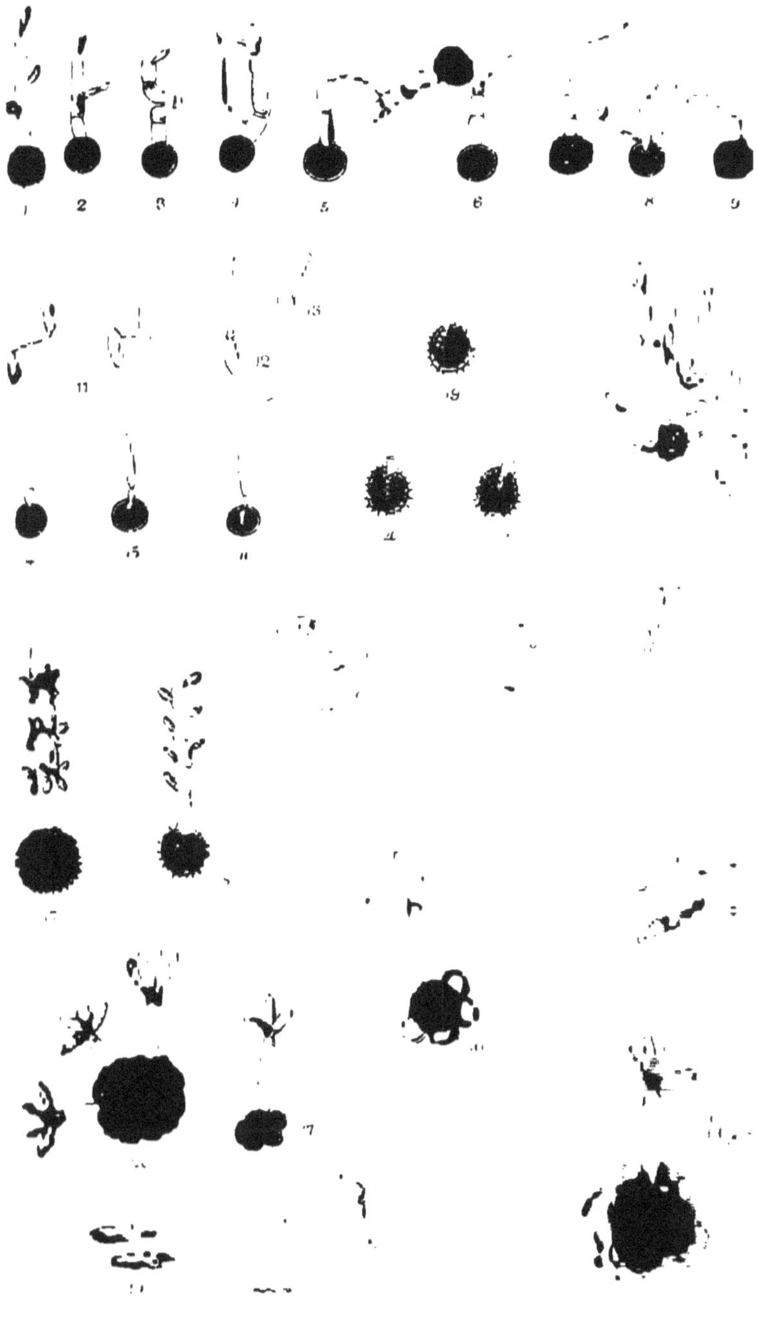

2
3
5
6
8
9
11
12

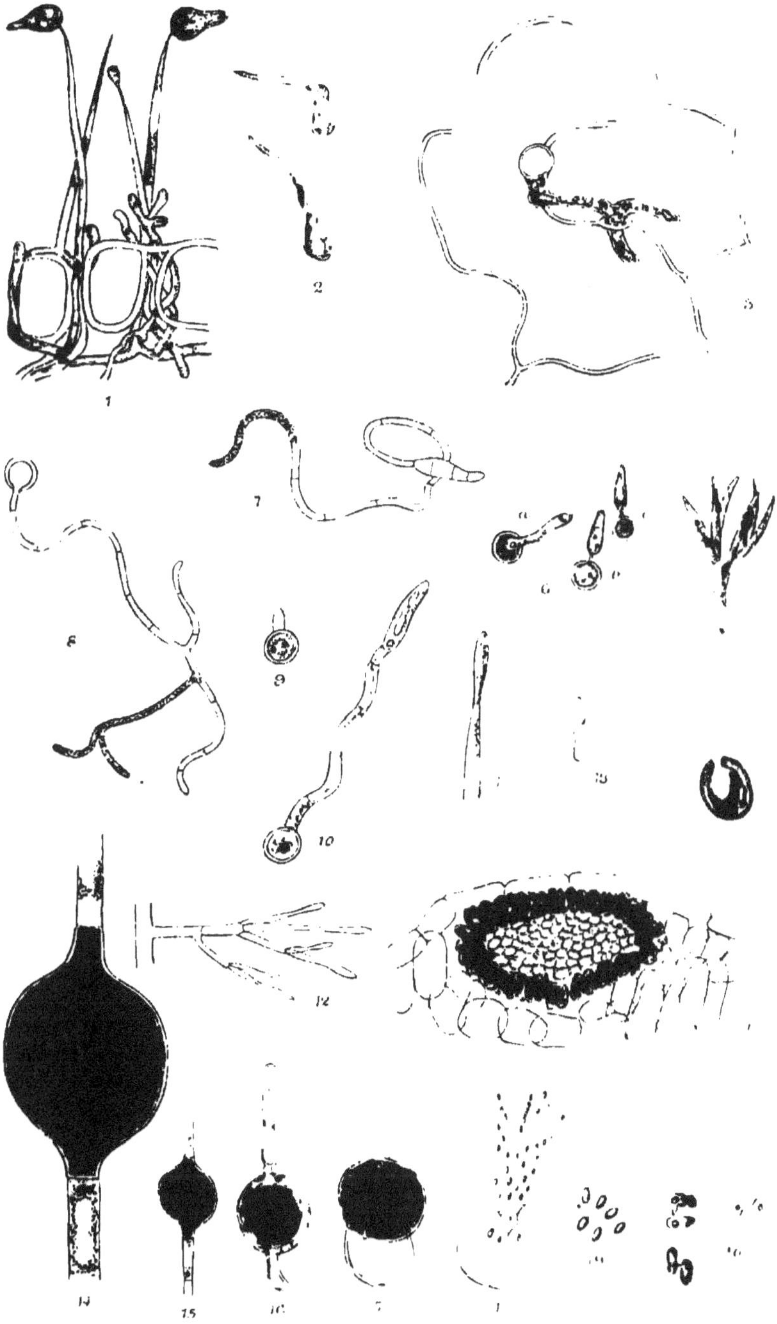

# INDEX OF HOST-PLANTS.

## UREDINEÆ.

## USTILAGINEÆ AND ALLIED SPECIES.

# BIOLOGICAL INDEX.

# INDEX OF SPECIES.

PRINTED BY WILLIAM CLOWES AND SONS, LIMITED, LONDON AND BECCLES.

*A LIST OF*

# *N PAUL, TRENCH, & CO.'S*

# *PUBLICATIONS.*

1 *Paternoster Square,*
*London.*

# A LIST OF
# KEGAN PAUL, TRENCH, & CO.'S
# PUBLICATIONS.

CONTENTS.



*A. K. H. B.*—From a Quiet Place. A New Volume of Sermons. Crown 8vo. 5*s*.

*AINSWORTH (F. W.)*—Personal Narrative of the Euphrates Expedition. 2 vols. 8vo. 30*s*.

*ALEXANDER (William, D.D., Bishop of Derry)*—The Great Question, and other Sermons. Crown 8vo. 6*s*.

*ALLIES (T. W.) M.A.*—Per Crucem ad Lucem. The Result of a Life. 2 vols. Demy 8vo. 25*s*.

A Life's Decision. Crown 8vo. 7*s*. 6*d*.

*AMHERST (Rev. W. J.)*—The History of Catholic Emancipation and the Progress of the Catholic Church in the British Isles (chiefy in England) from 1771–1820. 2 vols. Demy 8vo. 24*s*.

*AMOS (Prof. Sheldon)*—The History and Principles of the Civil Law of Rome. Demy 8vo. 16*s*.

*ARISTOTLE*—The Nicomachean Ethics of Aristotle. Translated by F. H. Peters, M.A. Third Edition. Crown 8vo. 6*s*.

*AUBERTIN (J. J.)*—A Flight to Mexico. With 7 full-page Illustrations and a Railway Map of Mexico. Crown 8vo. 7*s*. 6*d*.

Six Months in Cape Colony and Natal. With Illustrations and Map. Crown 8vo. 6*s*.

A Fight with Distances. With 8 Illustrations and 2 Maps. Crown 8vo. 7*s*. 6*d*.

Aucassin and Nicolette. Edited in Old French and rendered in Modern English by F. W. Bourdillon. Fcap. 8vo. 7*s*. 6*d*.

*AUCHMUTY (A. C.)*—Dives and Pauper, and other Sermons. Crown 8vo. 3*s*. 6*d*.

*AZARIAS (Brother)*—Aristotle and the Christian Church. Small crown 8vo. 3*s*. 6*d*.

*BADGER (George Percy) D.C.L.*—An English-Arabic Lexicon. In which the equivalents for English Words and Idiomatic Sentences are rendered into literary and colloquial Arabic. Royal 4to. 80*s*.

*BAGEHOT* (*Walter*)—THE ENGLISH CONSTITUTION. Fifth Edition. Crown 8vo. 7*s*. 6*d*.

LOMBARD STREET. A Description of the Money Market. Ninth Edition. Crown 8vo. 7*s*. 6*d*.

ESSAYS ON PARLIAMENTARY REFORM. Crown 8vo. 5*s*.

SOME ARTICLES ON THE DEPRECIATION OF SILVER, AND TOPICS CONNECTED WITH IT. Demy 8vo. 5*s*.

*BAGOT* (*Alan*) *C.E.*—ACCIDENTS IN MINES: Their Causes and Prevention. Crown 8vo. 6*s*.

THE PRINCIPLES OF COLLIERY VENTILATION. Second Edition, greatly enlarged, crown 8vo. 5*s*.

THE PRINCIPLES OF CIVIL ENGINEERING IN ESTATE MANAGEMENT. Crown 8vo. 7*s*. 6*d*.

*BAIRD* (*Henry M.*)—THE HUGUENOTS AND HENRY OF NAVARRE. 2 vols. 8vo. With Maps. 24*s*.

*BALDWIN* (*Capt. J. H.*)—THE LARGE AND SMALL GAME OF BENGAL AND THE NORTH-WESTERN PROVINCES OF INDIA. With 20 Illustrations. New and Cheaper Edition. Small 4to. 10*s*. 6*d*.

*BALLIN* (*Ada S. and F. L.*)—A HEBREW GRAMMAR. With Exercises selected from the Bible. Crown 8vo. 7*s*. 6*d*.

*BALL* (*John, F.R.S.*)—NOTES OF A NATURALIST IN SOUTH AMERICA. Crown 8vo. 8*s*. 6*d*.

*BARCLAY* (*Edgar*)—MOUNTAIN LIFE IN ALGERIA. Crown 4to. With numerous Illustrations by Photogravure. 16*s*.

*BASU* (*K. P.*) *M.A.*—STUDENTS' MATHEMATICAL COMPANION. Containing problems in Arithmetic, Algebra, Geometry, and Mensuration, for Students of the Indian Universities. Crown 8vo. 6*s*.

*BAUR* (*Ferdinand*) *Dr. Ph., Professor in Maulbronn.*—A PHILOLOGICAL INTRODUCTION TO GREEK AND LATIN FOR STUDENTS. Translated and adapted from the German by C. KEGAN PAUL, M.A., and the Rev. E. D. STONE, M.A. Third Edition. Crown 8vo. 6*s*.

*BENN* (*Alfred W.*)—THE GREEK PHILOSOPHERS. 2 vols. Demy 8vo. 28*s*.

*BENSON* (*A. C.*)—WILLIAM LAUD, SOMETIME ARCHBISHOP OF CANTERBURY. A Study. With Portrait. Crown 8vo. 6*s*.

*BIBLE FOLK-LORE.*—A STUDY IN COMPARATIVE MYTHOLOGY. Large crown 8vo. 10*s*. 6*d*.

*BIRD* (*Charles*) *F.G.S.*—HIGHER EDUCATION IN GERMANY AND ENGLAND: Being a Brief Practical Account of the Organisation and Curriculum of the German Higher Schools. With Critical Remarks and Suggestions with reference to those of England. Small crown 8vo. 2*s*. 6*d*.

BIRTH AND GROWTH OF RELIGION. A Book for Workers. Crown 8vo. cloth, 2*s*.; paper covers, 1*s*.

*BLACKBURN* (*Mrs. Hugh*)—BIBLE BEASTS AND BIRDS. A New Edition of 'Illustrations of Scripture by an Animal Painter.' With Twenty-two Plates, Photographed from the Originals, and Printed in Platinotype. 4to. cloth extra, gilt edges, 42*s*.

*COOPER (James Fenimore)*—LIFE. By T. R. LOUNDSBURY. With Portrait. Crown 8vo. 5*s.*

*CORY (William)*—A GUIDE TO MODERN ENGLISH HISTORY. Part I.—MDCCCXV.-MDCCCXXX. Demy 8vo. 9*s.* Part II.—MDCCCXXX.-MDCCCXXXV. 15*s.*

*COTTERILL (H. B.)*—AN INTRODUCTION TO THE STUDY OF POETRY. Crown 8vo. 7*s.* 6*d.*

*COTTON (H. J. S.)*—NEW INDIA, OR INDIA IN TRANSITION. Third Edition. Crown 8vo. 4*s.* 6*d.* Popular Edition, paper covers, 1*s.*

*COWIE (Right Rev. W. G.)*—OUR LAST YEAR IN NEW ZEALAND. 1887. Crown 8vo. 7*s.* 6*d.*

*COX (Rev. Sir George W.) M.A., Bart.*—THE MYTHOLOGY OF THE ARYAN NATIONS. New Edition. Demy 8vo. 16*s.*

TALES OF ANCIENT GREECE. New Edition. Small crown 8vo. 6*s.*

A MANUAL OF MYTHOLOGY IN THE FORM OF QUESTION AND ANSWER. New Edition. Fcp. 8vo. 3*s.*

AN INTRODUCTION TO THE SCIENCE OF COMPARATIVE MYTHOLOGY AND FOLK-LORE. Second Edition. Crown 8vo. 7*s.* 6*d.*

*COX (Rev. Sir G. W.) M.A., Bart., and JONES (Eustace Hinton)*—POPULAR ROMANCES OF THE MIDDLE AGES. Third Edition, in 1 vol. Crown 8vo. 6*s.*

*COX (Rev. Samuel) D.D.*—A COMMENTARY ON THE BOOK OF JOB. With a Translation. Second Edition. Demy 8vo. 15*s.*

SALVATOR MUNDI; or, Is Christ the Saviour of all Men? Eleventh Edition. Crown 8vo. 2*s.* 6*d.*

THE LARGER HOPE: a Sequel to 'SALVATOR MUNDI.' Second Edition. 16mo. 1*s.*

THE GENESIS OF EVIL, AND OTHER SERMONS, mainly expository. Third Edition. Crown 8vo. 6*s.*

BALAAM: An Exposition and a Study. Crown 8vo. 5*s.*

MIRACLES. An Argument and a Challenge. Crown 8vo. 2*s.* 6*d.*

*CRAVEN (Mrs.)*—A YEAR'S MEDITATIONS. Crown 8vo. 6*s.*

*CRAWFURD (Oswald)*—PORTUGAL, OLD AND NEW. With Illustrations and Maps. New and Cheaper Edition. Crown 8vo. 6*s.*

*CRUISE (F. R.) M.D.*—THOMAS À KEMPIS. Notes of a Visit to the Scenes in which his Life was spent, with some Account of the Examination of his Relics. Demy 8vo. Illustrated. 12*s.*

*CUNNINGHAM (W., B.D.)*—POLITICS AND ECONOMICS: An Essay on the Nature of the Principles of Political Economy, together with a Survey of Recent Legislation. Crown 8vo. 5*s.*

*DARMESTETER (Arsène)*—THE LIFE OF WORDS AS THE SYMBOLS OF IDEAS. Crown 8vo. 4*s.* 6*d.*

*DAVIDSON (Rev. Samuel) D.D., LL.D.*—CANON OF THE BIBLE: Its Formation, History, and Fluctuations. Third and revised Edition. Small crown 8vo. 5*s.*

THE DOCTRINE OF LAST THINGS, contained in the New Testament, compared with the Notions of the Jews and the Statements of Church Creeds. Small crown 8vo. 3*s.* 6*d.*

*DAWSON* (*Geo.*) *M.A.*—PRAYERS, WITH A DISCOURSE ON PRAYER. Edited by his Wife. First Series. New and Cheaper Edition. Crown 8vo. 3*s.* 6*d.*

PRAYERS, WITH A DISCOURSE ON PRAYER. Edited by GEORGE ST. CLAIR. Second Series. Crown 8vo. 6*s.*

SERMONS ON DISPUTED POINTS AND SPECIAL OCCASIONS. Edited by his Wife. Fourth Edition. Crown 8vo. 6*s.*

SERMONS ON DAILY LIFE AND DUTY. Edited by his Wife. Fifth Edition. Crown 8vo. 3*s.* 6*d.*

THE AUTHENTIC GOSPEL, and other Sermons. Edited by GEORGE ST. CLAIR. Third Edition. Crown 8vo. 6*s.*

EVERY-DAY COUNSELS. Edited by GEORGE ST. CLAIR, F.G.S. Crown 8vo. 6*s.*

BIOGRAPHICAL LECTURES. Edited by GEORGE ST. CLAIR, F.G.S. Second Edition. Large crown 8vo. 7*s.* 6*d.*

HAKESPEARE, and other Lectures. Edited by GEORGE ST. CLAIR, F.G.S. Large crown 8vo. 7*s.* 6*d.*

*DE BURY* (*Richard*)—THE PHILOBIBLON. Translated and Edited by ERNEST C. THOMAS.

*DE JONCOURT* (*Madame Marie*)—WHOLESOME COOKERY. Fourth Edition. Crown 8vo. cloth, 1*s.* 6*d.*; paper covers, 1*s.*

*DENT* (*H. C.*)—A YEAR IN BRAZIL. With Notes on Religion, Meteorology, Natural History, &c. Maps and Illustrations. Demy 8vo. 18*s.*

DOCTOR FAUST. The Old German Puppet Play, turned into English. with Introduction, etc., by T. C. H. HEDDERWICK. Large post 8vo. 7*s.* 6*d.*

*DOWDEN* (*Edward*) *LL.D.*—SHAKSPERE: a Critical Study of his Mind and Art. Eighth Edition. Post 8vo. 12*s.*

STUDIES IN LITERATURE, 1789–1877. Fourth Edition. Post 8vo. 6*s.*

TRANSCRIPTS AND STUDIES. Post 8vo. 12*s.*

*DRUMMOND* (*Thomas*)—LIFE. By R. BARRY O'BRIEN. 8vo. 14*s.*

DULCE DOMUM. Fcp. 8vo. 5*s.*

*DU MONCEL* (*Count*)—THE TELEPHONE, THE MICROPHONE, AND THE PHONOGRAPH. With 74 Illustrations. Third Edition. Small crown 8vo. 5*s.*

*DUNN* (*H. Percy*) *F.R.C.S.*—INFANT HEALTH. The Physiology and Hygiene of Early Life. Crown 8vo. 3*s.* 6*d.*

*DURUY* (*Victor*)—HISTORY OF ROME AND THE ROMAN PEOPLE. Edited by Professor MAHAFFY, with nearly 3,000 Illustrations. 4to. 6 Vols. in 12 Parts, 30*s.* each volume.

EDUCATION LIBRARY. Edited by Sir PHILIP MAGNUS:—

INDUSTRIAL EDUCATION. By Sir PHILIP MAGNUS.

AN INTRODUCTION TO THE HISTORY OF EDUCATIONAL THEORIES. By OSCAR BROWNING, M.A. Second Edition. 3*s.* 6*d.*

OLD GREEK EDUCATION. By the Rev. Prof. MAHAFFY, M.A. Second Edition. 3*s.* 6*d.*

SCHOOL MANAGEMENT; including a General View of the Work of Education. By JOSEPH LANDON. Sixth Edition. 6*s.*

*EDWARDES* (*Major-General Sir Herbert B.*)—MEMORIALS OF HIS LIFE. By his WIFE. With Portrait and Illustrations. 2 vols. 8vo. 36*s.*

EIGHTEENTH CENTURY ESSAYS. Selected and Edited by AUSTIN DOBSON. Fcp. 8vo. 1*s.* 6*d.*

*ELSDALE* (*Henry*)—STUDIES IN TENNYSON'S IDYLLS. Crown 8vo. 5*s.*

*EMERSON'S* (*Ralph Waldo*) LIFE. By OLIVER WENDELL HOLMES. [English Copyright Edition.] With Portrait. Crown 8vo. 6*s.*

*ERANUS.* A COLLECTION OF EXERCISES IN THE ALCAIC AND SAPPHIC METRES. Edited by F. W. CORNISH, Assistant Master at Eton. Second Edition. Crown 8vo. 2*s.*

FIVE O'CLOCK TEA. Containing Receipts for Cakes of every description, Savoury Sandwiches, Cooling Drinks, &c. Fcp. 8vo. 1*s.* 6*d.*, or 1*s.* sewed.

*FLINN* (*D. Edgar*)—IRELAND: its Health Resorts and Watering-Places. With Frontispiece and Maps. Demy 8vo. 5s.

*FORBES* (*Bishop*)—A MEMOIR, by the Rev. DONALD J. MACKEY. Portrait and Map. Crown 8vo. 7*s.* 6*d.*

*FORDYCE* (*John*)—THE NEW SOCIAL ORDER. Crown 8vo. 3*s.* 6*d.*

*FOTHERINGHAM* (*James*)—STUDIES IN THE POETRY OF ROBERT BROWNING. Crown 8vo. 6*s.*

*FRANKLIN* (*Benjamin*)—AS A MAN OF LETTERS. By J. B. MCMASTER. Crown 8vo. 5*s.*

FROM WORLD TO CLOISTER; or, My Novitiate. By BERNARD. Crown 8vo. 5*s.*

*GARDINER* (*Samuel R.*) *and J. BASS MULLINGER, M.A.*—INTRODUCTION TO THE STUDY OF ENGLISH HISTORY Second Edition. Large crown 8vo. 9*s.*

*GEORGE* (*Henry*)—PROGRESS AND POVERTY: an Inquiry into the Causes of Industrial Depressions, and of Increase of Want with Increase of Wealth. The Remedy. Library Edition. Post 8vo. 7*s.* 6*d.* Cabinet Edition, crown 8vo. 2*s.* 6*d.*

SOCIAL PROBLEMS. Crown 8vo. 5*s.*

PROTECTION, OR FREE TRADE. An Examination of the Tariff Question, with especial regard to the Interests of Labour. Second Edition. Crown 8vo. 5*s.*

*** Also Cheap Editions of each of the above, limp cloth, 1*s.* 6*d.*; paper covers, 1*s.*

*GILBERT* (*Mrs.*)—AUTOBIOGRAPHY, and other Memorials. Edited by JOSIAH GILBERT. Fifth Edition. Crown 8vo. 7*s.* 6*d.*

*GILLMORE* (*Col. Parker*)—DAYS AND NIGHTS BY THE DESERT. With numerous Illustrations. Demy 8vo. 10*s.* 6*d.*

*GLANVILL* (*Joseph*)—SCEPSIS SCIENTIFICA; or, Confest Ignorance, th Way to Science; in an Essay of the Vanity of Dogmatising and Confiden Opinion. Edited, with Introductory Essay, by JOHN OWEN. Elzevir 8vo. printed on hand-made paper, 6*s.*

*GLASS* (*Henry Alex.*)—THE STORY OF THE PSALTERS. Crown 8vo. 5*s.*

GLOSSARY OF TERMS AND PHRASES. Edited by the Rev. H. PERCY SMITH and others. Medium 8vo. 7*s.* 6*d.*

*GLOVER* (*F.*) *M.A.*—EXEMPLA LATINA. A First Construing Book, with Short Notes, Lexicon, and an Introduction to the Analysis of Sentences. Second Edition. Fcp. 8vo. 2*s.*

*GOODENOUGH* (*Commodore J. G.*)—MEMOIR OF, with Extracts from his Letters and Journals. Edited by his Widow. With Steel Engraved Portrait. Third Edition. Crown 8vo. 5*s.*

*GORDON (Major-Gen. C. G.)*—HIS JOURNALS AT KARTOUM. Printed from the Original MS. With Introduction and Notes by A. EGMONT HAKE. Portrait, 2 Maps, and 30 Illustrations. 2 vols. Demy 8vo. 21*s*. Also a Cheap Edition in 1 vol., 6*s*.

GORDON'S (GENERAL) LAST JOURNAL. A Facsimile of the last Journal received in England from General Gordon. Reproduced by Photo-lithography. Imperial 4to. £3. 3*s*.

EVENTS IN HIS LIFE. From the Day of his Birth to the Day of his Death. By Sir H. W. GORDON. With Maps and Illustrations. Demy 8vo. 7*s*. 6*d*.

*GOSSE (Edmund)* — SEVENTEENTH CENTURY STUDIES. A Contribution to the History of English Poetry. Demy 8vo. 10*s*. 6*d*.

*GOULD (Rev. S. Baring) M.A.*—GERMANY, PRESENT AND PAST. New and Cheaper Edition. Large crown 8vo. 7*s*. 6*d*.

THE VICAR OF MORWENSTOW: a Life of Robert Stephen Hawker, M.A. New and Cheaper Edition. Crown 8vo. 5*s*.

*GOWAN (Major Walter E.)* — A. IVANOFF'S RUSSIAN GRAMMAR. (16th Edition). Translated, enlarged, and arranged for use of Students of the Russian Language. Demy 8vo. 6*s*.

*GOWER (Lord Ronald)*—MY REMINISCENCES. Limp Parchment, Antique, with Etched Portrait, 10*s*. 6*d*.

BRIC-À-BRAC. Being some Photoprints illustrating Art objects at Gower Lodge, Windsor. Super royal 8vo. 15*s*.; Persian leather, 21*s*.

LAST DAYS OF MARY ANTOINETTE. An Historical Sketch. With Portrait and Facsimiles. Fcp. 4to. 10*s*. 6*d*.

NOTES OF A TOUR FROM BRINDISI TO YOKOHAMA, 1883–1884. Fcp. 8vo. 2*s*. 6*d*.

*GRAHAM (William) M.A.*—THE CREED OF SCIENCE, Religious, Moral, and Social. Second Edition, revised. Crown 8vo. 6*s*.

THE SOCIAL PROBLEM IN ITS ECONOMIC, MORAL, AND POLITICAL ASPECTS. Demy 8vo. 14*s*.

*GRIMLEY (Rev. H. N.) M.A.*—TREMADOC SERMONS, CHIEFLY ON THE SPIRITUAL BODY, THE UNSEEN WORLD, AND THE DIVINE HUMANITY. Fourth Edition. Crown 8vo. 6*s*.

THE TEMPLE OF HUMANITY, and other Sermons. Crown 8vo. 6*s*.

*GURNEY (Edmund)*—TERTIUM QUID: Chapters on various Disputed Questions. 2 vols. Crown 8vo. 12*s*.

*HADDON (Caroline)*—THE LARGER LIFE, STUDIES IN HINTON'S ETHICS. Crown 8vo. 5*s*.

*HAECKEL (Prof. Ernst)*—THE HISTORY OF CREATION. Translation revised by Professor E. RAY LANKESTER, M.A., F.R.S. With Coloured Plates and Genealogical Trees of the various groups of both plants and animals. 2 vols. Third Edition. Post 8vo. 32*s*.

THE HISTORY OF THE EVOLUTION OF MAN. With numerous Illustrations. 2 vols. Post 8vo. 32*s*.

A VISIT TO CEYLON. Post 8vo. 7*s*. 6*d*.

FREEDOM IN SCIENCE AND TEACHING. With a Prefatory Note by T. H. HUXLEY, F.R.S. Crown 8vo. 5*s*.

*HALCOMBE (J. J.)*—GOSPEL DIFFICULTIES DUE TO A DISPLACED SECTION OF ST. LUKE. Second Edition. Crown 8vo. 6*s*.

HAMILTON, MEMOIRS OF ARTHUR, B.A., of Trinity College, Cambridge. Crown 8vo. 6*s*.

HANDBOOK OF HOME RULE, being Articles on the Irish Question by Various Writers. Edited by JAMES BRYCE, M.P. Second Edition. Crown 8vo. 1*s*. sewed, or 1*s*. 6*d*. cloth.

*HART (Rev. J. W. T.)*—AUTOBIOGRAPHY OF JUDAS ISCARIOT. A Character-Study. Crown 8vo. 3*s*. 6*d*.

*HAWEIS (Rev. H. R.) M.A.*—CURRENT COIN. Materialism—The Devil — Crime — Drunkenness — Pauperism — Emotion — Recreation — The Sabbath. Fifth Edition. Crown 8vo. 5*s*.

ARROWS IN THE AIR. Fifth Edition. Crown 8vo. 5*s*.

SPEECH IN SEASON. Fifth Edition. Crown 8vo. 5*s*.

THOUGHTS FOR THE TIMES. Fourteenth Edition. Crown 8vo. 5*s*.

UNSECTARIAN FAMILY PRAYERS. New Edition. Fcp. 8vo. 1*s*. 6*d*.

*HAWTHORNE (Nathaniel)*—WORKS. Complete in 12 vols. Large post 8vo. each vol. 7*s*. 6*d*.

VOL. I. TWICE-TOLD TALES.
II. MOSSES FROM AN OLD MANSE.
III. THE HOUSE OF THE SEVEN GABLES, and THE SNOW IMAGE.
IV. THE WONDER BOOK, TANGLEWOOD TALES, and GRANDFATHER'S CHAIR.
V. THE SCARLET LETTER, and THE BLITHEDALE ROMANCE.
VI. THE MARBLE FAUN. (Transformation.)
VII. & VIII. OUR OLD HOME, and ENGLISH NOTE-BOOKS.
IX. AMERICAN NOTE-BOOKS.
X. FRENCH AND ITALIAN NOTE-BOOKS.
XI. SEPTIMIUS FELTON, THE DOLLIVER ROMANCE, FANSHAWE, and, in an appendix, THE ANCESTRAL FOOTSTEP.
XII. TALES AND ESSAYS, AND OTHER PAPERS, WITH A BIOGRAPHICAL SKETCH OF HAWTHORNE.

*HEATH (Francis George)*—AUTUMNAL LEAVES. Third and Cheaper Edition. Large crown 8vo. 6*s*.

SYLVAN WINTER. With 70 Illustrations. Large crown 8vo. 14*s*.

*HEGEL*—THE INTRODUCTION TO HEGEL'S PHILOSOPHY OF FINE ART. Translated from the German, with Notes and Prefatory Essay, by BERNARD BOSANQUET, M.A. Crown 8vo. 5*s*.

*HEIDENHAIN (Rudolph) M.D.*—HYPNOTISM; or Animal Magnetism. With Preface by G. J. ROMANES, F.R.S. Second Edition. Small crown 8vo. 2*s*. 6*d*.

*HENNESSY (Sir John Pope)*—RALEGH IN IRELAND, WITH HIS LETTERS ON IRISH AFFAIRS AND SOME CONTEMPORARY DOCUMENTS. Large crown 8vo. printed on hand-made paper, parchment, 10*s*. 6*d*.

*HENRY (Philip)*—DIARIES AND LETTERS. Edited by MATTHEW HENRY LEE, M.A. Large crown 8vo. 7*s*. 6*d*.

*HINTON (J.)*—THE MYSTERY OF PAIN. New Edition. Fcp. 8vo. 1*s*.

LIFE AND LETTERS. With an Introduction by Sir W. W. GULL, Bart., and Portrait engraved on Steel by C. H. JEENS. Fifth Edition. Crown 8vo. 8*s*. 6*d*.

*HINTON* (*J.*)—continued.

PHILOSOPHY AND RELIGION. Selections from the MSS. of the late JAMES HINTON. Edited by CAROLINE HADDON. Second Edition. Crown 8vo. 5*s.*

THE LAW BREAKER AND THE COMING OF THE LAW. Edited by MARGARET HINTON. Crown 8vo. 6*s.*

*HOOPER* (*Mary*)—LITTLE DINNERS: HOW TO SERVE THEM WITH ELEGANCE AND ECONOMY. Twentieth Edition. Crown 8vo. 2*s.* 6*d.*

COOKERY FOR INVALIDS, PERSONS OF DELICATE DIGESTION, AND CHILDREN. Fifth Edition. Crown 8vo. 2*s.* 6*d.*

EVERY-DAY MEALS. Being Economical and Wholesome Recipes for Breakfast, Luncheon, and Supper. Seventh Edition. Crown 8vo. 2*s.* 6*d.*

*HOPKINS* (*Ellice*)—WORK AMONGST WORKING MEN. Fifth Edition. Crown 8vo. 3*s.* 6*d.*

*HORNADAY* (*W. T.*)—TWO YEARS IN A JUNGLE. With Illustrations. Demy 8vo. 21*s.*

*HOSPITALIER* (*E.*)—THE MODERN APPLICATIONS OF ELECTRICITY. Translated and Enlarged by JULIUS MAIER, Ph.D. 2 vols. Second Edition, revised, with many additions and numerous Illustrations. Demy 8vo. 12*s.* 6*d.* each volume.

VOL. I.—Electric Generators, Electric Light.

II.—Telephone: Various Applications: Electrical Transmission of Energy.

*HOWARD* (*Robert*) *M.A.*—THE CHURCH OF ENGLAND AND OTHER RELIGIOUS COMMUNIONS. A Course of Lectures delivered in the Parish Church of Clapham. Crown 8vo. 7*s.* 6*d.*

HOW TO MAKE A SAINT; or, The Process of Canonisation in the Church of England. By The Prig. Fcp. 8vo. 3*s.* 6*d.*

*HYNDMAN* (*H. M.*)—THE HISTORICAL BASIS OF SOCIALISM IN ENGLAND. Large crown 8vo. 8*s.* 6*d.*

*IM THURN* (*Everard F.*)—AMONG THE INDIANS OF GUIANA. Being Sketches, chiefly Anthropologic, from the Interior of British Guiana. With 53 Illustrations and a Map. Demy 8vo. 18*s.*

IXORA: A Mystery. Crown 8vo. 6*s.*

*JACCOUD* (*Prof. S.*)—THE CURABILITY AND TREATMENT OF PULMONARY PHTHISIS. Translated and Edited by M. LUBBOCK, M.D. 8vo. 15*s.*

JAUNT IN A JUNK: A Ten Days' Cruise in Indian Seas. Large crown 8vo. 7*s.* 6*d.*

*JENKINS* (*E.*) *and RAYMOND* (*J.*)—THE ARCHITECT'S LEGAL HANDBOOK. Third Edition, Revised. Crown 8vo. 6*s.*

*JENKINS* (*Rev. Canon R. C.*)—HERALDRY: English and Foreign. With a Dictionary of Heraldic Terms and 156 Illustrations. Small crown 8vo. 3*s.* 6*d.*

STORY OF THE CARAFFA. Small crown 8vo. 3*s.* 6*d.*

*JEROME* (*Saint*)—LIFE, by Mrs. CHARLES MARTIN. Large cr. 8vo. 6*s.*

*JOEL* (*L.*)—A CONSUL'S MANUAL AND SHIPOWNER'S AND SHIPMASTER'S PRACTICAL GUIDE IN THEIR TRANSACTIONS ABROAD. With Definitions of Nautical, Mercantile, and Legal Terms; a Glossary of Mercantile Terms in English, French, German, Italian, and Spanish; Tables of the Money, Weights, and Measures of the Principal Commercial Nations and their Equivalents in British Standards; and Forms of Consular and Notarial Acts. Demy 8vo. 12*s.*

*JORDAN* (*Furneaux*) *F.R.C.S.*—ANATOMY AND PHYSIOLOGY IN CHARACTER. Crown 8vo. 5*s.*

*KAUFMANN (Rev. M.) M.A.*—SOCIALISM: its Nature, its Dangers, and its Remedies considered. Crown 8vo. 7*s.* 6*d.*

UTOPIAS; or, Schemes of Social Improvement, from Sir Thomas More to Karl Marx. Crown 8vo. 5*s.*

CHRISTIAN SOCIALISM. Crown 8vo. 4*s.* 6*d.*

*KAY (David)*—EDUCATION AND EDUCATORS. Crown 8vo. 7*s.* 6*d.*

MEMORY: What it is, and how to improve it. Crown 8vo. 6*s.*

*KAY (Joseph)*—FREE TRADE IN LAND. Edited by his Widow. With Preface by the Right Hon. JOHN BRIGHT, M.P. Seventh Edition. Crown 8vo. 5*s.*

*** Also a cheaper edition, without the Appendix, but with a Review of Recent Changes in the Land Laws of England, by the Right Hon. G. OSBORNE MORGAN, Q.C., M.P. Cloth, 1*s.* 6*d.*; Paper covers, 1*s.*

*KELKE (W. H. H.)*—AN EPITOME OF ENGLISH GRAMMAR FOR THE USE OF STUDENTS. Adapted to the London Matriculation Course and Similar Examinations. Crown 8vo. 4*s.* 6*d.*

*KEMPIS (Thomas à)*—OF THE IMITATION OF CHRIST. Parchment Library Edition, parchment or cloth, 6*s.*; vellum, 7*s.* 6*d.* The Red Line Edition, fcp. 8vo. cloth extra, 2*s.* 6*d.* The Cabinet Edition, small 8vo. cloth limp, 1*s.*; or cloth boards, red edges, 1*s.* 6*d.* The Miniature Edition, 32mo. cloth limp, 1*s.*; or with red lines, 1*s.* 6*d.*

*** All the above Editions may be had in various extra bindings.

*KENNARD (Rev. H. B.)*—MANUAL OF CONFIRMATION. 16mo. cloth, 1*s.* Sewed, 3*d.*

*KENDALL (Henry)*—THE KINSHIP OF MEN: Genealogy viewed as a Science. Crown 8vo. 5*s.*

*KETTLEWELL (Rev. S.) M.A.*—THOMAS À KEMPIS AND THE BROTHERS OF COMMON LIFE. 2 vols. With Frontispieces. Demy 8vo. 30*s.*

*** Also an Abridged Edition in 1 vol. With Portrait. Crown 8vo. 7*s.* 6*d.*

*KIDD (Joseph) M.D.*—THE LAWS OF THERAPEUTICS; or, the Science and Art of Medicine. Second Edition. Crown 8vo. 6*s.*

*KINGSFORD (Anna) M.D.*—THE PERFECT WAY IN DIET. A Treatise advocating a Return to the Natural and Ancient Food of Race. Small crown 8vo. 2*s.*

*KINGSLEY (Charles) M.A.*—LETTERS AND MEMORIES OF HIS LIFE. Edited by his WIFE. With Two Steel Engraved Portraits and Vignettes. Sixteenth Cabinet Edition, in 2 vols. Crown 8vo. 12*s.*

*** Also a People's Edition in 1 vol. With Portrait. Crown 8vo. 6*s.*

ALL SAINTS' DAY, and other Sermons. Edited by the Rev. W. HARRISON. Third Edition. Crown 8vo. 7*s.* 6*d.*

TRUE WORDS FOR BRAVE MEN. A Book for Soldiers' and Sailors' Libraries. Fourteenth Edition. Crown 8vo. 2*s.* 6*d.*

*KNOX (Alexander A.)*—THE NEW PLAYGROUND; or, Wanderings in Algeria. New and Cheaper Edition. Large crown 8vo. 6*s.*

*LAMARTINE (Alphonse de).* By Lady MARGARET DOMVILE. Large crown 8vo., with Portrait, 7*s.* 6*d.*

LAND CONCENTRATION AND IRRESPONSIBILITY OF POLITICAL POWER, as causing the Anomaly of a Widespread State of Want by the Side of the Vast Supplies of Nature. Crown 8vo. 5*s.*

*LANDON (Joseph)*—SCHOOL MANAGEMENT; including a General View of the Work of Education, Organisation, and Discipline. Sixth Edition. Crown 8vo. 6*s.*

*LAURIE (S. S.)*—LECTURES ON THE RISE AND EARLY CONSTITUTION OF UNIVERSITIES. With a Survey of Mediæval Education. Crown 8vo. 6*s.*

*LEE (Rev. F. G.) D.C.L.*—THE OTHER WORLD; or, Glimpses of the Supernatural. 2 vols. A New Edition. Crown 8vo. 15*s.*

*LEFEVRE (Right Hon. G. Shaw)*—PEEL AND O'CONNELL. Demy 8vo. 10*s.* 6*d.*

INCIDENTS OF COERCION. A Journal of Visits to Ireland. Crown 8vo. 1*s.*

LETTERS FROM AN UNKNOWN FRIEND. By the Author of 'Charles Lowder.' With a Preface by the Rev. W. H. Cleaver. Fcp. 8vo. 1*s.*

*LEWARD (Frank)*—Edited by CHAS. BAMPTON. Crown 8vo. 7*s.* 6*d.*

LIFE OF A PRIG. By ONE. Third Edition. Fcp. 8vo. 3*s.* 6*d.*

*LILLIE (Arthur) M.R.A.S.*—THE POPULAR LIFE OF BUDDHA. Containing an Answer to the Hibbert Lectures of 1881. With Illustrations. Crown 8vo. 6*s.*

BUDDHISM IN CHRISTENDOM; or, Jesus, the Essene. Demy 8vo. with Illustrations. 15*s.*

*LOCHER (Carl)*—EXPLANATION OF THE ORGAN STOPS, with Hints for Effective Combinations. Illustrated. Demy 8vo. 5*s.*

*LONGFELLOW (H. Wadsworth)*—LIFE. By his Brother, SAMUEL LONGFELLOW. With Portraits and Illustrations. 3 vols. Demy 8vo. 42*s.*

*LONSDALE (Margaret)*—SISTER DORA: a Biography. With Portrait. Cheap Edition. Crown 8vo. 2*s.* 6*d.*

GEORGE ELIOT: Thoughts upon her Life, her Books, and Herself. Second Edition. Small crown 8vo. 1*s.* 6*d.*

*LOWDER (Charles)*—A BIOGRAPHY. By the Author of 'St. Teresa.' New and Cheaper Edition. Crown 8vo. With Portrait. 3*s.* 6*d.*

*LÜCKES (Eva C. E.)*—LECTURES ON GENERAL NURSING, delivered to the Probationers of the London Hospital Training School for Nurses. Second Edition. Crown 8vo. 2*s.* 6*d.*

*LYTTON (Edward Bulwer, Lord)*—LIFE, LETTERS, AND LITERARY REMAINS. By his Son the EARL OF LYTTON. With Portraits, Illustrations, and Facsimiles. Demy 8vo. cloth. Vols. I. and II. 32*s.*

*MACHIAVELLI (Niccolò)*—HIS LIFE AND TIMES. By Prof. VILLARI. Translated by LINDA VILLARI. 4 vols. Large post 8vo. 48*s.*

DISCOURSES ON THE FIRST DECADE OF TITUS LIVIUS. Translated from the Italian by NINIAN HILL THOMSON, M.A. Large crown 8vo. 12*s.*

THE PRINCE. Translated from the Italian by N. H. T. Small crown 8vo. printed on hand-made paper, bevelled boards, 6*s.*

*MACNEILL (J. G. Swift)*—HOW THE UNION WAS CARRIED. Crown 8vo. cloth, 1*s.* 6*d.*; paper covers, 1*s.*

*MAGNUS (Lady)*—ABOUT THE JEWS SINCE BIBLE TIMES. From the Babylonian Exile till the English Exodus. Small crown 8vo. 6*s.*

*MAGUIRE* (*Thomas*)—LECTURES ON PHILOSOPHY. Demy 8vo. 9*s.*

*MAINTENON* (*Madame de*). By EMILY BOWLES. With Portrait. Large crown 8vo. 7*s.* 6*d.*

MANY VOICES.—Extracts from Religious Writers, from the First to the Sixteenth Century. With Biographical Sketches. Crown 8vo. cloth extra, 6*s.*

*MARKHAM* (*Capt. Albert Hastings*) *R.N.*—THE GREAT FROZEN SEA: a Personal Narrative of the Voyage of the *Alert* during the Arctic Expedition of 1875-6. With 33 Illustrations and Two Maps. Sixth Edition. Crown 8vo. 6*s.*

*MARTINEAU* (*Gertrude*)—OUTLINE LESSONS ON MORALS. Small crown 8vo. 3*s.* 6*d.*

*MASON* (*Charlotte M.*)—HOME EDUCATION. A Course of Lectures to Ladies, delivered in Bradford in the winter of 1885-1886. Crown 8vo. 3*s.* 6*d.*

MATTER AND ENERGY: An Examination of the Fundamental Conceptions of Physical Force. By B. L. L. Small crown 8vo. 2*s.*

*MATUCE* (*H. Ogram*)—A WANDERER. Crown 8vo. 5*s.*

*MAUDSLEY* (*H.*) *M.D.*—BODY AND WILL. Being an Essay Concerning Will, in its Metaphysical, Physiological, and Pathological Aspects. 8vo. 12*s.*

NATURAL CAUSES AND SUPERNATURAL SEEMINGS. Second Edition. Crown 8vo. 6*s.*

*McGRATH* (*Terence*)—PICTURES FROM IRELAND. New and Cheaper Edition. Crown 8vo. 2*s.*

*McKINNEY* (*S. B. G.*)—THE SCIENCE AND ART OF RELIGION. Crown 8vo. 8*s.* 6*d.*

*MILLER* (*Edward*)—THE HISTORY AND DOCTRINES OF IRVINGISM; or, the so-called Catholic and Apostolic Church. 2 vols. Large post 8vo. 15*s.*

THE CHURCH IN RELATION TO THE STATE. Large crown 8vo. 4*s.*

*MILLS* (*Herbert*)—POVERTY AND THE STATE; or, Work for the Unemployed. An Enquiry into the Causes and Extent of Enforced Idleness. Cr. 8vo. 6*s.*

*MINTON* (*Rev. Francis*)—CAPITAL AND WAGES. 8vo. 15*s.*

*MITCHELL* (*John*)—LIFE. By WILLIAM DILLON. With Portrait. Demy 8vo.

*MITCHELL* (*Lucy M.*)—A HISTORY OF ANCIENT SCULPTURE. With numerous Illustrations, including six Plates in Phototype. Super royal, 42*s.*

SELECTIONS FROM ANCIENT SCULPTURE. Being a Portfolio containing Reproductions in Phototype of 36 Masterpieces of Ancient Art, to illustrate Mrs. MITCHELL'S 'History of Ancient Sculpture.' 18*s.*

*MIVART* (*St. George*)—ON TRUTH. 8vo. 16*s.*

*MOCKLER* (*E.*)—A GRAMMAR OF THE BALOOCHEE LANGUAGE, as it is spoken in Makran (Ancient Gedrosia), in the Persia-Arabic and Roman characters. Fcp. 8vo. 5*s.*

*MOHL* (*Julius and Mary*)—LETTERS AND RECOLLECTIONS OF. By M. C. M. SIMPSON. With Portraits and Two Illustrations. Demy 8vo. 15*s.*

*MOLESWORTH* (*W. Nassau*)—HISTORY OF THE CHURCH OF ENGLAND FROM 1660. Large crown 8vo. 7*s.* 6*d.*

*MORELL (J. R.)*—EUCLID SIMPLIFIED IN METHOD AND LANGUAGE. Being a Manual of Geometry. Compiled from the most important French Works, approved by the University of Paris and the Minister of Public Instruction. Fcp. 8vo. 2*s.* 6*d.*

*MORISON (James Cotter)*—THE SERVICE OF MAN. An Essay towards the Religion of the Future. Demy 8vo. 10*s.* 6*d.*; Cheap Edition, crown 8vo. 5*s.*

*MORSE (E. S.) Ph.D.*—FIRST BOOK OF ZOOLOGY. With numerous Illustrations. New and Cheaper Edition. Crown 8vo. 2*s.* 6*d.*

MY LAWYER: A Concise Abridgment of the Laws of England. By a Barrister-at-Law. Crown 8vo. 6*s.* 6*d.*

*NELSON (J. H.) M.A.*—A PROSPECTUS OF THE SCIENTIFIC STUDY OF THE HINDÛ LAW. Demy 8vo. 9*s.*

INDIAN USAGE AND JUDGE-MADE LAW IN MADRAS. Demy 8vo. 12*s.*

NEW SOCIAL TEACHINGS. By POLITICUS. Small crown 8vo. 5*s.*

*NEWMAN (Cardinal)*—CHARACTERISTICS FROM THE WRITINGS OF. Being Selections from his various Works. Arranged with the Author's personal Approval. Seventh Edition. With Portrait. Crown 8vo. 6*s.*

*** A Portrait of Cardinal Newman, mounted for framing, can be had, 2*s.* 6*d.*

*NEWMAN (Francis William)*—ESSAYS ON DIET. Small crown 8vo. 2*s.*

MISCELLANIES. Vol. II.: Essays, Tracts, and Addresses, Moral and Religious. Demy 8vo. 12*s.*

REMINISCENCES OF TWO EXILES AND TWO WARS. Crown 8vo. 3*s.* 6*d.*

*NICOLS (Arthur) F.G.S., F.R.G.S.*—CHAPTERS FROM THE PHYSICAL HISTORY OF THE EARTH: an Introduction to Geology and Palæontology. With numerous Illustrations. Crown 8vo. 5*s.*

*NIHILL (Rev. H. D.)*—THE SISTERS OF ST. MARY AT THE CROSS: SISTERS OF THE POOR AND THEIR WORK. Crown 8vo. 2*s.* 6*d.*

*NOEL (The Hon. Roden)*—ESSAYS ON POETRY AND POETS. Demy 8vo. 12*s.*

*NOPS (Marianne)*—CLASS LESSONS ON EUCLID. Part I. containing the First Two Books of the Elements. Crown 8vo. 2*s.* 6*d.*

NUCES: EXERCISES ON THE SYNTAX OF THE PUBLIC SCHOOL LATIN PRIMER. New Edition in Three Parts. Crown 8vo. each 1*s.*

*** The Three Parts can also be had bound together in cloth, 3*s.*

*OATES (Frank) F.R.G.S.*—MATABELE LAND AND THE VICTORIA FALLS. A Naturalist's Wanderings in the Interior of South Africa. Edited by C. G. OATES, B.A. With numerous Illustrations and 4 Maps. Demy 8vo. 21*s.*

*O'BRIEN (R. Barry)*—IRISH WRONGS AND ENGLISH REMEDIES, with other Essays. Crown 8vo. 5*s.*

*OGLE (W.) M.D., F.R.C.P.*—ARISTOTLE ON THE PARTS OF ANIMALS. Translated, with Introduction and Notes. Royal 8vo. 12*s.* 6*d.*

*OLIVER (Robert)*—UNNOTICED ANALOGIES. A Talk on the Irish Question. Crown 8vo. 3*s.* 6*d.*

*O'MEARA (Kathleen)*—HENRI PERREYVE AND HIS COUNSELS TO THE SICK. Small crown 8vo. 5*s.*

ONE AND A HALF IN NORWAY. A Chronicle of Small Beer. By Either and Both. Small crown 8vo. 3*s.* 6*d.*

*O'NEIL (The late Rev. Lord)*.—SERMONS. With Memoir and Portrait. Crown 8vo. 6*s.*

ESSAYS AND ADDRESSES. Crown 8vo. 5*s.*

*OTTLEY* (*Henry Bickersteth*)—The Great Dilemma : Christ His own Witness or His own Accuser. Six Lectures. Second Edition. Crown 8vo. 3*s*. 6*d*.

Our Priests and their Tithes. By a Priest of the Province of Canterbury. Crown 8vo. 5*s*.

Our Public Schools—Eton, Harrow, Winchester, Rugby, Westminster, Marlborough, The Charterhouse. Crown 8vo. 6*s*.

*OWEN* (*F. M.*)—John Keats : a Study. Crown 8vo. 6*s*.

*PADGHAM* (*Richard*)—In the Midst of Life we are in Death. Crown 8vo. 5*s*.

*PALMER* (*the late William*)—Notes of a Visit to Russia in 1840–41. Selected and arranged by John H. Cardinal Newman. With Portrait. Crown 8vo. 8*s*. 6*d*.

Early Christian Symbolism. A series of Compositions from Fresco-Paintings, Glasses, and Sculptured Sarcophagi. Edited by the Rev. Provost Northcote, D.D., and the Rev. Canon Brownlow, M.A. With Coloured Plates, folio, 42*s*. ; or with plain plates, folio, 25*s*.

Parchment Library. Choicely printed on hand-made paper, limp parchment antique or cloth, 6*s*. ; vellum, 7*s*. 6*d*. each volume.

Carlyle's Sartor Resartus.

Milton's Poetical Works. 2 vols.

Chaucer's Canterbury Tales. 2 vols. Edited by Alfred W. Pollard.

Selections from the Prose Writings of Jonathan Swift. With a Preface and Notes by Stanley Lane-Poole, and Portrait.

English Sacred Lyrics.

Sir Joshua Reynolds' Discourses. Edited by Edmund Gosse.

Selections from Milton's Prose Writings. Edited by Ernest Myers.

The Book of Psalms. Translated by the Rev. Canon Cheyne, D.D.

The Vicar of Wakefield. With Preface and Notes by Austin Dobson.

English Comic Dramatists. Edited by Oswald Crawfurd.

English Lyrics.

The Sonnets of John Milton. Edited by Mark Pattison. With Portrait after Vertue.

French Lyrics. Selected and Annotated by George Saintsbury. With miniature Frontispiece, designed and etched by H. G. Glindoni.

Fables by Mr. John Gay. With Memoir by Austin Dobson, and an etched Portrait from an unfinished Oil-sketch by Sir Godfrey Kneller.

Select Letters of Percy Bysshe Shelley. Edited, with an Introtion, by Richard Garnett.

The Christian Year; Thoughts in Verse for the Sundays and Holy Days throughout the Year. With etched Portrait of the Rev. J. Keble, after the Drawing by G. Richmond, R.A.

Shakspere's Works. Complete in Twelve Volumes.

Eighteenth Century Essays. Selected and Edited by Austin Dobson. With a Miniature Frontispiece by R. Caldecott.

PARCHMENT LIBRARY—continued.

Q. HORATI FLACCI OPERA. Edited by F. A. CORNISH, Assistant Master at Eton. With a Frontispiece after a design by L. ALMA TADEMA. Etched by LEOPOLD LOWENSTAM.

EDGAR ALLAN POE'S POEMS. With an Essay on his Poetry by ANDREW LANG, and a Frontispiece by Linley Sambourne.

SHAKSPERE'S SONNETS. Edited by EDWARD DOWDEN. With a Frontispiece etched by Leopold Lowenstam, after the Death Mask.

ENGLISH ODES. Selected by EDMUND GOSSE. With Frontispiece on India paper by Hamo Thornycroft, A.R.A.

OF THE IMITATION OF CHRIST. By THOMAS À KEMPIS. A revised Translation. With Frontispiece on India paper, from a Design by W. B. Richmond.

POEMS: Selected from PERCY BYSSHE SHELLEY. Dedicated to Lady Shelley. With Preface by RICHARD GARNETT and a Miniature Frontispiece.

LETTERS AND JOURNALS OF JONATHAN SWIFT. Selected and edited, with a Commentary and Notes, by STANLEY LANE POOLE.

DE QUINCEY'S CONFESSIONS OF AN ENGLISH OPIUM EATER. Reprinted from the First Edition. Edited by RICHARD GARNETT.

THE GOSPEL ACCORDING TO MATTHEW, MARK, AND LUKE.

*PARSLOE (Joseph)* — OUR RAILWAYS. Sketches, Historical and Descriptive. With Practical Information as to Fares and Rates, &c., and a Chapter on Railway Reform. Crown 8vo. 6*s.*

*PASCAL (Blaise)*—THE THOUGHTS OF. Translated from the Text of AUGUSTE MOLINIER by C. KEGAN PAUL. Large crown 8vo. with Frontispiece, printed on hand-made paper, parchment antique, or cloth, 12*s.*; vellum, 15*s.* New Edition, crown 8vo. 6*s.*

*PATON (W. A.)*—DOWN THE ISLANDS; a Voyage to the Caribbees. Illustrations. Demy 8vo. 16*s.*

*PAUL (C. Kegan)*—BIOGRAPHICAL SKETCHES. Printed on hand-made paper, bound in buckram. Second Edition. Crown 8vo. 7*s.* 6*d.*

*PEARSON (Rev. S.)*—WEEK-DAY LIVING. A Book for Young Men and Women. Second Edition. Crown 8vo. 5*s.*

*PENRICE (Major J.)*—ARABIC AND ENGLISH DICTIONARY OF THE KORAN. 4to. 21*s.*

*PESCHEL (Dr. Oscar)*—THE RACES OF MAN AND THEIR GEOGRAPHICAL DISTRIBUTION. Second Edition, large crown 8vo. 9*s.*

*PETERS (F. H.)*—THE NICOMACHEAN ETHICS OF ARISTOTLE. Translated by. Crown 8vo. 6*s.*

*PIDGEON (D.)*—AN ENGINEER'S HOLIDAY; or, Notes of a Round Trip from Long. 0° to 0°. New and Cheaper Edition. Large crown 8vo. 7*s.* 6*d.*

OLD WORLD QUESTIONS AND NEW WORLD ANSWERS. Large crown 8vo. 7*s.* 6*d.*

PLAIN THOUGHTS FOR MEN. Eight Lectures delivered at the Foresters' Hall, Clerkenwell, during the London Mission, 1884. Crown 8vo. 1*s.* 6*d.*; paper covers, 1*s.*

*PLOWRIGHT (C. B.)*—THE BRITISH UREDINEÆ AND USTILAGINEÆ. With Illustrations. Demy 8vo. 10*s.* 6*d.*

*POE* (*Edgar Allan*)—WORKS OF. With an Introduction and a Memoir by RICHARD HENRY STODDARD. In 6 vols. with Frontispieces and Vignettes. Large crown 8vo. 6*s*. each vol.

*PRICE* (*Prof. Bonamy*)—CHAPTERS ON PRACTICAL POLITICAL ECONOMY. Being the Substance of Lectures delivered before the University of Oxford. New and Cheaper Edition. Large post 8vo. 5*s*.

PRIG'S BEDE: The Venerable Bede Expurgated, Expounded, and Exposed. By the PRIG, Author of 'The Life of a Prig.' Fcp. 8vo. 3*s*. 6*d*.

PRIGMENT (THE). A Collection of 'The Prig' Books. Crown 8vo. 6*s*.

PULPIT COMMENTARY (THE). Old Testament Series. Edited by the Rev. J. S. EXELL and the Very Rev. Dean H. D. M. SPENCE.

GENESIS. By Rev. T. WHITELAW, M.A. With Homilies by the Very Rev. J. F. MONTGOMERY, D.D., Rev. Prof. R. A. REDFORD, M.A., LL.B., Rev. F. HASTINGS, Rev. W. ROBERTS, M.A.; an Introduction to the Study of the Old Testament by the Venerable Archdeacon FARRAR, D.D., F.R.S.; and Introductions to the Pentateuch by the Right Rev. H. COTTERILL, D.D., and Rev. T. WHITELAW, M.A. Eighth Edition. One vol. 15*s*.

EXODUS. By the Rev. Canon RAWLINSON. With Homilies by Rev. J. ORR, Rev. D. YOUNG, Rev. C. A. GOODHART, Rev. J. URQUHART, and Rev. H. T. ROBJOHNS. Fourth Edition. Two vols. each 9*s*.

LEVITICUS. By the Rev. Prebendary MEYRICK, M.A. With Introductions by Rev. R. COLLINS, Rev. Professor A. CAVE, and Homilies by Rev. Prof. REDFORD, LL.B., Rev. J. A. MACDONALD, Rev. W. CLARKSON, Rev. S. R. ALDRIDGE, LL.B., and Rev. MCCHEYNE EDGAR. Fourth Edition. 15*s*.

NUMBERS. By the Rev R. WINTERBOTHAM, LL.B. With Homilies by the Rev. Professor W. BINNIE, D.D., Rev. E. S. PROUT, M.A., Rev. D. YOUNG, Rev. J. WAITE; and an Introduction by the Rev. THOMAS WHITELAW, M.A. Fifth Edition. 15*s*.

DEUTERONOMY. By Rev. W. L. ALEXANDER, D.D. With Homilies by Rev. D. DAVIES, M.A., Rev. C. CLEMANCE, D.D., Rev. J. ORR, B.D., and Rev. R. M. EDGAR, M.A. Fourth Edition. 15*s*.

JOSHUA. By Rev. J. J. LIAS, M.A. With Homilies by Rev. S. R. ALDRIDGE, LL.B., Rev. R. GLOVER, Rev. E. DE PRESSENSÉ, D.D., Rev. J. WAITE, B.A. Rev. W. F. ADENEY, M.A.; and an Introduction by the Rev. A. PLUMMER, M.A. Fifth Edition. 12*s*. 6*d*.

JUDGES AND RUTH. By the Bishop of Bath and Wells and Rev. J. MORISON, D.D. With Homilies by Rev. A. F. MUIR, M.A., Rev. W. F. ADENEY, M.A., Rev. W. M. STATHAM, and Rev. Professor J. THOMSON, M.A. Fifth Edition. 10*s*. 6*d*.

1 and 2 SAMUEL. By the Very Rev. R. P. SMITH, D.D. With Homilies by Rev. DONALD FRASER, D.D., Rev. Prof. CHAPMAN, Rev. B. DALE, and Rev G. WOOD. Vol. I. Sixth Edition, 15*s*. Vol. II. 15*s*.

1 KINGS. By the Rev. JOSEPH HAMMOND, LL.B. With Homilies by the Rev. E DE PRESSENSÉ, D.D., Rev. J. WAITE, B.A., Rev. A. ROWLAND, LL.B., Rev. J. A. MACDONALD, and Rev. J. URQUHART. Fifth Edition. 15*s*.

1 CHRONICLES. By the Rev. Prof. P. C. BARKER, M.A., LL.B. With Homilies by Rev. Prof. J. R. THOMSON, M.A., Rev. R. TUCK, B.A., Rev. W. CLARKSON, B.A., Rev. F. WHITFIELD, M A., and Rev. RICHARD GLOVER. 15*s*.

PULPIT COMMENTARY (THE). Old Testament Series—continued.

EZRA, NEHEMIAH, AND ESTHER. By Rev. Canon G. RAWLINSON, M.A. With Homilies by Rev. Prof. J. R. THOMSON, M.A., Rev. Prof. R. A. REDFORD, LL.B., M.A., Rev. W. S. LEWIS, M.A., Rev. J. A. MACDONALD, Rev. A. MACKENNAL, B.A., Rev. W. CLARKSON, B.A., Rev. F. HASTINGS, Rev. W. DINWIDDIE, LL.B., Rev. Prof. ROWLANDS, B.A., Rev. G. WOOD, B.A., Rev. Prof. P. C. BARKER, LL.B., M.A., and Rev. J. S. EXELL, M.A. Sixth Edition. One vol. 12*s.* 6*d.*

ISAIAH. By the Rev. Canon G. RAWLINSON, M.A. With Homilies by Rev. Prof. E. JOHNSON, M.A., Rev. W. CLARKSON, B.A., Rev. W. M. STATHAM, and Rev. R. TUCK, B.A. Second Edition. 2 vols. each 15*s.*

JEREMIAH (Vol. I.). By the Rev. Canon CHEYNE, D.D. With Homilies by the Rev W. F. ADENEY, M.A., Rev. A. F. MUIR, M.A., Rev. S. CONWAY, B.A., Rev. J. WAITE, B.A., and Rev. D. YOUNG, B.A. Third Edition. 15*s.*

JEREMIAH (Vol. II.), AND LAMENTATIONS. By the Rev. Canon CHEYNE, D.D. With Homilies by Rev. Prof. J. R. THOMSON, M.A., Rev. W. F. ADENEY, M.A., Rev. A. F. MUIR, M.A., Rev. S. CONWAY, B.A., Rev. D. YOUNG, B.A. 15*s.*

HOSEA AND JOEL. By the Rev. Prof. J. J. GIVEN, Ph.D., D.D. With Homilies by the Rev. Prof. J. R. THOMSON, M.A., Rev. A. ROWLAND, B.A., LL.B., Rev. C. JERDAN, M.A., LL.B., Rev. J. ORR, M.A., B.D., and Rev. D. THOMAS, D.D. 15*s.*

PULPIT COMMENTARY (THE). New Testament Series.

ST. MARK. By the Very Rev. E. BICKERSTETH, D.D., Dean of Lichfield. With Homilies by the Rev. Prof. THOMSON, M.A., Rev. Prof. GIVEN, M.A., Rev. Prof. JOHNSON, M.A., Rev. A. ROWLAND, LL.B., Rev. A. MUIR, M.A., and Rev. R. GREEN. Fifth Edition. 2 vols. each 10*s.* 6*d.*

ST. JOHN. By the Rev. Prof. H. R. REYNOLDS, D.D. With Homilies by Rev. Prof. T. CROSKERY, D.D., Rev. Prof. J. R. THOMSON, Rev. D. YOUNG, Rev. B. THOMAS, and Rev. G. BROWN. 2 vols. each 15*s.*

THE ACTS OF THE APOSTLES. By the Bishop of BATH AND WELLS. With Homilies by Rev. Prof. P. C. BARKER, M.A., Rev. Prof. E. JOHNSON, M.A., Rev. Prof. R. A. REDFORD, M.A., Rev. R. TUCK, B.A., Rev. W. CLARKSON, B.A. Fourth Edition. Two vols. each 10*s.* 6*d.*

I CORINTHIANS. By the Ven. Archdeacon FARRAR, D.D. With Homilies by Rev. Ex-Chancellor LIPSCOMB, LL.D., Rev. DAVID THOMAS, D.D., Rev. DONALD FRASER, D.D., Rev. Prof. J. R. THOMSON, M.A., Rev. R. TUCK, B.A., Rev. E. HURNDALL, M.A., Rev. J. WAITE, B.A., Rev. H. BREMNER, B.D. Third Edition. 15*s.*

II CORINTHIANS AND GALATIANS. By the Ven. Archdeacon FARRAR, D.D., and Rev. Preb. E. HUXTABLE. With Homilies by Rev. Ex-Chancellor LIPSCOMB, LL.D., Rev. DAVID THOMAS, D.D., Rev. DONALD FRASER, D.D., Rev. R. TUCK, B.A., Rev. E. HURNDALL, M.A., Rev. Prof. J. R. THOMSON, M.A., Rev. R. FINLAYSON, B.A., Rev. W. F. ADENEY, M.A., Rev. R. M. EDGAR, M.A., and Rev. T. CROSKERY, D.D. 21*s.*

EPHESIANS, PHILIPPIANS, AND COLOSSIANS. By the Rev. Prof. W. G. BLAIKIE, D.D., Rev. B. C. CAFFIN, M.A., and Rev. G. G. FINDLAY, B.A. With Homilies by Rev. D. THOMAS, D.D., Rev. R. M. EDGAR, M.A., Rev. R. FINLAYSON, B.A., Rev. W. F. ADENEY, M.A., Rev. Prof. T. CROSKERY, D.D., Rev. E. S. PROUT, M.A., Rev. Canon VERNON HUTTON, and Rev. U. R. THOMAS, D.D. Second Edition. 21*s.*

PULPIT COMMENTARY (THE). New Testament Series—continued.

THESSALONIANS, TIMOTHY, TITUS, AND PHILEMON. By the BISHOP OF BATH AND WELLS, Rev. Dr. GLOAG, and Rev. Dr. EALES. With Homilies by the Rev. B. C. CAFFIN, M.A., Rev. R. FINLAYSON, B.A., Rev. Prof. T. CROSKERY, D.D., Rev. W. F. ADENEY, M.A., Rev. W. M. STATHAM, and Rev. D. THOMAS, D.D. 15*s*.

HEBREWS AND JAMES. By the Rev. J. BARMBY, D.D., and Rev. Prebendary E. C. S. GIBSON, M.A. With Homiletics by the Rev. C. JERDAN, M.A., LL.B., and Rev. Prebendary E. C. S. GIBSON. And Homilies by the Rev. W. JONES, Rev. C. NEW, Rev. D. YOUNG, B.A., Rev. J. S. BRIGHT, Rev. T. F. LOCKYER, B.A., and Rev. C. JERDAN, M.A., LL.B. Second Edition. Price 15*s*.

*PUSEY (Dr.)*—SERMONS FOR THE CHURCH'S SEASONS FROM ADVENT TO TRINITY. Selected from the published Sermons of the late EDWARD BOUVERIE PUSEY, D.D. Crown 8vo. 5*s*.

*QUEKETT (Rev. William)*—MY SAYINGS AND DOINGS, WITH REMINISCENCES OF MY LIFE. Demy 8vo. 18*s*.

*RANKE (Leopold von)*—UNIVERSAL HISTORY. The Oldest Historical Group of Nations and the Greeks. Edited by G. W. PROTHERO. Demy 8vo. 16*s*.

*RENDELL (J. M.)*—CONCISE HANDBOOK OF THE ISLAND OF MADEIRA. With Plan of Funchal and Map of the Island. Fcp. 8vo. 1*s*. 6*d*.

*REYNOLDS (Rev. J. W.)*—THE SUPERNATURAL IN NATURE. A Verification by Free Use of Science. Third Edition, revised and enlarged. Demy 8vo. 14*s*.

THE MYSTERY OF MIRACLES. Third and Enlarged Edition. Crown 8vo. 6*s*.

THE MYSTERY OF THE UNIVERSE: Our Common Faith. Demy 8vo. 14*s*.

THE WORLD TO COME: Immortality a Physical Fact. Crown 8vo. 6*s*.

*RIBOT (Prof. Th.)*—HEREDITY: a Psychological Study on its Phenomena, its Laws, its Causes, and its Consequences. Second Edition. Large crown 8vo. 9*s*.

*RIVINGTON (Luke)*—AUTHORITY, OR A PLAIN REASON FOR JOINING THE CHURCH OF ROME. Crown 8vo. 3*s*. 6*d*.

*ROBERTSON (The late Rev. F. W.) M.A.*—LIFE AND LETTERS OF. Edited by the Rev. Stopford Brooke, M.A.

I. Two vols., uniform with the Sermons. With Steel Portrait. Crown 8vo. 7*s*. 6*d*.

II. Library Edition, in demy 8vo. with Portrait. 12*s*.

III. A Popular Edition, in 1 vol. Crown 8vo. 6*s*.

SERMONS. Four Series. Small crown 8vo. 3*s*. 6*d*. each.

THE HUMAN RACE, and other Sermons. Preached at Cheltenham, Oxford, and Brighton. New and Cheaper Edition. Small crown 8vo. 3*s*. 6*d*.

NOTES ON GENESIS. New and Cheaper Edition. Small crown 8vo. 3*s*. 6*d*.

EXPOSITORY LECTURES ON ST. PAUL'S EPISTLES TO THE CORINTHIANS. A New Edition. Small crown 8vo. 5*s*.

LECTURES AND ADDRESSES, with other Literary Remains. A New Edition. Small crown 8vo. 5*s*.

AN ANALYSIS OF TENNYSON'S 'IN MEMORIAM.' (Dedicated by Permission to the Poet-Laureate.) Fcp. 8vo. 2*s*.

THE EDUCATION OF THE HUMAN RACE. Translated from the German of Gotthold Ephraim Lessing. Fcp. 8vo. 2*s*. 6*d*.

*** A Portrait of the late Rev. F. W. Robertson, mounted for framing, can be had, 2*s*. 6*d*.

*ROGERS (William)*—REMINISCENCES. Compiled by R. H. HADDEN. With Portrait. Third Edition. Crown 8vo. 6*s*.

ROMANCE OF THE RECUSANTS. By the Author of 'Life of a Prig.' Cr. 8vo. 5*s*.

*ROMANES (G. J.)*—MENTAL EVOLUTION IN ANIMALS. With a Posthumous Essay on Instinct, by CHARLES DARWIN, F.R.S. Demy 8vo. 12*s*.

MENTAL EVOLUTION IN MAN. Vol. I. 8vo. 14*s*.

*ROSMINI SERBATI (A.) Founder of the Institute of Charity*—LIFE. By FATHER LOCKHART. 2 vols. Crown 8vo. 12*s*.

ROSMINI'S ORIGIN OF IDEAS. Translated from the Fifth Italian Edition of the Nuovo Saggio. *Sull' origine delle idee.* 3 vols. Demy 8vo. 10*s*. 6*d*. each.

ROSMINI'S PSYCHOLOGY. 3 vols. Demy 8vo. [Vols. I. & II. now ready, 10*s*. 6*d*. each.

*ROSS (Janet)*—ITALIAN SKETCHES. With 14 full-page Illustrations. Crown 8vo. 7*s*. 6*d*.

*RULE (Martin) M.A.*—THE LIFE AND TIMES OF ST. ANSELM, ARCHBISHOP OF CANTERBURY AND PRIMATE OF THE BRITAINS. 2 vols. Demy 8vo. 32*s*.

*SAMUEL (Sydney M.)*—JEWISH LIFE IN THE EAST. Small crown 8vo. 3*s*. 6*d*.

*SAYCE (Rev. Archibald Henry)*—INTRODUCTION TO THE SCIENCE OF LANGUAGE. 2 vols. Second Edition. Large post 8vo. 21*s*.

*SCOONES (W. Baptiste)*—FOUR CENTURIES OF ENGLISH LETTERS: A Selection of 350 Letters by 150 Writers, from the Period of the Paston Letters to the Present Time. Third Edition. Large crown 8vo. 6*s*.

*SÉE (Prof. Germain)*—BACILLARY PHTHISIS OF THE LUNGS. Translated and Edited for English Practitioners, by WILLIAM HENRY WEDDELL, M.R.C.S. Demy 8vo. 10*s*. 6*d*.

*SELWYN (Augustus) D.D.*—LIFE. By Canon G. H. CURTEIS. Crown 8vo. 6*s*.

*SEYMOUR (W. Digby)*—HOME RULE AND STATE SUPREMACY. Crown 8vo. 3*s*. 6*d*.

*SHAKSPERE*—WORKS. The Avon Edition, 12 vols. fcp. 8vo. cloth, 18*s*.; in cloth box, 21*s*.; bound in 6 vols., cloth, 15*s*.

*SHAKSPERE*—WORKS (An Index to). By EVANGELINE O'CONNOR. Crown 8vo. 5*s*.

*SHELLEY (Percy Bysshe)*.—LIFE. By EDWARD DOWDEN, LL.D. With Portraits and Illustrations, 2 vols., demy 8vo. 36*s*.

*SHILLITO (Rev. Joseph)*—WOMANHOOD: its Duties, Temptations, and Privileges. A Book for Young Women. Third Edition. Crown 8vo. 3*s*. 6*d*.

SHOOTING, PRACTICAL HINTS ON. Being a Treatise on the Shot Gun and its Management. By '20-Bore.' With 55 Illustrations. Demy 8vo. 12*s*.

SISTER AUGUSTINE, Superior of the Sisters of Charity at the St. Johannis Hospital at Bonn. Cheap Edition. Large crown 8vo. 4*s*. 6*d*.

SKINNER (JAMES). A Memoir. By the Author of 'Charles Lowder.' With a Preface by Canon CARTER, and Portrait. Large crown 8vo. 7*s*. 6*d*.

*** Also a Cheap Edition, with Portrait. Crown 8vo. 3*s*. 6*d*.

*SMEATON (Donald)*.—THE LOYAL KARENS OF BURMAH. Crown 8vo. 4*s*. 6*d*.

*SMITH (Edward) M.D., LL.B., F.R.S.*—TUBERCULAR CONSUMPTION IN ITS EARLY AND REMEDIABLE STAGES. Second Edition. Crown 8vo. 6*s*.

*SMITH (L. A.)*—MUSIC OF THE WATERS: Sailors' Chanties, or Working Songs of the Sea of all Maritime Nations. Demy 8vo. 12*s.*

*SMITH (Sir W. Cusack, Bart.)*—OUR WAR SHIPS. A Naval Essay. Crown 8vo. 5*s.*

SPANISH MYSTICS. By the Editor of 'Many Voices.' Crown 8vo. 5*s.*

SPECIMENS OF ENGLISH PROSE STYLE FROM MALORY TO MACAULAY. Selected and Annotated, with an Introductory Essay, by GEORGE SAINTSBURY. Large crown 8vo., printed on hand-made paper, parchment antique, or cloth, 12*s.*; vellum, 15*s.*

*SPEDDING (James)*—REVIEWS AND DISCUSSIONS, LITERARY, POLITICAL, AND HISTORICAL NOT RELATING TO BACON. Demy 8vo. 12*s.* 6*d.*

EVENINGS WITH A REVIEWER; or, Bacon and Macaulay. With a Prefatory Notice by G. S. VENABLES, Q.C. 2 vols. Demy 8vo. 18*s.*

*STRACHEY (Sir John)*—LECTURES ON INDIA. 8vo. 15*s.*

STRAY PAPERS ON EDUCATION AND SCENES FROM SCHOOL LIFE. By B. H. Second Edition. Small crown 8vo. 3*s.* 6*d.*

*STREATFEILD (Rev. G. S.) M.A.*—LINCOLNSHIRE AND THE DANES. Large crown 8vo. 7*s.* 6*d.*

*STRECKER-WISLICENUS*—ORGANIC CHEMISTRY. Translated and Edited, with Extensive Additions, by W. R. HODGKINSON, Ph.D., and A. J. GREENAWAY, F.I.C. Demy 8vo. 12*s.* 6*d.*

SUAKIN, 1885; being a Sketch of the Campaign of this Year. By an Officer who was there. Second Edition. Crown 8vo. 2*s.* 6*d.*

*SULLY (James) M.A.*—PESSIMISM: a History and a Criticism. Second Edition. Demy 8vo. 14*s.*

*TARRING (Charles James) M.A.*—A PRACTICAL ELEMENTARY TURKISH GRAMMAR. Crown 8vo. 6*s.*

*TAYLOR (Hugh)*—THE MORALITY OF NATIONS. A Study in the Evolution of Ethics. Crown 8vo. 6*s.*

*TAYLOR (Rev. Isaac)*—THE ALPHABET. An Account of the Origin and Development of Letters. Numerous Tables and Facsimiles. 2 vols. 8vo. 36*s.*

LEAVES FROM AN EGYPTIAN NOTE-BOOK. Crown 8vo. 5*s.*

*TAYLOR (Reynell) C.B., C.S.I.*—A BIOGRAPHY. By E. GAMBIER PARRY. With Portrait and Map. Demy 8vo. 14*s.*

*THOM (John Hamilton)*—LAWS OF LIFE AFTER THE MIND OF CHRIST. Two Series. Crown 8vo. 7*s.* 6*d.* each.

*THOMPSON (Sir H.)*—DIET IN RELATION TO AGE AND ACTIVITY. Fcp. 8vo. cloth, 1*s.* 6*d.*; Paper covers, 1*s.*

*TIDMAN (Paul F.)*—GOLD AND SILVER MONEY. Part I.—A Plain Statement. Part II.—Objections Answered. Third Edition. Crown 8vo. 1*s.*

MONEY AND LABOUR. 1*s.* 6*d.*

*TODHUNTER (Dr. J.)*—A STUDY OF SHELLEY. Crown 8vo. 7*s.*

*TOLSTOI (Count Leo)*—CHRIST'S CHRISTIANITY. Translated from the Russian. Large crown 8vo. 7*s.* 6*d.*

*TRANT (William)*—TRADE UNIONS; Their Origin and Objects, Influence and Efficacy. Small crown 8vo. 1*s.* 6*d.*; paper covers, 1*s.*

*TRENCH (The late R. C., Archbishop)*—LETTERS AND MEMORIALS. Edited by the Author of 'Charles Lowder, a Biography,' &c. With two Portraits. 2 vols. demy 8vo. 21*s.*

SERMONS NEW AND OLD. Crown 8vo. 6*s.*

WESTMINSTER AND DUBLIN SERMONS. Crown 8vo. 6*s.*

NOTES ON THE PARABLES OF OUR LORD. Fourteenth Edition. 8vo. 12*s.*; Popular Edition, crown 8vo. 7*s.* 6*d.*

NOTES ON THE MIRACLES OF OUR LORD. Twelfth Edition. 8vo. 12*s.*; Popular Edition, crown 8vo. 7*s.* 6*d.*

STUDIES IN THE GOSPELS. Fifth Edition, Revised. 8vo. 10*s.* 6*d.*

BRIEF THOUGHTS AND MEDITATIONS ON SOME PASSAGES IN HOLY Scripture. Third Edition. Crown 8vo. 3*s.* 6*d.*

SYNONYMS OF THE NEW TESTAMENT. Tenth Edition, Enlarged. 8vo. 12*s.*

ON THE AUTHORISED VERSION OF THE NEW TESTAMENT. Second Edition. 8vo. 7*s.*

COMMENTARY ON THE EPISTLE TO THE SEVEN CHURCHES IN ASIA. Fourth Edition, Revised. 8vo. 8*s.* 6*d.*

THE SERMON ON THE MOUNT. An Exposition drawn from the Writings of St. Augustine, with an Essay on his Merits as an Interpreter of Holy Scripture. Fourth Edition, Enlarged. 8vo. 10*s.* 6*d.*

SHIPWRECKS OF FAITH. Three Sermons preached before the University of Cambridge in May 1867. Fcp. 8vo. 2*s.* 6*d.*

LECTURES ON MEDIÆVAL CHURCH HISTORY. Being the Substance of Lectures delivered at Queen's College, London. Second Edition. 8vo. 12*s.*

ENGLISH, PAST AND PRESENT. Thirteenth Edition, Revised and Improved. Fcp. 8vo. 5*s.*

ON THE STUDY OF WORDS. Nineteenth Edition, Revised. Fcp. 8vo. 5*s.*

SELECT GLOSSARY OF ENGLISH WORDS USED FORMERLY IN SENSES DIFFERENT FROM THE PRESENT. Sixth Edition, Revised and Enlarged. Fcp. 8vo. 5*s.*

PROVERBS AND THEIR LESSONS. Seventh Edition, Enlarged. Fcp. 8vo. 4*s.*

POEMS. Collected and Arranged Anew. Ninth Edition. Fcp. 8vo. 7*s.* 6*d.*

POEMS. Library Edition. 2 vols. Small crown 8vo. 10*s.*

SACRED LATIN POETRY. Chiefly Lyrical, Selected and Arranged for Use. Third Edition, Corrected and Improved. Fcp. 8vo. 7*s.*

A HOUSEHOLD BOOK OF ENGLISH POETRY. Selected and Arranged, with Notes. Fourth Edition, Revised. Extra fcp. 8vo. 5*s.* 6*d.*

AN ESSAY ON THE LIFE AND GENIUS OF CALDERON. With Translations from his 'Life's a Dream' and 'Great Theatre of the World.' Second Edition, Revised and Improved. Extra fcp. 8vo. 5*s.* 6*d.*

GUSTAVUS ADOLPHUS IN GERMANY, AND OTHER LECTURES ON THE THIRTY YEARS' WAR. Third Edition, Enlarged. Fcp. 8vo. 4*s.*

PLUTARCH: HIS LIFE, HIS LIVES, AND HIS MORALS. Second Edition, Enlarged. Fcap. 8vo. 3*s.* 6*d.*

REMAINS OF THE LATE MRS. RICHARD TRENCH. Being Selections from her Journals, Letters, and other Papers. New and Cheaper Issue. With Portrait. 8vo. 6*s.*

*TUTHILL (C. A. H.)*—ORIGIN AND DEVELOPMENT OF CHRISTIAN DOGMA. Crown 8vo. 3*s.* 6*d.*

*TWINING (Louisa)*—WORKHOUSE VISITING AND MANAGEMENT DURING TWENTY-FIVE YEARS. Small crown 8vo. 2*s.*

TWO CENTURIES OF IRISH HISTORY. Edited by JAMES BRYCE, M.P. 8vo. 16*s.*

*UMLAUFT (F.)*—THE ALPS. Illustrated. 8vo.

*VAL D'EREMAO (J. P.) D.D.*—THE SERPENT OF EDEN. Crown 8vo. 4*s.* 6*d.*

*VAUGHAN (H. Halford)*—NEW READINGS AND RENDERINGS OF SHAKESPEARE'S TRAGEDIES. 3 vols. Demy 8vo. 12*s.* 6*d.* each.

*VICARY (J. Fulford)*—SAGA TIME. With Illustrations. Cr. 8vo. 7*s.* 6*d.*

*VOLCKXSOM (E. W. v.)*—CATECHISM OF ELEMENTARY MODERN CHEMISTRY. Small crown 8vo. 3*s.*

*WALPOLE (Chas. George)*—A SHORT HISTORY OF IRELAND FROM THE EARLIEST TIMES TO THE UNION WITH GREAT BRITAIN. With 5 Maps and Appendices. Third Edition. Crown 8vo. 6*s.*

*WARD (William George) Ph.D.*—ESSAYS ON THE PHILOSOPHY OF THEISM. Edited, with an Introduction, by WILFRID WARD. 2 vols. demy 8vo. 21*s.*

*WARD (Wilfrid)*—THE WISH TO BELIEVE: A Discussion concerning the Temper of Mind in which a reasonable Man should undertake Religious Inquiry. Small crown 8vo. 5*s.*

*WARNER (Francis) M.D.*—LECTURES ON THE ANATOMY OF MOVEMENT. Crown 8vo. 4*s.* 6*d.*

*WARTER (J. W.)*—AN OLD SHROPSHIRE OAK. 2 vols. demy 8vo. 28*s.*

*WEDMORE (Frederick)*—THE MASTERS OF GENRE PAINTING. With Sixteen Illustrations. Post 8vo. 7*s.* 6*d.*

*WHIBLEY (Charles)*—CAMBRIDGE ANECDOTES. Crown 8vo. 7*s.* 6*d.*

*WHITMAN (Sidney)*—CONVENTIONAL CANT: Its Results and Remedy. Crown 8vo. 6*s.*

*WHITNEY (Prof. William Dwight)*—ESSENTIALS OF ENGLISH GRAMMAR, for the Use of Schools. Second Edition, crown 8vo. 3*s.* 6*d.*

*WHITWORTH (George Clifford)*—AN ANGLO-INDIAN DICTIONARY: a Glossary of Indian Terms used in English. Demy 8vo. cloth, 12*s.*

*WILBERFORCE (Samuel) D.D.*—LIFE. By R. G. WILBERFORCE. Crown 8vo. 6*s.*

*WILSON (Mrs. R. F.)*—THE CHRISTIAN BROTHERS: THEIR ORIGIN AND WORK. Crown 8vo. 6*s.*

*WOLTMANN (Dr. Alfred), and WOERMANN (Dr. Karl)*—HISTORY OF PAINTING. Vol. I. Ancient, Early, Christian, and Mediæval Painting. With numerous Illustrations. Super-royal 8vo. 28*s.*; bevelled boards, gilt leaves, 30*s.* Vol. II. The Painting of the Renascence. Cloth, 42*s.*; cloth extra, bevelled boards, 45*s.*

WORDS OF JESUS CHRIST TAKEN FROM THE GOSPELS. Small crown 8vo. 2*s.* 6*d.*

*YOUMANS (Eliza A.)*—FIRST BOOK OF BOTANY. Designed to cultivate the Observing Powers of Children. With 300 Engravings. New and Cheaper Edition. Crown 8vo. 2*s.* 6*d.*

*YOUMANS (Edward L.) M.D.*—A CLASS BOOK OF CHEMISTRY, on the Basis of the New System. With 200 Illustrations. Crown 8vo. 5*s.*

*YOUNG (Arthur).*—AXIAL POLARITY OF MAN'S WORD-EMBODIED IDEAS, AND ITS TEACHING. Demy 4to. 15*s.*

# THE INTERNATIONAL SCIENTIFIC SERIES.

I. FORMS OF WATER: a Familiar Exposition of the Origin and Phenomena of Glaciers. By J. Tyndall, LL.D., F.R.S. With 25 Illustrations. Ninth Edition. Crown 8vo. 5*s*.

II. PHYSICS AND POLITICS; or, Thoughts on the Application of the Principles of 'Natural Selection' and 'Inheritance' to Political Society. By Walter Bagehot. Eighth Edition. Crown 8vo. 5*s*.

III. FOODS. By Edward Smith, M.D., LL.B., F.R.S. With numerous Illustrations. Ninth Edition. Crown 8vo. 5*s*.

IV. MIND AND BODY: the Theories and their Relation. By Alexander Bain, LL.D. With Four Illustrations. Eighth Edition. Crown 8vo. 5*s*.

V. THE STUDY OF SOCIOLOGY. By Herbert Spencer. Thirteenth Edition. Crown 8vo. 5*s*.

VI. ON THE CONSERVATION OF ENERGY. By Balfour Stewart, M.A., LL.D., F.R.S. With 14 Illustrations. Seventh Edition. Crown 8vo. 5*s*.

VII. ANIMAL LOCOMOTION; or, Walking, Swimming, and Flying. By J. B. Pettigrew, M.D., F.R.S., &c. With 130 Illustrations. Third Edition. Crown 8vo. 5*s*.

VIII. RESPONSIBILITY IN MENTAL DISEASE. By Henry Maudsley, M.D. Fourth Edition. Crown 8vo. 5*s*.

IX. THE NEW CHEMISTRY. By Professor J. P. Cooke. With 31 Illustrations. Ninth Edition, remodelled and enlarged. Crown 8vo. 5*s*.

X. THE SCIENCE OF LAW. By Professor Sheldon Amos. Sixth Edition. Crown 8vo. 5*s*.

XI. ANIMAL MECHANISM: a Treatise on Terrestrial and Aerial Locomotion. By Professor E. J. Marey. With 117 Illustrations. Third Edition. Crown 8vo. 5*s*.

XII. THE DOCTRINE OF DESCENT AND DARWINISM. By Professor Oscar Schmidt. With 26 Illustrations. Seventh Edition. Crown 8vo. 5*s*.

XIII. THE HISTORY OF THE CONFLICT BETWEEN RELIGION AND SCIENCE. By J. W. Draper, M.D., LL.D. Twentieth Edition. Crown 8vo. 5*s*.

XIV. FUNGI: their Nature, Influences, Uses, &c. By M. C. Cooke, M.D., LL.D. Edited by the Rev. M. J. Berkeley, M.A., F.L.S. With numerous Illustrations. Fourth Edition. Crown 8vo. 5*s*.

XV. THE CHEMICAL EFFECTS OF LIGHT AND PHOTOGRAPHY. By Dr. Hermann Vogel. Translation thoroughly revised. With 100 Illustrations. Fifth Edition. Crown 8vo. 5*s*.

XVI. THE LIFE AND GROWTH OF LANGUAGE. By Professor William Dwight Whitney. Fifth Edition. Crown 8vo. 5*s*.

XVII. MONEY AND THE MECHANISM OF EXCHANGE. By W. Stanley Jevons, M.A., F.R.S. Eighth Edition. Crown 8vo. 5*s*.

XVIII. THE NATURE OF LIGHT. With a General Account of Physical Optics. By Dr. Eugene Lommel. With 188 Illustrations and a Table of Spectra in Chromo-lithography. Fourth Edit. Crown 8vo. 5*s*.

XIX. ANIMAL PARASITES AND MESSMATES. By P. J. Van Beneden. With 83 Illustrations. Third Edition. Crown 8vo. 5*s*.

XX. FERMENTATION. By Professor Schützenberger. With 28 Illustrations. Fourth Edition. Crown 8vo. 5*s*.

XXI. THE FIVE SENSES OF MAN. By Professor Bernstein. With 91 Illustrations. Fifth Edition. Crown 8vo. 5*s*.

XXII. THE THEORY OF SOUND IN ITS RELATION TO MUSIC. By Professor Pietro Blaserna. With numerous Illustrations. Third Edition. Crown 8vo. 5*s*.

XXIII. STUDIES IN SPECTRUM ANALYSIS. By J. Norman Lockyer, F.R.S. Fourth Edition. With six Photographic Illustrations of Spectra, and numerous Engravings on Wood. Crown 8vo. 6*s*. 6*d*.

XXIV. A HISTORY OF THE GROWTH OF THE STEAM ENGINE. By Professor R. H. Thurston. With numerous Illustrations. Fourth Edition. Crown 8vo. 5*s*.

XXV. EDUCATION AS A SCIENCE. By Alexander Bain, LL.D. Sixth Edition. Crown 8vo. 5*s*.

XXVI. THE HUMAN SPECIES. By Prof. A. De Quatrefages. Fourth Edition. Crown 8vo. 5*s*.

XXVII. MODERN CHROMATICS. With Applications to Art and Industry. By Ogden N. Rood. With 130 original Illustrations. Second Edition. Crown 8vo. 5*s*.

XXVIII. THE CRAYFISH: an Introduction to the Study of Zoology. By Professor T. H. Huxley. With 82 Illustrations. Fourth Edition. Crown 8vo. 5*s*.

XXIX. THE BRAIN AS AN ORGAN OF MIND. By H. Charlton Bastian, M.D. With numerous Illustrations. Third Edition. Crown 8vo. 5*s*.

XXX. THE ATOMIC THEORY. By Prof. Wurtz. Translated by G. Cleminshaw, F.C.S. Fifth Edition. Crown 8vo. 5*s*.

XXXI. THE NATURAL CONDITIONS OF EXISTENCE AS THEY AFFECT ANIMAL LIFE. By Karl Semper. With 2 Maps and 106 Woodcuts. Third Edition. Crown 8vo. 5*s*.

XXXII. GENERAL PHYSIOLOGY OF MUSCLES AND NERVES. By Prof. J. Rosenthal. Third Edition. With Illustrations. Crown 8vo. 5*s*.

XXXIII. SIGHT: an Exposition of the Principles of Monocular and Binocular Vision. By Joseph Le Conte, LL.D. Second Edition. With 132 Illustrations. Crown 8vo. 5*s*.

XXXIV. ILLUSIONS: a Psychological Study. By James Sully. Third Edition. Crown 8vo. 5*s*.

XXXV. VOLCANOES: WHAT THEY ARE AND WHAT THEY TEACH. By Professor J. W. Judd, F.R.S. With 92 Illustrations on Wood. Fourth Edition. Crown 8vo. 5*s*.

XXXVI. SUICIDE: an Essay on Comparative Moral Statistics. By Prof. H. Morselli. Second Edition. With Diagrams. Crown 8vo. 5*s*.

XXXVII. THE BRAIN AND ITS FUNCTIONS. By J. Luys. Second Edition. With Illustrations. Crown 8vo. 5*s*.

XXXVIII. MYTH AND SCIENCE: an Essay. By Tito Vignoli. Third Edition. Crown 8vo. 5*s*.

XXXIX. THE SUN. By Professor Young. With Illustrations. Third Edition. Crown 8vo. 5*s*.

XL. ANTS, BEES, AND WASPS: a Record of Observations on the Habits of the Social Hymenoptera. By Sir John Lubbock, Bart., M.P. With 5 Chromo-lithographic Illustrations. Ninth Edition. Crown 8vo 5*s*.

XLI. ANIMAL INTELLIGENCE. By G. J. Romanes, LL.D., F.R.S. Fourth Edition. Crown 8vo. 5*s*.

XLII. THE CONCEPTS AND THEORIES OF MODERN PHYSICS. By J. B. Stallo. Third Edition. Crown 8vo. 5*s*.

XLIII. DISEASES OF MEMORY: an Essay in the Positive Pyschology. By Prof. Th. Ribot. Third Edition. Crown 8vo. 5*s*.

XLIV. MAN BEFORE METALS. By N. Joly. Fourth Edition. Crown 8vo. 5*s*.

XLV. THE SCIENCE OF POLITICS. By Prof. Sheldon Amos. Third Edit. Crown. 8vo. 5*s*.

XLVI. ELEMENTARY METEOROLOGY. By Robert H. Scott. Fourth Edition. With numerous Illustrations. Crown 8vo. 5*s*.

XLVII. THE ORGANS OF SPEECH AND THEIR APPLICATION IN THE FORMATION OF ARTICULATE SOUNDS By Georg Hermann von Meyer. With 47 Woodcuts. Crown 8vo. 5*s*.

XLVIII. FALLACIES: a View of Logic from the Practical Side. By Alfred Sidgwick. Second Edition. Crown 8vo. 5*s*.

XLIX. ORIGIN OF CULTIVATED PLANTS. By Alphonse de Candolle. Second Edition. Crown 8vo. 5*s*.

L. JELLY FISH, STAR FISH, AND SEA URCHINS. Being a Research on Primitive Nervous Systems. By G. J. Romanes. Crown 8vo. 5*s*.

LI. THE COMMON SENSE OF THE EXACT SCIENCES. By the late William Kingdon Clifford. Second Edition. With 100 Figures. 5*s*.

LII. PHYSICAL EXPRESSION: ITS MODES AND PRINCIPLES. By Francis Warner, M.D., F.R.C.P. With 50 Illustrations. 5*s.*

LIII. ANTHROPOID APES. By Robert Hartmann. With 63 Illustrations. 5*s.*

LIV. THE MAMMALIA IN THEIR RELATION TO PRIMEVAL TIMES. By Oscar Schmidt. With 51 Woodcuts. 5*s.*

LV. COMPARATIVE LITERATURE. By H. Macaulay Posnett, LL.D. 5*s.*

LVI. EARTHQUAKES AND OTHER EARTH MOVEMENTS. By Prof. JOHN MILNE. With 38 Figures. Second Edition. 5*s.*

LVII. MICROBES, FERMENTS, AND MOULDS. By E. L. TROUESSART. With 107 Illustrations. 5*s.*

LVIII. GEOGRAPHICAL AND GEOLOGICAL DISTRIBUTION OF ANIMALS. By Professor A. Heilprin. With Frontispiece. 5*s.*

LIX. WEATHER. A Popular Exposition of the Nature of Weather Changes from Day to Day. By the Hon. Ralph Abercromby. With 96 Illustrations. Second Edition. 5*s.*

LX. ANIMAL MAGNETISM. By Alfred Binet and Charles Féré. 5*s.*

LXI. MANUAL OF BRITISH DISCOMYCETES, with descriptions of all the Species of Fungi hitherto found in Britain included in the Family, and Illustrations of the Genera. By William Phillips, F.L.S. 5*s.*

LXII. INTERNATIONAL LAW. With Materials for a Code of International Law. By Professor Leone Levi. 5*s.*

LXIII. THE GEOLOGICAL HISTORY OF PLANTS. By Sir J. William Dawson. With 80 Illustrations. 5*s.*

LXIV. THE ORIGIN OF FLORAL STRUCTURES THROUGH INSECT AND OTHER AGENCIES. By Professor G. Henslow.

LXV. ON THE SENSES, INSTINCTS, AND INTELLIGENCE OF ANIMALS. With special Reference to Insects. By Sir John Lubbock, Bart., M.P. 100 Illustrations. 5*s.*

## MILITARY WORKS.

*BARRINGTON (Capt. J. T.)*—ENGLAND ON THE DEFENSIVE; or, the Problem of Invasion Critically Examined. Large crown 8vo. with Map, 7*s.* 6*d.*

*BRACKENBURY (Col. C. B.) R.A.*—MILITARY HANDBOOKS FOR REGIMENTAL OFFICERS:

I. MILITARY SKETCHING AND RECONNAISSANCE. By Colonel F. J. Hutchison and Major H. G. MacGregor. Fifth Edition. With 15 Plates. Small crown 8vo. 4*s.*

II. THE ELEMENTS OF MODERN TACTICS PRACTICALLY APPLIED TO ENGLISH FORMATIONS. By Lieut.-Col. Wilkinson Shaw. Sixth Edit. With 25 Plates and Maps. Small crown 8vo. 9*s.*

III. FIELD ARTILLERY: its Equipment, Organisation, and Tactics. By Major Sisson C. Pratt, R.A. With 12 Plates. Third Edition. Small crown 8vo. 6*s.*

IV. THE ELEMENTS OF MILITARY ADMINISTRATION. First Part: Permanent System of Administration. By Major J. W. Buxton. Small crown 8vo. 7*s.* 6*d.*

*BRACKENBURY (Col. C. B.) R.A.*—continued.

V. MILITARY LAW: its Procedure and Practice. By Major Sisson C. Pratt, R.A. Third Edition. Small crown 8vo. 4*s.* 6d.

VI. CAVALRY IN MODERN WAR. By Major-General F. Chenevix Trench. Small crown 8vo. 6*s.*

VII. FIELD WORKS. Their Technical Construction and Tactical Application. By the Editor, Col. C. B. Brackenbury, R.A. Small crown 8vo.

*BROOKE (Major C. K.)*—A SYSTEM OF FIELD TRAINING. Small crown 8vo. 2*s.*

*CLERY (Col. C. Francis) C.B.*—MINOR TACTICS. With 26 Maps and Plans. Eighth Edition. Crown 8vo. 9*s.*

*COLVILE (Lieut.-Col. C. F.)*—MILITARY TRIBUNALS. Sewed, 2*s.* 6*d.*

*CRAUFURD (Capt. H. J.)*—SUGGESTIONS FOR THE MILITARY TRAINING OF A COMPANY OF INFANTRY. Crown 8vo. 1*s.* 6*d.*

*HAMILTON* (*Capt. Ian*) *A.D.C.*—THE FIGHTING OF THE FUTURE. 1*s*.

*HARRISON* (*Lieut.-Col. R.*) — THE OFFICER'S MEMORANDUM BOOK FOR PEACE AND WAR. Fourth Edition. Oblong 32mo. roan, with pencil, 3*s*. 6*d*.

NOTES ON CAVALRY TACTICS, ORGANISATION, &c. By a Cavalry Officer. With Diagrams. Demy 8vo. 12*s*.

*PARR* (*Col. H. Hallam*) *C.M.G.*—THE DRESS, HORSES, AND EQUIPMENT OF INFANTRY AND STAFF OFFICERS. Crown 8vo. 1*s*.

FURTHER TRAINING AND EQUIPMENT OF MOUNTED INFANTRY. Crown 8vo. 1*s*.

*SCHAW* (*Col. H.*)—THE DEFENCE AND ATTACK OF POSITIONS AND LOCALITIES. Third Edition, revised and corrected. Crown 8vo. 3*s*. 6*d*.

*STONE* (*Capt. F. Gleadowe*) *R.A.*—TACTICAL STUDIES FROM THE FRANCO-GERMAN WAR OF 1870–71. With 22 Lithographic Sketches and Maps. Demy 8vo. 10*s*. 6*d*.

THE CAMPAIGN OF FREDERICKSBURG, November–December, 1862: a Study for Officers of Volunteers. By a Line Officer. Second Edition. Crown 8vo. With Five Maps and Plans. 5*s*.

*WILKINSON* (*H. Spenser*) *Capt. 20th Lancashire R.V.*—CITIZEN SOLDIERS. Essays towards the Improvement of the Volunteer Force. Cr. 8vo. 2*s*. 6*d*.

## POETRY.

*ADAM OF ST. VICTOR*—THE LITURGICAL POETRY OF ADAM OF ST. VICTOR. From the text of Gautier. With Translations into English in the Original Metres, and Short Explanatory Notes. By Digby S. Wrangham, M.A. 3 vols. Crown 8vo. printed on hand-made paper, boards, 21*s*.

*ALEXANDER* (*William*) *D.D.*, *Bishop of Derry*—ST. AUGUSTINE'S HOLIDAY, and other Poems. Crown 8vo. 6*s*.

*AUCHMUTY* (*A. C.*)—POEMS OF ENGLISH HEROISM: From Brunanburgh to Lucknow; from Athelstan to Albert. Small crown 8vo. 1*s*. 6*d*.

*BARNES* (*William*)—POEMS OF RURAL LIFE, IN THE DORSET DIALECT. New Edition, complete in one vol. Crown 8vo. 6*s*.

*BAYNES* (*Rev. Canon H. R.*)—HOME SONGS FOR QUIET HOURS. Fourth and cheaper Edition. Fcp. 8vo. 2*s*. 6*d*.

*BEVINGTON* (*L. S.*)—KEY NOTES. Small crown 8vo. 5*s*.

*BLUNT* (*Wilfrid Scawen*)—THE WIND AND THE WHIRLWIND. Demy 8vo. 1*s*. 6*d*.

THE LOVE SONNETS OF PROTEUS. Fifth Edition. 18mo. cloth extra, gilt top, 5*s*.

*BOWEN* (*H. C.*) *M.A.*—SIMPLE ENGLISH POEMS. English Literature for Junior Classes. In Four Parts. Parts I. II. and III. 6*d*. each, and Part IV. 1*s*., complete 3*s*.

*BRYANT* (*W. C.*) — POEMS. Cheap Edition, with Frontispiece. Small crown 8vo. 3*s*. 6*d*.

CALDERON'S DRAMAS: the Wonder-working Magician—Life is a Dream—the Purgatory of St. Patrick. Translated by Denis Florence MacCarthy. Post 8vo. 10*s*.

*CAMPBELL* (*Lewis*)—SOPHOCLES. The Seven Plays in English Verse. Crown 8vo. 7*s*. 6*d*.

*CERVANTES.* — JOURNEY TO PARNASSUS. Spanish Text, with Translation into English Tercets, Preface, and Illustrative Notes, by JAMES Y. GIBSON. Crown 8vo. 12*s*.

NUMANTIA; a Tragedy. Translated from the Spanish, with Introduction and Notes, by JAMES Y. GIBSON. Crown 8vo., printed on hand-made paper, 5*s*.

CID BALLADS, and other Poems. Translated from Spanish and German by J. Y. Gibson. 2 vols. Crown 8vo. 12*s*.

*CHRISTIE* (*A. J.*)—THE END OF MAN. Fourth Edition. Fcp. 8vo. 2*s*. 6*d*.

*COXHEAD* (*Ethel*)—BIRDS AND BABIES. Imp. 16mo. With 33 Illustrations. 1*s*.

*DANTE*—THE DIVINA COMMEDIA OF DANTE ALIGHIERI. Translated, line for line, in the 'Terza Rima' of the original, with Notes, by FREDERICK K. H. HASELFOOT, M.A. Demy 8vo. 16*s*.

*DE BERANGER.*—A SELECTION FROM HIS SONGS. In English Verse. By WILLIAM TOYNBEE. Small crown 8vo. 2*s.* 6*d.*

*DENNIS (J.)* — ENGLISH SONNETS. Collected and Arranged by. Small crown 8vo. 2*s.* 6*d.*

*DE VERE (Aubrey)*—POETICAL WORKS:
I. THE SEARCH AFTER PROSERPINE, &c. 6*s.*
II. THE LEGENDS OF ST. PATRICK, &c. 6*s.*
III. ALEXANDER THE GREAT, &c. 6*s.*

THE FORAY OF QUEEN MEAVE, and other Legends of Ireland's Heroic Age. Small crown 8vo. 5*s.*

LEGENDS OF THE SAXON SAINTS. Small crown 8vo. 6*s.*

LEGENDS AND RECORDS OF THE CHURCH AND THE EMPIRE. Small crown 8vo. 6*s.*

*DOBSON (Austin)*—OLD WORLD IDYLLS, and other Verses. Eighth Edition. Elzevir 8vo. cloth extra, gilt tops, 6*s.*

AT THE SIGN OF THE LYRE. Fifth Edition. Elzevir 8vo., gilt top, 6*s.*

*DOWDEN (Edward) LL.D.*—SHAKSPERE'S SONNETS. With Introduction and Notes. Large post 8vo. 7*s.* 6*d.*

*DUTT (Toru)*—A SHEAF GLEANED IN FRENCH FIELDS. New Edition. Demy 8vo. 10*s.* 6*d.*

ANCIENT BALLADS AND LEGENDS OF HINDUSTAN. With an Introductory Memoir by EDMUND GOSSE. Second Edition. 18mo. Cloth extra, gilt top, 5*s.*

*ELLIOTT (Ebenezer), The Corn Law Rhymer*—POEMS. Edited by his Son, the Rev. Edwin Elliott, of St. John's, Antigua. 2 vols. crown 8vo. 18*s.*

ENGLISH VERSE. Edited by W. J. LINTON and R. H. STODDARD. In 5 vols. Crown 8vo. each 5*s.*
1. CHAUCER TO BURNS.
2. TRANSLATIONS.
3. LYRICS OF THE NINETEENTH CENTURY.
4. DRAMATIC SCENES AND CHARACTERS.
5. BALLADS AND ROMANCES.

*EVANS (Anne)*—POEMS AND MUSIC. With Memorial Preface by ANN THACKERAY RITCHIE. Large crown 8vo. 7*s.*

*GOSSE (Edmund W.)*—NEW POEMS. Crown 8vo. 7*s.* 6*d.*

FIRDAUSI IN EXILE, and other Poems. Elzevir 8vo. gilt top, 6*s.*

*GURNEY (Rev. Alfred)*—THE VISION OF THE EUCHARIST, and other Poems. Crown 8vo. 5*s.*

A CHRISTMAS FAGGOT. Small crown 8vo. 5*s.*

*HARRISON (Clifford)*—IN HOURS OF LEISURE. Second Edition. Crown 8vo. 5*s.*

*KEATS (John)* — POETICAL WORKS. Edited by W. T. ARNOLD. Large crown 8vo. choicely printed on hand-made paper, with Portrait in *eau forte*. Parchment, or cloth, 12*s.*; vellum, 15*s.*

Also, a smaller Edition. Crown 8vo. 3*s.* 6*d.*

*KING (Mrs. Hamilton)*—THE DISCIPLES. Eighth Edition, with Portrait and Notes. Crown 8vo. 5*s.* Elzevir Edition, 6*s.*

A BOOK OF DREAMS. Third Edition. Crown 8vo. 3*s.* 6*d.*

THE SERMON IN THE HOSPITAL. Reprinted from 'The Disciples.' Fcp. 8vo. 1*s.* Cheap Edition, 3*d.*, or 20*s.* per 100.

*KNOX (The Hon. Mrs. O. N.)*—FOUR PICTURES FROM A LIFE, and other Poems. Small crown 8vo. 3*s.* 6*d.*

*LANG (A.)*—XXXII BALLADES IN BLUE CHINA. Elzevir 8vo. parchment, or cloth, 5*s.*

RHYMES À LA MODE. With Frontispiece by E. A. Abbey. Elzevir 8vo. cloth extra, gilt top, 5*s.*

*LAWSON (Right Hon. Mr. Justice)*—HYMNI USITATI LATINE REDDITI, with other Verses. Small 8vo. parchment, 5*s.*

LIVING ENGLISH POETS. MDCCCLXXXII. With Frontispiece by Walter Crane. Second Edition. Large crown 8vo. printed on hand-made paper. Parchment, or cloth, 12*s.*; vellum, 15*s.*

*LOCKER (F.)*—LONDON LYRICS. New Edition, with Portrait. 18mo. cloth extra, gilt tops, 5*s.*

LOVE IN IDLENESS. A Volume of Poems. With an etching by W. B. Scott. Small crown 8vo. 5*s.*

*LUMSDEN* (*Lieut.-Col. H. W.*)—BEOWULF: an Old English Poem. Translated into Modern Rhymes. Second and revised Edition. Small crown 8vo. 5*s*.

*MAGNUSSON* (*Eirikr*) *M.A.*, *and PALMER* (*E. H.*) *M.A.*—JOHAN LUDVIG RUNEBERG'S LYRICAL SONGS, IDYLLS, AND EPIGRAMS. Fcp. 8vo. 5*s*.

*MEREDITH* (*Owen*) [*The Earl of Lytton*]—LUCILE. New Edition. With 32 Illustrations. 16mo. 3*s*. 6*d*.; cloth extra, gilt edges, 4*s*. 6*d*.

*MORRIS* (*Lewis*)—POETICAL WORKS. New and Cheaper Editions, with Portrait, complete in 4 vols. 5*s*. each.

Vol. I. contains Songs of Two Worlds. Twelfth Edition.

Vol. II. contains The Epic of Hades. Twenty-second Edition.

Vol. III. contains Gwen and the Ode of Life. Seventh Edition.

Vol. IV. contains Songs Unsung and Gycia. Fifth Edition.

SONGS OF BRITAIN. Third Edition. Fcp. 8vo. 5*s*.

THE EPIC OF HADES. With 16 Autotype Illustrations after the drawings by the late George R. Chapman. 4to. cloth extra, gilt leaves, 21*s*.

THE EPIC OF HADES. Presentation Edit. 4to. cloth extra, gilt leaves, 10*s*. 6*d*.

THE LEWIS MORRIS BIRTHDAY BOOK. Edited by S. S. Copeman. With Frontispiece after a design by the late George R. Chapman. 32mo. cloth extra, gilt edges, 2*s*.; cloth limp, 1*s*. 6*d*.

*MORSHEAD* (*E. D. A.*)—THE HOUSE OF ATREUS. Being the Agamemnon, Libation-Bearers, and Furies of Æschylus. Translated into English Verse. Crown 8vo. 7*s*.

THE SUPPLIANT MAIDENS OF ÆSCHYLUS. Crown 8vo. 3*s*. 6*d*.

*MULHOLLAND* (*Rosa*).—VAGRANT VERSES. Small crown 8vo. 5*s*.

*NADEN* (*Constance C. W.*)—A MODERN APOSTLE, and other Poems. Small crown 8vo. 5*s*.

*NOEL* (*The Hon. Roden*)—A LITTLE CHILD'S MONUMENT. Third Edition. Small crown 8vo. 3*s*. 6*d*.

THE RED FLAG, and other Poems. New Edition. Small crown 8vo. 6*s*.

THE HOUSE OF RAVENSBURG. New Edition. Small crown 8vo. 6*s*.

*NOEL* (*The Hon. Roden*)—continued.

SONGS OF THE HEIGHTS AND DEEPS. Crown 8vo. 6*s*.

A MODERN FAUST. Small crown 8vo.

*O'BRIEN* (*Charlotte Grace*)—LYRICS. Small crown 8vo. 3*s*. 6*d*.

*O'HAGAN* (*John*)—THE SONG OF ROLAND. Translated into English Verse. New and Cheaper Edition. Crown 8vo. 5*s*.

*PFEIFFER* (*Emily*)—THE RHYME OF THE LADY OF THE ROCK AND HOW IT GREW. Small crown 8vo. 3*s*. 6*d*.

GERARD'S MONUMENT, and other Poems. Second Edition. Crown 8vo. 6*s*.

UNDER THE ASPENS: Lyrical and Dramatic. With Portrait. Crown 8vo. 6*s*.

*PIATT* (*J. J.*)—IDYLS AND LYRICS OF THE OHIO VALLEY. Crown 8vo. 5*s*.

*PIATT* (*Sarah M. B.*)—A VOYAGE TO THE FORTUNATE ISLES, and other Poems. 1 vol. Small crown 8vo. gilt top, 5*s*.

IN PRIMROSE TIME. A New Irish Garland. Small crown 8vo. 2*s*. 6*d*.

RARE POEMS OF THE 16TH AND 17TH CENTURIES. Edited by W. J. Linton. Crown 8vo. 5*s*.

*RHOADES* (*James*)—THE GEORGICS OF VIRGIL. Translated into English Verse. Small crown 8vo. 5*s*.

*ROBINSON* (*A. Mary F.*)—A HANDFUL OF HONEYSUCKLE. Fcp. 8vo. 3*s*. 6*d*.

THE CROWNED HIPPOLYTUS. Translated from Euripides. With New Poems. Small crown 8vo. cloth, 5*s*.

SHAKSPERE'S WORKS. The Avon Edition, 12 vols. fcp. 8vo. cloth, 18*s*.; and in box, 21*s*.; bound in 6 vols. cloth, 15*s*.

SOPHOCLES: The Seven Plays in English Verse. Translated by Lewis Campbell. Crown 8vo. 7*s*. 6*d*.

*SYMONDS* (*John Addington*)—VAGABUNDULI LIBELLUS. Crown 8vo. 6*s*.

*TAYLOR* (*Sir H.*)—Works Complete in Five Volumes. Crown 8vo. 30*s*.

PHILIP VAN ARTEVELDE. Fcp. 8vo. 3*s*. 6*d*.

THE VIRGIN WIDOW, &c. Fcp. 8vo. 3*s*. 6*d*.

THE STATESMAN. Fcp. 8vo. 3*s*. 6*d*.

*TODHUNTER (Dr. J.)*—LAURELLA, and other Poems. Crown 8vo. 6*s.* 6*d.*

FOREST SONGS. Small crown 8vo. 3*s.* 6*d.*

THE TRUE TRAGEDY OF RIENZI: a Drama. Crown 8vo. 3*s.* 6*d.*

ALCESTIS: a Dramatic Poem. Extra fcp. 8vo. 5*s.*

HELENA IN TROAS. Small crown 8vo. 2*s.* 6*d.*

*TYNAN (Katherine)*—LOUISE DE LA VALLIERE, and other Poems. Small crown 8vo. 3*s.* 6*d.*

SHAMROCKS. Small crown 8vo. 5*s.*

VICTORIAN HYMNS: English Sacred Songs of Fifty Years. Dedicated to the Queen. Large post 8vo. 10*s.* 6*d.*

*WATTS (Alaric Alfred and Emma Mary Howitt)*—AURORA: a Medley of Verse. Fcp. 8vo. 5*s.*

*WORDSWORTH*—SELECTIONS. By Members of the Wordsworth Society. Large crown 8vo parchment, 12*s.*; vellum, 15*s.* Also, cr. 8vo. cl. 4*s.* 6*d.*

WORDSWORTH BIRTHDAY BOOK, THE. Edited by ADELAIDE and VIOLET WORDSWORTH. 32mo. limp cloth, 1*s.* 6*d.*; cloth extra, 2*s.*

## WORKS OF FICTION.

'ALL BUT:' a Chronicle of Laxenford Life. By PEN OLIVER, F.R.C.S. With 20 Illustrations. Second Edit. Crown 8vo. 6*s.*

*BANKS (Mrs. G. L.)*—GOD'S PROVIDENCE HOUSE. New Edition. Crown 8vo. 6*s.*

*CHICHELE (Mary)*—DOING AND UNDOING: a Story. Crown 8vo. 4*s.* 6*d.*

*CRAWFURD (Oswald)*—SYLVIA ARDEN. Crown 8vo. 6*s.*

*GARDINER (Linda)*—HIS HERITAGE. Crown 8vo. 6*s.*

*GRAY (Maxwell)*—THE SILENCE OF DEAN MAITLAND. Fourth Edition. Crown 8vo. 6*s.*

*GREY (Rowland)*—BY VIRTUE OF HIS OFFICE. Crown 8vo. 6*s.*

IN SUNNY SWITZERLAND. Small crown 8vo. 5*s.*

LINDENBLUMEN, and other Stories. Small crown 8vo. 5*s.*

*HUNTER (Hay)*—CRIME OF CHRISTMAS DAY. A Tale of the Latin Quarter. By the Author of 'My Ducats and My Daughter.' 1*s.*

*HUNTER (Hay) and WHYTE (Walter)* MY DUCATS AND MY DAUGHTER. New and Cheaper Edition. With Frontispiece. Crown 8vo. 6*s.*

*INGELOW (Jean)*—OFF THE SKELLIGS. A Novel. With Frontispiece. Second Edition. Crown 8vo. 6*s.*

IXORA. A Mystery. Crown 8vo. 6*s.*

*JENKINS (Edward)*—A SECRET OF TWO LIVES. Crown 8vo. 2*s.* 6*d.*

*KIELLAND (Alexander L.)*—GARMAN AND WORSE. A Norwegian Novel. Authorised Translation by W. W. Kettlewell. Crown 8vo. 6*s.*

*LANG (Andrew)*—IN THE WRONG PARADISE, and other Stories. Crown 8vo. 6*s.*

*MACDONALD (G.)*—DONAL GRANT. Crown 8vo. 6*s.*

CASTLE WARLOCK. Crown 8vo. 6*s.*

MALCOLM. With Portrait of the Author engraved on Steel. Crown 8vo. 6*s.*

THE MARQUIS OF LOSSIE. Crown 8vo. 6*s.*

ST. GEORGE AND ST. MICHAEL. Crown 8vo. 6*s.*

PAUL FABER, SURGEON. Crown 8vo. 6*s.*

THOMAS WINGFOLD, CURATE. Crown 8vo. 6*s.*

WHAT'S MINE'S MINE. Second Edition. Crown 8vo. 6*s.*

ANNALS OF A QUIET NEIGHBOURHOOD. Crown 8vo. 6*s.*

THE SEABOARD PARISH: a Sequel to 'Annals of a Quiet Neighbourhood.' Crown 8vo. 6*s.*

WILFRED CUMBERMEDE. An Autobiographical Story. Crown 8vo. 6*s.*

THE ELECT LADY. Crown 8vo. 6*s.*

*MALET (Lucas)*—COLONEL ENDERBY'S WIFE. Crown 8vo. 6*s.*

A COUNSEL OF PERFECTION. Crown 8vo. 6*s.*

*MULHOLLAND (Rosa)*—MARCELLA GRACE. An Irish Novel. Crown 8vo. 6*s.*

A FAIR EMIGRANT. Crown 8vo. 6*s.*

*OGLE (A. C.) ('Ashford Owen.')* A LOST LOVE. Small crown 8vo. 2*s.* 6*d.*

*PALGRAVE* (*W. Gifford*)—HERMANN AGHA: an Eastern Narrative. Third Edition. Crown 8vo. 6*s*.

*SEVERNE* (*Mrs.*)—THE PILLAR HOUSE. With Frontispiece. Crown 8vo. 6*s*.

*SHAW* (*Flora L.*)—CASTLE BLAIR; a Story of Youthful Days. New and Cheaper Edition. Crown 8vo. 3*s*. 6*d*.

*STRETTON* (*Hesba*) — THROUGH A NEEDLE'S EYE. A Story. New and Cheaper Edition, with Frontispiece. Crown 8vo. 6*s*.

*TAYLOR* (*Col. Meadows*) *C.S.I., M.R.I.A.* SEETA. A Novel. New and Cheaper Edition. With Frontispiece. Crown 8vo. 6*s*.

*TAYLOR* (*Col. Meadows*) *C.S.I., M.R.I.A.*—continued.

TIPPOO SULTAUN: a Tale of the Mysore War. New Edition, with Frontispiece. Crown 8vo. 6*s*.

RALPH DARNELL. New and Cheaper Edition. With Frontispiece. Crown 8vo. 6*s*.

A NOBLE QUEEN. New and Cheaper Edition. With Frontispiece. Crown 8vo. 6*s*.

THE CONFESSIONS OF A THUG. Crown 8vo. 6*s*.

TARA: a Mahratta Tale. Crown 8vo. 6*s*.

WITHIN SOUND OF THE SEA. New and Cheaper Edition, with Frontispiece. Crown 8vo. 6*s*.

## BOOKS FOR THE YOUNG.

BRAVE MEN'S FOOTSTEPS. A Book of Example and Anecdote for Young People. By the Editor of 'Men who have Risen.' With Four Illustrations by C. Doyle. Eighth Edition. Crown 8vo. 2*s*. 6*d*.

*COXHEAD* (*Ethel*)—BIRDS AND BABIES. With 33 Illustrations. Imp. 16mo. cloth gilt, 1*s*.

*DAVIES* (*G. Christopher*) — RAMBLES AND ADVENTURES OF OUR SCHOOL FIELD CLUB. With Four Illustrations. New and Cheaper Edition. Crown 8vo. 3*s*. 6*d*.

*EDMONDS* (*Herbert*) — WELL-SPENT LIVES: a Series of Modern Biographies. New and Cheaper Edition. Crown 8vo. 3*s*. 6*d*.

*EVANS* (*Mark*)—THE STORY OF OUR FATHER'S LOVE, told to Children. Sixth and Cheaper Edition of Theology for Children. With Four Illustrations. Fcp. 8vo. 1*s*. 6*d*.

*MAC KENNA* (*S. J.*)—PLUCKY FELLOWS. A Book for Boys. With Six Illustrations. Fifth Edition. Crown 8vo. 3*s*. 6*d*.

*MALET* (*Lucas*)—LITTLE PETER. A Christmas Morality for Children of any Age. With numerous Illustrations. 5*s*.

*REANEY* (*Mrs. G. S.*)—WAKING AND WORKING; or, From Girlhood to Womanhood. New and Cheaper Edition. With a Frontispiece. Cr. 8vo. 3*s*. 6*d*.

BLESSING AND BLESSED: a Sketch of Girl Life. New and Cheaper Edition. Crown 8vo. 3*s*. 6*d*.

ROSE GURNEY'S DISCOVERY. A Book for Girls. Dedicated to their Mothers. Crown 8vo. 3*s*. 6*d*.

ENGLISH GIRLS: Their Place and Power. With Preface by the Rev. R. W. Dale. Fourth Edition. Fcp. 8vo. 2*s*. 6*d*.

JUST ANYONE, and other Stories. Three Illustrations. Royal 16mo. 1*s*. 6*d*.

SUNBEAM WILLIE, and other Stories. Three Illustrations. Royal 16mo. 1*s*. 6*d*.

SUNSHINE JENNY, and other Stories. Three Illustrations. Royal 16mo. 1*s*. 6*d*.

*STORR* (*Francis*) *and TURNER* (*Hawes*). CANTERBURY CHIMES; or, Chaucer Tales Re-told to Children. With Six Illustrations from the Ellesmere MS. Third Edition. Fcp. 8vo. 3*s*. 6*d*.

*STRETTON* (*Hesba*)—DAVID LLOYD'S LAST WILL. With Four Illustrations. New Edition. Royal 16mo. 2*s*. 6*d*.

*WHITAKER* (*Florence*)—CHRISTY'S INHERITANCE: A London Story. Illustrated. Royal 16mo. 1*s*. 6*d*.

*Spottiswoode & Co. Printers, New-street Square, London.*

www.ingramcontent.com/pod-product-compliance
Lightning Source LLC
Chambersburg PA
CBHW020101020526
44112CB00032B/797